SCIENCE AND THE RAJ

'Looking through the window of history of British colonialism in India,
one would truly admire Kumar for all the variegated sights
he brings before us.'
—*Nature*

'Kumar has been one of the pioneers of the social and administrative
history of science in India.'
—*British Journal for the History of Science*

'...empirically and conceptually well-grounded...combines in-depth case
studies with a wider analytical perspective...a must for historians of
science and medicine.'
—*Social History of Medicine*

'This work is far from a simple panorama of case studies of science in India
during the colonial period; it proposes, on the contrary, a critical synthesis
of the facts through a serious theoretical reflection.'
—*Science, Technology and Society*

SCIENCE AND THE RAJ

A Study of British India

SECOND EDITION

DEEPAK KUMAR

OXFORD

UNIVERSITY PRESS

OXFORD
UNIVERSITY PRESS

Oxford University Press is a department of the University of Oxford.
It furthers the University's objective of excellence in research, scholarship,
and education by publishing worldwide. Oxford is a registered trademark of
Oxford University Press in the UK and in certain other countries

Published in India by
Oxford University Press
22 Workspace, 2nd Floor, 1/22 Asaf Ali Road, New Delhi 110 002

First published 1995
Oxford India Paperbacks 1997
Second edition 2006
Oxford India Paperbacks (Second edition) 2006
28th impression 2025

ISBN-13: 978-0-19-568714-9
ISBN-10: 0-19-568714-0

Typeset in Agaramond 10.5/12.5
by Sai Graphic Design, New Delhi 110 055
Printed in India by Manipal Technologies Limited, Manipal

To my teachers

S. N. Sahu, V. N. Datta,
R. L. Shukla and R. MacLeod

Contents

Preface to the Second Edition

An opportunity to revisit one's work is always welcome. Full revision was not possible. Earlier the focus was on Victorian India; now I have taken the narrative up to 1947 in a new chapter. I hope the readers would enjoy at least the footnotes! For this chapter 'Reconstructing India' new sources were consulted. These are from the National Archives of India – Sanitary and Education Proceedings (1911–19) and Education, Health and Land Proceedings (1932–5). Information in the Native Newspaper Reports for Madras, Bengal and Bombay (1893–1923) are truly fascinating and give us a glimpse of local response on several issues. I have consciously tried to supplement official documents with private papers which are so well-preserved at the different libraries and archives of UK. Significant among these are: (a) India Office Collections, British Library, London—Cornelia Sorabjee Papers, Mss. Eur. D.1103; H. Butler Papers, Mss. Eur. f.111, 654-55; Recordings of E.J. Beer, Mss. Eur. D551/30; S.K. Brown Papers, Mss. Eur. F235/46; (b) Wellcome Library for History of Medicine, London—Family Planning Association Papers; Champneys Papers; Leonard Rogers Papers, GOR/A.5511-110; Marie Stopes Papers; (c) Rockefeller Archive Centre (USA)—International Health Board and International Health Department Papers; Diaries of J.B. Grant and Dr. Heiser; (d) Centre for South Asian Studies, Cambridge —E.J. Somerset Papers; G.F. Heany Papers; (e) Churchill College Archive, Cambridge—A.V. Hill Papers (AVH); (f) Royal Society, London—A.C. Egerton Papers; A.V. Hill Paper (MDA); (g) Cambridge University Library—J.D. Bernal Papers.

In addition to this several contemporary publications and journals like *Science and Culture* (1935–50), *Harijan* (1934–46), *Young India* (1921–7), *The Indian Medical Gazette* (1866–1930) were consulted. Since the publication of this volume a decade ago, several new books have come up. Scholars like Richard Grove, Dhruv Raina, S. Irfan Habib, Gyan Prakash, Anil Kumar, Mark Harrison, Mridula Ramanna, Subrata Dasgupta, Pratik Chakrabarti, J. Lourdoswamy, Benjamin Zachariah, V.V. Krishna, and Manu Goswami have enriched the theme immensely. I have enjoyed reading them as other readers would have. I wish these new insights could be reflected adequately in the new edition. But I hope my arguments and citations still remain valid! Few minor corrections have been made for which I am grateful to Daniel Headrick, Patrick Petitjean, Santanu Chakravarty, and Dhrub Kumar Singh.

November 2005   DEEPAK KUMAR
New Delhi

Preface to the Paperback Edition

It is a nice feeling to see an affordable Paperback Edition. Few minor corrections have been made for which I am grateful to Daniel Headrick, Patrick Petitjean and Santanu Chacravarty. Some critical reviews appeared and these are likely to benefit my future work.

DEEPAK KUMAR

Preface to the First Edition

The present work seeks to explore the development of science in a colonial situation, its social implications and its economic ramifications. The object is to reveal the nature and working of the relationship between the techno-scientific imperatives and colonial requirements. The history of India during the last century spectacularly illustrates a close link between science and the Raj and I have endeavoured to explore the contours, course and significance of this link.

The entire gamut of the relationship between science and colonization can be precisely expressed by the term *colonial science*. No other word so aptly sums up the state of science, and its limitations, triumphs and failures in a colony, and Victorian India furnishes the best example. In some ways, however, colonial science did represent an advance over pre-colonial science. It was far more systematic, methodical, penetrative and pervasive. But how did it emerge? A western historian of science explains: 'In addition to greed for riches and domination, the white man became possessed suddenly of a strange spirit of adventure, of an insatiable intellectual curiosity.'[1] Is the phenomenon so simple as to be explained by 'curiosity' and the sudden 'strange spirit of adventure' the scholar has referred to? Or did there exist a pattern, a design and a 'policy'?

More than two hundred years ago, Robert Kyd, while pleading for a botanical garden in Calcutta, defined policy as 'the common sense of government'.[2] And how right he was; the common sense of an alien power could scarcely have thought of anything other than profit. But can this be the final verdict? The more one studies the archival and contemporary sources, the more one finds various shades surfacing at different levels of the official hierarchy as well as at non-official forums. Several gaps, interruptions, incoherence and inconsistencies raise doubts as to whether or not the Raj had a science policy at all. One can argue

---

[1] K. Mendelssohn, *Science and Western Domination*, London, 1976, p. 17.
[2] Kyd Papers, IOL, Mss. Eur. F.95/1. Interestingly enough, T.H. Huxley later found science itself as nothing but 'trained and organized common sense', T.H. Huxley, *Science and Education*, London, 1893, p.45.

that even the absence of an articulate policy itself often points to the existence of some sort of negative policy. The government had set certain goals, certain objectives—administrative as well as economic—though its activities often seem ad hoc, as if they were spontaneous responses to the sudden requirements of the day, while in reality a cause and effect relationship did exist between them. Certain implicit strands of their science policy can therefore be gleaned from the great mass of informative tit-bits that still survive in contemporary accounts and documents.

Modern science has traditionally been viewed as benevolent, apolitical and value-neutral. Is this really so? If not, then several questions emerge that demand an explanation. Can there be an imperialist side to the core of natural knowledge? Should one talk of colonial medicine or tropical medicine and, in this context, tropical environment or colonialism, which was the main determinant? What was the shape that 'modern' and 'universal' science took in a colony? What was the colonial posture in science and to what extent were scientific discourses used to achieve political and cultural goals? No less important is to glean how the recipient culture sought to appropriate or redefine the metropolitan ideology of science. How was the indigenous scientific tradition perceived and how did the latter react to the introduction of the former? Was a synthesis possible? Finally, could the integration of technological and scientific traditions have taken place as part of the natural evolution of the Indian society had colonization not intervened?

Clear cut answers are difficult to attempt, for colonialism was no monolith and it left several facts and questions open which can be interpreted either way. The concept of colonial science has been discussed by several scholars, a review of which, followed by a brief discussion on pre-colonial science and technology, appears in Chapter I. These provide a backdrop against which the changes that were to occur in the nineteenth century are looked into. The subsequent chapters deal with the early exploratory activities, problems in science administration, science education, scientific research works and, of course, the Indian response in these areas. Taken together, I hope, they make a useful study.

A clarification will be in order here. Terms like 'science' and 'technology', 'modern science' and 'western science' have sometimes been used interchangeably. Moreover, the term 'science' in this volume refers more to the physical and biological sciences which in the good old days went by the name of 'natural history'! As for the concept of western science, I believe that there exists a mainstream of scientific activity to which different countries in different periods contributed in various way.[3] The term science is thus used in a broad sense and in relation to institutions (both official and non-official), groups, individuals, application, etc. Similarly, the term 'native' has been used to keep the suggestive flavour of the colonial writings. The present work, though dealing primarily with the state and its *policy*, does not have an apex-centred approach. Hundreds of characters appear who may be taken as subalterns of science. Unfortunately, I could not gather biographical details on them and this remains a serious lacuna in my work.

No one can deny the importance of concepts, for they infuse life into long-forgotten and almost dead data; but the latter are equally important because it is they who impart a distinct shape and weight to the former. So care has been taken to draw mostly upon the primary sources, particularly archival documents and private papers. But I must confess my limitations. Marathi, Tamil and Urdu sources could not be consulted. Only a few Bengali tracts have been seen and, I am sure, much more still remains untapped. The archival materials are mostly the Government of India's 'A' proceedings and, unless otherwise indicated, are from the National Archives of India. Some work was also done at the West Bengal State Archives, Bihar State Archives, Tamilnadu State Archives and the Maharashtra State Archives. Contemporary tracts, reports and journals were consulted in Calcutta at the National Library, and the libraries of the Asiatic Society, Bangiya Sahitya Parishad, J.C. Bose Trust, AHSI, BSI, GSI, ZSI, St. Xavier's College and the Calcutta Medical College. Important references were also found at the Haffkine Institute, VJTI, Grant Medical College, BNHS and the Asiatic Society in Bombay; the Meteorological Department in Poona; the Forest Research Institute at Dehra Dun; and the Khuda Bux Library and the Sinha Library in Patna. Some Ph.D. theses and recent works were consulted at the Smithsonian Institution Library in Washington and at the NMML and the NISTADS Library in New Delhi. The libraries in UK are stacked with private papers and the details are given in the bibliography. I thank the staff of all these institutions for help and co-operation.

[3] As Joseph Needham wrote, 'We speak of the unity of science . . . there can be no doubt at all that the work of Asian people was at least as important in the history of sciences and technology as that of the Europeans, until the time of the Renaissance'. Mansel, D. (ed.), *A Selection from the Writings of Joseph Needham*, Lewes, 1990, p. 142.

I feel a deep debt of gratitude to my teachers. This work is dedicated to my school-teacher, S.N. Sahu, who made me a student of history, R.L. Shukla, who nursed my interest in the subject at Patna College, V.N. Datta from whom I learnt the tools of historical research and to Roy MacLeod, whose papers introduced me to the theme of this volume.

New Delhi   DEEPAK KUMAR

Acknowledgements

The theme of this book required extensive archive research and this would not have been possible without generous financial assistance from the Indian Council of Historical Research, the Council of Scientific and Industrial Research, the Indian Council of Social Science Research, Sir Dorabjee Tata Trust (Bombay), St. John's College (Cambridge), the British Academy, the Royal Society, the Institute of Commonwealth Studies (London), Rockefeller Archive Centre (New York), and the Smithsonian Institution (Washington D.C.). I thank them for supporting me in different stages of the work.

This volume is the revised version of a doctoral exercise at Delhi University. I am grateful to V.N. Datta and Barun De for help in developing the theme, and to Aparna Basu and Sumit Sarkar for affectionate supervision. Roy MacLeod, Michael Worboys and Ian Inkster read the entire manuscript and made valuable suggestions, for which I am sincerely grateful. I am deeply indebted to S. Bhattacharya, S. Ambirajan, Lewis Pyenson and A.K. Bagchi for comments and constant encouragement. John Iliffe and Nathan Reingold made my work at Cambridge and Washington both possible and memorable. I am grateful to A. Rahman and Ashok Jain for evincing a keen interest in my work and for providing excellent institutional facilities. To my editor at the Oxford University Press I owe a special thanks for help and care in publication.

At NISTADS, I have benefited from the writings of Satpal Sangwan, Dhruv Raina, S. Irfan Habib, V.V. Krishna and Amitabha Ghosh. We have worked closely for so long that words fail me in adequately registering my debt and gratitude. Works of Richard Grove simply inspire and his friendship a treasure. In the late 1970s Anil Kumar and Satpal Sangwan were the first with whom I shared views. They have not only been generous to me but have enriched the theme in their own way. Two friendly couples, Kiran-Ravindra and Meera-Anil, took care of me during my early work at Calcutta and Delhi. In my lean hours I received soothing encouragement from K.L. Tuteja, P.K. Shukla, Amod Mishra, Pradosh Nath, J.N. Sinha and of course, my childhood friends, Shambhoo and xvii *Acknowledgements*

Babloo. All my labours where neatly put to print by Naresh and Rajender.
I am beholden to all of them for their concern and friendship. My parents
and my wife, Sunnu, helped me in so many ways and suffered. To thank
them would be like thanking my own self!

Abbreviations

Agri.	Agriculture
AHSI	Agriculture and Horticulture Society of India
AMD	Army Medical Department
BAAS	British Association for the Advancement of Science
BM	British Museum
BNHS	Bombay Natural History Society
BSA	Board of Scientific Advice
BSAP	Bihar State Archives, Patna
BSI	Botanical Survey of India
Court	Court of Directors
DG	Director-General
DPI	Director Public Instruction
Edu.	Education
FRI	Forest Research Institute (Dehra Dun)
GG	Governor-General
GOI	Government of India
GSI	Geological Survey of India
GTSI	Great Trigonometrical Survey of India
IAC	Indian Advisory Committee
IACS	Indian Association for Cultivation of Science
IESHR	Indian Economic and Social History Review
IG	Inspector-General
IHR	Indian Historical Review
IJHS	Indian Journal of History of Science
IMS	Indian Medical Service
INC	Indian National Congress
IOL	Indian Office Library
IOR	Indian Office Records

Jr.	Journal
JASB	Journal f the Asiatic Society of Bengal
JRSA	Journal of Royal Society of Arts
MSA	Maharashtra State Archives
Misc.	Miscellaneous
NAI	National Archives of India
NISTADS	National Institute of Science, Technology and Development Studies
n.d.	Not dated
n.p.	Not paginated
PWD	Public Works Department
Rev.	Revenue
VJTI	Victoria Jubilee Technical Institute
WBSA	West Bengal State Archives
ZSI	Zoological Survey of India

CHAPTER 1

Science in a Colony: Concept and Contours

Colonization was an extremely important historical process with wide-ranging results. We do talk about colonial polity, colonial society, colonial economy, colonial legacy, etc. Can the ethos and function of colonialism be extended to the realm of science and technology? Can the latter be spotted at the centre of the colonial whirlpool? Is it possible to talk of colonial science? Of course, few would deny the universality, rationality, and utility of science. But colonialism has also affected our 'universe' no less profoundly than, say, the Scientific or Industrial Revolution. Many claimed it to be 'rational' as well, and some still harp on its 'utilitarian' ideology.

Colonial expansion was not a disjointed phenomenon. Changes in technology played a vital role and entailed the movement and transfer of ideas. Sometimes this transfer was a two-way traffic but mostly it remained an exercise in dominance. 'Imperialism is the spirit of rule, ascendancy or predominance',[1] and this is how it was perceived in its heyday. Domination by one meant subordination of another, and this non-binary equation existed at both the centre and circumference of colonialism. Colonial science can be defined as 'a dependent science wherein the result-oriented research in applied science heavily supersedes the curiosity-oriented research in pure science'.[2] I realize how insufficient this description is. The distinctions between pure and applied science appeared only in the late nineteenth century and are still unclear. They cannot be applied to earlier times. One could have replaced 'result-oriented' with 'power-oriented' and 'curiosity' with 'knowledge'. But are power and knowledge not two sides of the same coin? Foucault presents

[1] Lord Rosebery quoted in J.G. Godard, *Racial Supremacy Being Studies in Imperialism*, Edinburgh, 1905, p. 5.

[2] Deepak Kumar, 'Patterns of Colonial Science in India', presented at the Indian History Congress, Hyderabad session, 1978. Later published in *Indian Journal of History of Science*, 15, 1 May 1980, pp. 105–13.

power/knowledge as a facet of power/culture. Together they make for dominance. This dominance is, however, not always monolithic nor does it remain uncontested. The exercise of even absolute power has within it some element of disempowerment. Colonialism has both unity and contradictions, direction and ambivalence, power and weaknesses. It involved a set of structures as well as a set of discourses which, when taken together, may provide an imbricated view of domination in its myriad forms. It affected every conceivable aspect of human existence—the body and mind of both the colonized and the colonizer. Science, now routinely studied as a social institution at all levels, could not be an exception and so we have the phenomenon of colonial science. Like colonialism itself, colonial science is more than a set of institutions or structures; it is an economic as well as a cultural intervention.

The contours of this discourse appear when one studies how work on this topic has developed over the years. Perhaps the earliest detailed study was a doctoral dissertation in 1941 by Charles Forman on *Science for Empire* covering the years 1895–1940.[3] During the 1960s and '70s certain articles appeared which not only drew attention to this theme but also sparked off lively debates and new researches, leading to a flood of articles and several monographs during the '80s. In the late '50s Dupree and Bernard Cohn wrote about America as a source of science for Europe.[4] Later Donald Fleming extended the American perceptions to Canada and Australia and probed their similarities and dissimilarities. One common denominator he found was that the 'reconnaissance of natural history' provided the leitmotif of the scientific enterprise in the nineteenth century. These investigations 'flowed without a break from the combination of practical and speculative motives that triggered the colonial enterprise itself'.[5] After all, the environment had to be 'mastered' and its economic potentialities properly 'canvassed'. Fleming notices two distinct features: (1) European scientists (e.g. Linnaeus or Banks) behaved like 'absentee landlords, ceaselessly commanding and receiving tribute (data) from the ends of the earth'; (2) those based in colonies—the

[3] Charles Forman, *Science for Empire: Britain's Development of the Empire through Scientific Research, 1895–1940*, unpublished Ph.D. thesis, University of Wisconsin, 1941.

[4] B. Cohen, 'The New World as a Source of Science for Europe', *Actes du IX Congress International d'Histoire des Science*, Barcelona, 1959.

[5] Donald Fleming, 'Science in Australia, Canada and the United States: Some Comparative Remarks', *Proceedings of the X International Congress of History of Science*, Ithaca, 1964, pp. 180–1.

colonial scientists themselves preferred a subservient role because *inter alia* it saved them from the 'more perilous enterprise of theoretical constructions'. The result was that 'the natural history man' looked like a variant of 'the economic man',[6] and not much different from an explorer, planter or colonizer.

Models

These dependency perspectives were converted into a diffusionist model by George Basalla, who, in 1967, presented a three-phase linear evolutionary framework to explain the 'spread' of western science in non-European areas.[7] During Phase I, the non-scientific society provides a source for European science; Phase II is termed colonial science; and Phase III completes the process of transplantation with a struggle to achieve an independent scientific tradition or culture. The first phase is basically explorative. The term 'non-scientific' refers to the absence of western science, and Basalla takes care not to offend the indigenous knowledge systems of ancient civilizations. Similarly, he does not use colonial science as a pejorative term implying 'the existence of some sort of scientific imperialism whereby science in the non-European nation is suppressed or maintained in a servile state by an imperial power'.[8] Moreover, the area of operation may not be a formal European colony (e.g. Japan or the USA). The determining factor is dependence upon the established scientific culture of western Europe. The real difficulty comes when Phase II gives way to Phase III. Basalla puts several conditions before Phase III can be taken to be fully operational. They are the removal of philosophical or religious resistance (like Confucianism) to science; achievement of social approval and full government support; organized science education; creation of a technological base; and greater professionalization through local scientific organizations, journals, international recognition, etc.

Basalla generalizes with a sweeping flourish. In a single linear scheme he tries to compress a complex phenomenon occurring in different culture areas at different points of time. He is cavalier with both time and space. He assumes uniformity and homogeneity where none exist, whether it be western scientific beliefs and institutions, or metropolis–

---

[6] Ibid., pp. 182–3.
[7] George Basalla, 'The Spread of Western Science', *Science*, 156, 5 May 1967, pp. 611–22.
[8] Ibid., p. 613.

colony and colony–colony relationships.[9] Even as a diffusionist model, it tends to be passive. Phase I, for example, is a one-way traffic. Here knowledge flows from the colony to Europe. So how did diffusion take place? In Phase II the colonial scientist is shown as dependent upon an external scientific culture, yet he is 'not a fully participating member of that culture'. This is not true. If one looks into the private papers of metropolitan savants (Joseph Banks, John Lindley, W.J. Hooker) one does not find that they ever undervalued the scientific potential of their peripheral colleagues.[10] Inkster shows how Australian scientists secured a place within the 'invisible college'.[11] And in Phase III, many conditions, mostly structural, have to be fulfilled before the embryo conceived in Phase II is granted an independent existence. The umbilical chord never gets snapped (even on maturity) in a Eurocentric model. This may be true or relevant for a white-settler colony like Australia or Canada. But what about India or Egypt? These could never have the same degree of identi-fication which the former could easily forge with the metropoles. Some sort of a cultural hiatus remained in their imperial relations all along and even grew with the passage of time. They were, moreover, no *tabula rasa* when the colonizers came; they had a fairly long techno-scientific tradition and a rich cultural heritage. When the phase of colonial science is in full swing, Basalla, unlike Fleming, does not account for the socio-psychological tensions that exist between the professionals on both sides of the fence and within their own local settings.[12] Fleming talks of the 'psychological combat between Australian intellectuals and the society around them',[13] though both shared the same ethnic and cultural heritage. The tension and its impact must be quite different and varied when two antipodal civilizations encounter and interact, as in South Asia. Basalla, nevertheless, provoked an intense debate and several counter-models (some counterfeits) appeared in response.

A decade later, Michael Worboys wrote *Science and British Colonial*

[9] D.W. Chambers, 'Period and Process in Colonial and National Science', in N. Reingold and M. Rothenberg (eds), *Scientific Colonialism*, Washington, 1987, p. 312.

[10] Satpal Sangwan, 'Natural History in Colonial Context: Profit or Pursuit? British Botanical Enterprise in India, 1778–1820, in P. Petitjean et al. (eds), *Science and Empires*, Dordrecht, 1992, p. 290.

[11] Ian Inkster, 'Scientific Enterprise and the Colonial Model: Observations on Australian Experience in Historical Context', *Social Studies of Science*, xv, 1985, p. 685.

[12] Ibid., p. 688.

[13] Donald Fleming, p. 186, note 5.

Imperialism.[14] Unlike Forman, he concentrates on policy issues and provides details and analyses with the clarity of a historian. Making a few exceptions, such as astronomy, he treats colonial science as applied science, i.e. 'science applied to production of systematic knowledge, the provision of material benefits, and the solution of practical problems'.[15] His colonial scientists were affected at all levels by colonialism. But he does not ignore the 'relative autonomy' that they enjoyed which made them sometimes oppose certain politico-economic policies. For example, the long-term ecological approach of scientists to soil management led to conflicts over land use and it was the colonial scientists who first complained of the neglect of 'native' food production.[16] Worboys also gives an excellent analysis of the influence colonial science had on certain British scientific institutions which had not been done earlier. But his approach is metropolitan. He concentrates on events in Britain or on events in the dependent empire as seen from the Colonial Office in London. He does not touch upon the results of colonial scientific 'development' efforts and the establishment of 'regional', later national, scientific communities. This aspect has been addressed by Roy MacLeod, and more specifically by Ian Inkster, in the context of Australia.

'Metropolis' Revisited

For MacLeod the central issue is not 'science in imperial history but science *as* imperial history'.[17] This is an important statement. Imperialism itself has been the subject of an almost unending debate. On this is superimposed the history of the natural sciences and technologies 'that served different functions at different times and that underwent quite distinct patterns of development'. MacLeod is aware of the hazards, yet he cannot resist the temptation of creating new categories. He 'visualizes' five stages in the evolution of the so-called British imperial science (during 1780–1939).[18] They are:

[14] Michael Worboys, *Science and British Colonial Imperialism, 1890–1940*, unpublished Ph.D. thesis, Univ. of Sussex, 1979.

[15] Ibid., p. 400.

[16] Michael Worboys, 'Science and the Colonial Empire', in Deepak Kumar (ed.) *Science and Empire*, Delhi, 1991, p. 21.

[17] Roy MacLeod, 'On Visiting the Moving Metropolis: Reflections on the Architecture of Imperial Science', in N. Reingold and M. Rothenberg (eds), p. 219, note 9.

[18] Ibid., p. 230.

I. Metropolitan science, characterized by an explorative, Banksian and systematist approach, leading to expansion of maritime trade, discovery of raw materials and new markets.

II. Colonial science, characterized by metropolitan dominance, autochthonous societies, individual research and scientific services with emphasis on primary products, adaptive technology and local markets.

III. Federative science characterized by co-operative, imperial research, higher education and professional legitimation, resulting in improved technology and greater participation in world markets.

IV. 'Efficient' imperial science, characterized by codified science and specialization of disciplines, and dominated by experts and sound management of land, resources and industry.

V. Empire or commonwealth science, characterized by a metropolitan trusteeship, co-ordinated fundamental research and state encouragement of applied science.

In all the above categories the binding thread is that of imperial science, marked by metropolitan dominance or 'trusteeship' (in the Gandhian sense?). On one occasion MacLeod defines imperial science as 'a set of structures', on another as 'an expression of a will and a purpose, a mission, a vocation, often inarticulate, but enormously powerful'.[19] Taken together it shows science *as* imperial history. This seems to perhaps be an overstatement. Was science ever organized or practised on an empire-wide basis? Worboys rather argues that there was no imperial science. In fact science followed the wider political and administrative divisions with separate activities and policies in the dominions, India and other colonies.[20] The 'purpose' or 'mission' was certainly there but its direction and intensity varied with the locale and time. Similarly, the 'structures' differed. Special features of a region or period should not be lost sight of when general patterns are discussed and vice-versa.

Moreover, the questions raised by MacLeod may get similar answers if science is treated as a factor in imperial history and not as imperial history itself. Science and empire (which MacLeod himself uses as a 'political expression in cultural terms') may appear to be a more appropriate description.[21] It grants relative autonomy to both and yet brings into focus the nexus between the two.

---

[19] Ibid., pp. 219–20.
[20] Michael Worboys, p. 13, note 16.
[21] Roy MacLeod, p. 244, note 17.

Returning to the categories, the five fine distinctions that MacLeod has made, overlap and intrude into each other in both content and time-frame. For example, when does the metropolitan phase end, and why should it be confined only to the explorative-Banksian era when the metropolitan science itself has been defined as 'a way of doing science'?[22] The argument is that the metropolis is not static, it moves to the periphery if certain conditions are fulfilled. Yet the fact remains that the metropolis maintained its overbearing presence in the content and context of science, at least in the first three phases. 'Intellectual deference', an important hallmark of colonial science, is visible even in the last phase of the so-called trusteeship. Only the emphasis on 'practical service' gets diluted when the colonies begin to acquire greater autonomy. MacLeod, however, does not describe colonial science as a 'low science' indulged in by 'lesser minds'; rather he shows how, by virtue of the new data it brought forth, colonial science held 'a vital, not a subordinate position in institutions at home'.[23] But this is done again in relation to the metropolitan imperatives (e.g. the use made of colonial science by the mother country). The periphery thereby gets peripheral treatment.

Even the federative phase has 'scientific soldiers' like Ronald Ross carrying on with his 'individual' research on mosquitoes. Then how does this phase differ from the earlier phase of colonial science? Only 'co-operative research' and 'professional legitimation' are not sufficient to distinguish between the two phases. Who determined the agenda is more important. And even if it was done on the periphery, it was most likely the result of some sort of a 'struggle'.[24] This struggle gets subsumed under terms like federative or commonwealth which of course emphasize co-operation, co-ordination, trusteeship, etc. These healthy values are not always visible either in the politics of nations or in the politics of science. The most intriguing is the phase of 'efficient' imperialism. The phrase itself is inane. The high-noon of imperialism is presented as 'efficient' imperialism and perhaps rightly so, for had it not been 'efficient' it would not have lasted so long. The rhetoric of a Chamberlain or a Ramsay was

[22] Ibid., p. 220.
[23] MacLeod writes: 'It (colonial science) meant different things: to those at home . . . it meant derivative science done by lesser minds. . . . It was, *looked at from the metropolis*, low science identified with fact gathering.' Ibid., p. 221 (emphasis added to show that it was the opinion of the 'colonizing' metropolis).
[24] The 'struggle' aspect is lucidly discussed in I. Habib and D. Raina, 'Copernicus, Columbus, Colonialism and the Role of Science in Nineteenth-Century India', *Social Scientist*, no. 190–1, March 1989, pp. 52–66.

impressive but the results were nought.[25] The British Association of Advancement of Science held its sessions in Australia, Canada and South Africa and it is often cited as federative bonhomie. But the Association never visited India. Imperial science was selective. In India and several other colonies, the phase of colonial science lasted longer than suggested by the models.

MacLeod does not confine the metropolis only to western Europe. Calcutta, Sydney and Montreal emerge as metropoles in their own right and create their own peripheries. Circles emerge within circles. A metropolitan 'aperture' is unable to capture the shifts and tensions in all its details and niceties. Perhaps no model can. Yet the remarkable thing about the 'Moving Metropolis' is its open-endedness, i.e. the recognition that changing socio-economic, political and cultural circumstances alter both the environment and perception of science.

A distinct 'peripheral' perspective is provided by Ian Inkster.[26] A truer 'colonial model', he argues, will be forged only in work which accounts for both the local and the imperial factors. Unlike the earlier models, he differentiates between the areas of recent settlement (like Canada and Australia) and the areas of relative economic backwardness (like India and Japan). He then provides a pyramidical model with (1) a large socio-economic base supporting (2) an all-important 'cultural-institutional infrastructure' which mediates and strengthens (3) the 'scientific super-structure' of research programmes. Here the first and second are most significant because they sustain the scientific superstructure which, in turn, works in tandem with the metropolitan or imperial centres.

This description fits the areas of recent (white) settlements but not other areas. For example, India and Japan had a 'traditional' socio-economic base and an 'infrastructure' of their own which the western culture, science and technique had to encounter and 'penetrate'. Here it was *not* like a quick transition from the stone-age to the steam-age. It was more in the nature of an encounter, whereas in Australia it was in the nature of an interaction. But in both cases, even having entered Basalla's Phase III, the dependency or 'colonial' mentality remains. A scientist on the periphery was spatially away from the centre, but could, at the same time, be 'mentally divorced' from the local setting in which he lived. Intriguingly, Inkster calls it 'cosmopolitan'.[27] Indian scientists were never 'cosmopolitan' in this sense. The local setting and colonial

[25] Michael Worboys, p. 14, note 16.
[26] Ian Inkster, pp. 677–703, note 11.
[27] Ibid., p. 689.

control weighed heavily on them. This, as he argues elsewhere, was because of the 'loss' of political and economic sovereignty.[28] Japan and Australia had retained their sovereignty, and this made all the difference. Japan could easily assimilate western science, thanks to her independent existence and a liberal set-up, particularly after the Meiji Restoration. In India, on the other hand, the requirements of a colonial government made science dependent and greatly limited in its scope.[29]

The Indian experience in Basalla's Phase II and Phase III has recently been illuminated by S. Sangwan[30] and V.V. Krishna.[31] Sangwan wishes to rescue the colonial scientists from the 'stigma' of colonialism. With numerous examples, he explains how, despite being sandwiched between a profit-oriented local government and research-oriented metropolitan savants, they successfully maintained their identity. Sangwan appeals for their work to not be dismissed as 'rubbish' or 'low science' (though so far no historian has done so).[32] Their work was certainly conditioned by colonialism which in itself is no 'stigma' but only a fact of history. Sangwan, nevertheless, makes a very valid point, i.e. to turn to the 'internal' organization of colonial science. To this V.V. Krishna has given some thought. He enquires as to who a colonial scientist actually is and places them into three categories:[33]

I. The 'gate-keepers' who helped to keep science dependent.
II. The 'scientific-soldiers' who merely executed their occupational roles.
III. 'National' scientists who struggled to cultivate modern science in the framework of emerging nationalism.

All these key terms, such as gate-keepers and scientific-soldiers, Krishna borrows from Roy MacLeod but puts them to different use.[34] He argues

[28] Ian Inkster, 'Prometheus Bound: Technology and Industrialization in Japan, China and India prior to 1914: A Political Economy Approach', *Annals of Science* 45, 1988, pp. 399–426.
[29] Deepak Kumar, p. 111, note 2.
[30] S. Sangwan, pp. 281–98, note 10; idem, 'From Gentlemen Amateurs to Professionals: Natural History Tradition in Colonial India 1780-1840', paper presented at an International Conference on Ecological History in South and South East Asia, New Delhi, 18–21 Feb. 1992.
[31] V.V. Krishna, 'The Colonial Model and the Emergence of National Science in India, 1876–1920, in P. Petitjean et al., pp. 57–72, note 10.
[32] Here again to MacLeod is attributed the epithet 'low science'. See note 23.
[33] V.V. Krishna, note 31.
[34] Roy MacLeod, pp. 221, 224, 239, note 17. The term 'gate-keepers' figures also

that the third category was responsible for the emergence of national science and that the other two categories were not. Thus he questions Basalla's hypothesis that Phase III (independent science in its embryonic form) is contained in Phase II (colonial science). But at the same time he emphasizes the fallacy of situating colonial science 'delinked from its colonial context—namely, colonialism and rising nationalism'. Once both are taken together, how can one deny at least the embryonic links between Phase II and Phase III? The constitution of national science might be outside the structures of the colonial scientific enterprise, but definitely not outside the colonial milieu. The advocates of national science were after all the products of a colonial educational system, and they also looked to western models and laurels, sometimes of course on their own terms. Krishna ignores the overlaps among his categories. They are not water-tight compartments. He, however, says that the national science was 'stemming' out of its embryonic form by 1920. The colonial grip over India by then had begun to loosen. Working on Mexican examples, D.W. Chambers prefers to view these changes as a complex of ideas and processes, not just staged development. He argues, not without reason, that 'colonial science may remain colonial even in politically independent nations'.[35] In a recent paper he tries to minimize the use of the words 'country' and 'nation' and instead uses the term 'localities', focusing on local sites of knowledge production. 'In the Basalla model the existence of a scientific centre in Europe was seen to be non-problematic; in the locality model the very existence of the centre/periphery divide is the thing to be explained, along with the emplacement of science in every locality and the observed similarities and variations.'[36]

Tools vs. Reason

The centre-periphery models mostly deal with scientific exchange, mutual relationships, professionalization, institutions, etc. But there are two specific questions which have simultaneously received considerable attention. They relate to the role and place of the 'tools' or technology and 'pure' science (reason?) in the history of imperialism. It is well-known that

---

in H.M.C. Vessuri, 'The Implantation of Modern Science in Venezuela', in J.J. Saldana (ed.), *Cross Cultural Diffusion of Science in Latin America*, Mexico, 1987, p. 108.

[35] D.W. Chambers, p. 314, note 9 (emphasis added).

[36] D.W. Chambers, 'Locality and Science: Myths of Centre and Periphery', in Lafuente and Elena (eds), *Mundializacion de la Ciencia*, Madrid, 1992 pp. 605–17.

technologies appear in a certain socio-economic context. But can there be a technological context in which social and all other events take place? To what extent did the technological context determine the course of European expansion and what were the consequences?[37] Diametrically opposed to this is another set of questions revolving round the pristine purity of 'pure' science amidst 'the slaughter and squalor of imperialist adventures'. Did the imperialist scientists manipulate learning or did the natural sciences change their form in the colonies? How can one separate the economic motive from the cultural (or more sublime) motives?[38]

In 1981 D.R. Headrick made a case for technological determinants in his *Tools of Empire*.[39] Imperialism, he argued, was not the result of mere western superiority, but of the unleashing of overwhelming force at minimal cost. Technological changes affected the timing and location of the European conquests and thus determined the economic relations of colonialism. It made European expansion swift, thorough and cheap. A ready example is the penetration of Africa as a result of the steamer, quinine and the quick-firing gun. In a subsequent volume entitled *The Tentacles of Progress*,[40] his emphasis is on consolidation and its consequences. The title is a bit intriguing. How do the tentacles of imperialism become 'the tentacles of progress'?

Headrick tries to show, and perhaps account for, 'the contrast between the successful relocation of European technologies under colonialism and the delays and failures in spreading the corresponding culture'. Here the question arises as to whether the relocation of European technologies was really successful. Perhaps not, partly because they were conceived of, and remained, as mere technology projects (not technology diffusion), and partly because 'the local requirements' and 'local knowledge' were seldom taken into consideration. Headrick gives an excellent analysis of the causes in relation to British India. Of raw materials there was no shortage; the 'natives' were not averse to innovations; and capital could have been generated. 'What was lacking was a consistent attitude on the part of the government.' But the colonial interests always put the blame on the oriental 'other-worldliness'. They consistently argued that the 'natives' had an in-built, cultural bias against manual work and technology. This

[37] D.R. Headrick, '*The Tentacles of Progress: Technology Transfer in the Age of Imperialism. 1850–1940*, New York, 1988, pp. I–X.
[38] Lewis Pyenson, *Cultural Imperialism and Exact Sciences: German Expansion Overseas 1900–1930*, New York, 1985, pp. 11–12.
[39] D.R. Headrick, *The Tools of Empire*, New York, 1981.
[40] D.R. Headrick, note 37.

was 'not an explanation but an excuse, or even a weapon'.[41] The rulers, moreover, educated their subjects only up to a point. Beyond that they withheld the culture of technology. Another way for technological diffusion could have been through enterprises and experience. Railways were enclavists and could not contribute to diffusion. Non-Europeans had to remain content with lower jobs. Entrepreneurship was not lacking but whoever dared to venture forth (like J.N. Tata) soon realized that the road to development lay through politics. The technologies, like trade, came 'wrapped in flags' and so politics was the most important factor.[42] The result was that the tropics experienced growth but little *development*. It is difficult to differ with this analysis.

In a work less prone to economic and technological 'bias', Michael Adas looks into the criteria by which the colonizers 'measured' the achievements and potentialities of non-western people.[43] He bases his surveys on hundreds of travelogues and contemporary European accounts. He argues that science and technology were central to the European sense of what it meant to be civilized. He agrees that 'race crops up more in the literature than comparisons of tools and concepts of the cosmos'. Yet he calls for 'a move away from racist reductionism' and asks to give more serious consideration to other standards, like science and technology, by which non-western people were judged.[44] Adas is interested more in measurement *per se* than in exploring how these standards became tools of domination. His concerns are 'the attitudes and ideologies' of the Europeans and not the processes of diffusion. Nor does he try to explore how the non-Europeans perceived the 'other'. For example, while rationalizing 'the civilizing mission' ideology, he does not discuss what the colonized thought about it.[45] He works on a huge geographical canvas but his great reliance on a particular category of sources limits his analysis to only one perspective.

Both Headrick and Adas tend to treat scientific dependence and technological dependence as two sides of the same coin which is largely true in the Afro-Asian context. But this was not so in the case of Australia or Canada. In a significant work on the Australian experience, Jan Todd

[41] Ibid., p. 309.

[42] Loss of political and economic sovereignty is what Inkster also holds as 'of greater importance than the supposed retardative impact of deeply held cultural traits'. Ian Inkster, 'Prometheus Bound', pp. 422–3, note 28.

[43] Michael Adas, *Machines as the Measure of Men*, Ithaca, 1989.

[44] Ibid., p. 341.

[45] Ibid., p. 268.

shows how the scientific enterprise functioned 'more as an intellectual resource facilitating the local assessment, selection, entry and assimilation of foreign frontier technology than in providing the intellectual content of new, local technological systems'.[46] The creation of any 'new' knowledge was thus not a necessary concomitant. There was hardly any 'direct' transfer from the scientific superstructure to the productive base. The result was that any boost to the scientific enterprise came from the 'bottom up' rather than 'top down', and in between there were several institutional conduits to facilitate the process.[47] In sharp contrast, British India presented the spectacle of a 'top-heavy' administration encouraging a 'top down' development.

Technological 'relocations' and the natural history 'taxonomies' do have an open and a strong economic content; but can the same be said of the exact sciences? Lewis Pyenson claims a 'distinct' status for the exact sciences. He argues that their *discourse* is insular, and unlike art and religion, pure science is *not* culture-bound. His case studies of physics and astronomy in the German and Dutch imperial settings exhibit no sign of socio-political determination.[48,49] He agrees that the imperialists spent money on physicists and astronomers in anticipation of receiving a return on their investment. But pure science did not oblige them in this way; it only served to confer prestige on the imperial powers in distant lands and, by extension, in rival imperial capitals. The scientific discourse itself did not undergo any 'transmutation' to meet colonial requirements. It was neither 'perverted' nor 'trivialized'. 'Those who achieved distinction overseas in the exact sciences did so by staying beyond the reach of the metropolitan octopus that moved men across oceans in order to satisfy metropolitan appetites.'[50] Based on this hypothesis, Pyenson formulates a model with the following three orthogonal axes:

[46] Jan Todd, 'Science at the Periphery: An Interpretation of Australian Scientific and Technological Dependency and Development prior to 1914', *Annals of Science*, 50, 1993, pp. 33–58.

[47] Ibid., pp. 55–6, see also J.H. Todd, *Transfer and Dependence: Aspects of Change in Australian Science and Technology 1880–1916*, Ph.D. thesis, University of NSW, 1991. For science and dependency in Canada, see V.M.G. de Vecchi, *Science and Government in Nineteenth-Century Canada*, Ph.D. thesis, University of Toronto, 1978.

[48] Lewis Pyenson, note 38.

[49] Lewis Pyenson, *The Empire of Reason: Exact Science in Indonesia, 1850–1950*, Leiden, 1989.

[50] Lewis Pyenson, 'Why Science May Serve Political Ends: Cultural Imperialism and the Mission to Civilize', *Berichte zur Wissenschaftsaeschichte*, 13, 1990, p. 77.

I. The functionary axis under which the colonial scientist is first and foremost a functionary of the empire.

II. The research axis wherein the research ethic remains paramount.

III. The mercantilist axis which has the scientist serving business interests.

Citing several examples, Pyenson locates the French on the functionary axis;[51] the Germans on the research axis; the British on both the research and mercantilist axes; and the Dutch are found to be strong on all axes! Pyenson's account illuminates the scientific edge of colonial science. The installation of a telescope in a remote colonial observatory does not relate *directly* to making money or exploiting a colony in a lucrative way and it definitely cannot be compared to an agricultural test station. Undoubtedly agriculture or medicine are quite different constructions from astrophysics. But such observatories and meteorological stations in the colonies did serve the navy which, in turn, was there to protect political and economic interests. In 1890, at the University of Buenos Aires, an Argentine civil engineer occupied the Chair of Astronomy and his students were themselves civil engineers. Those branches of physics were mostly encouraged which had direct applications, for example, electro-technology.[52] Pyenson tends to underestimate this. It is impossible, perhaps undesirable, to separate the economic motive from the political or cultural one; they are all very closely linked. Colonial scientific institutions do have 'a special character' and this is what colonial science is.

Of course, there would be local variations. But do such variations alter the basic character of expansion? In 1903, Emil Wiechert, a geophysicist at Gottingen, asked for a network of seismological stations. His argument was that Germany would fall behind in considering 'the great questions of research' just as it had in the acquisition of colonies unless such a network were established. 'We have to seek fulcrums in our colonies not only politically and economically but also scientifically.'[53] This shows that there did exist a singularity of purpose and motive. After all, as Pyenson himself points out, 'The German authorities did not send their musicians

---

[51] As late as in 1937, the French Academy of Sciences is quoted advocating 'a division of labour' between research pursued in the colonies and those in the metropolitan laboratories, ibid.

[52] Electricity generation in Argentina was totally in the hands of foreign-owned companies and German professors were brought in to create an indigenous corps of skilled electrical engineers, Lewis Pyenson, p. 150, note 38.

[53] Ibid., p. 38.

and artists to Samoa and Buenos Aires'. For them natural knowledge was an adjunct to political power.

Scientific works in the Dutch or German colonial outposts no doubt 'lit the wilderness for metropolitan travellers' but whether they 'illuminated local residents with the light of superior learning' is doubtful. These outposts mostly remained enclaves—superimposed and isolated. The recipient groups (except in India and Argentina) were largely passive or sidelined. The fact that no 'transmutation' took place in such outposts refers rather to the negative aspects of colonial science, be it the Dutch or the German variety.

At the Core

Returning to where I began, I consider colonial science to be inextricably woven into the whole fabric of colonialism. Several colonial scientists felt uncomfortable, yet they had to perform a dual role—to serve the colonial state and to serve science. This state claimed superiority in terms of structure, power, race, etc. Science claimed superiority in terms of knowledge and *inter alia* helped the colonial state dismiss 'other' epistemologies. Both needed each other and became mutually dependent.

Colonial science lacked sovereignty. Its contours were of course drawn on the colonial terrain, but it enjoyed a rather limited autonomy which was further reduced as the colonial grip tightened. Its dependence on the state and the metropolitan institutions kept growing with the passage of time. In fact, the very concept of a 'state scientist' emerged in the colonies.[54] But this does not mean that it was always derivative. There are several instances when original work was done under the most trying circumstances, and was widely recognized. Quite often it was more than data gathering and sometimes their work was (mis)appropriated by the metropolis.[55] Yet in the initial stages, a colonial scientist was, to a large extent, the master of his agenda; and a whole new world of flora, fauna and minerals was open to him. But as the colonial arteries hardened,

[54] Richard Grove, 'Colonial Conservation and Popular Resistance', in J. MacKenzie (ed.), *Imperialism and the Natural World*, Manchester, 1990, p. 22.

[55] A recent study shows how certain original observations and specimens sent by a colonial scientist during the 1830s were 'appropriated' by Richard Owen of the Royal Society. E.D. Newland, 'Dr George Bennett and Sir Richard Owen: A Case Study of the Colonization of Early Australian Science', in R.W. Home and S.G. Kohlstedt (eds), *International Science and National Scientific Identity: Australia between Britain and America*, Dordrecht, 1991, pp. 55–73.

science came under the purview of official knowledge with its official hierarchies, rituals, etc.[56] The state involvement made colonial science more 'utility-oriented'. In Britain itself there had occurred a gradual shift from science-as-avocation to science-as-enterprise. In the wake of the industrial revolution there had developed what Berman calls an entrepreneurial ideology of science.[57] Bacon and Bentham both coalesce on the point that the men who mattered began to look for utility or result-oriented science.

By focussing on the state and its economic imperatives, the intention is not to portray colonial science as a result or consequence of European political or market forces. Economic considerations were definitely an extremely important determinant, but they did not operate in neutral conditions. These, as well as social and cultural factors, cannot be studied in isolation. The social historians of science do take this view. But they seem to have a special fascination for the centre-periphery explanations. In fact all the models discussed earlier have this relationship at their centre. Chambers describes the centre as 'an infrastructure that links knowledge to power through the control of the processes of knowledge construction and knowledge communication'.[58] Hancock thinks of it not as 'an inert slab of geography' but as mobile, flexible and dispersed. 'Following the moving metropolis is a (serious) game that one might play in other societies than their own.'[59] What did it mean to the 'other' players?

This takes us into the realm of cultural encounters; its nuances and its psyche are more important than the political strategies of hegemonization and counter-hegemonization. Robinson projects imperialism more as 'a function of Afro-Asian politics than of European politics and economics'.[60] Was colonial science a function of its victims' collaboration or non-collaboration? Or did new groups from the periphery cause ruptures in the metropolitan tradition? In spite of (not because of) the all-pervasive and powerful influence of the imperium, one may argue, the generation,

[56] J.J. Saldana, p. 51, note 34.

[57] M. Berman, *Social Change and Scientific Organization: The Royal Institution, 1799–1844*, Ithaca,1978, p. 93.

[58] D.W. Chambers, 'Does Distance Tyrannize Science?'; in R.W. Home and S.G. Kohlstedt (eds), p. 32, note 55.

[59] Keith Hancock, 'The Moving Metropolis', in A.R. Lewis and T.F. McGann (eds), *The New World Looks at its History*, Texas, 1963, pp. 140–1.

[60] R. Robinson, 'Non-European Foundations of European Imperialism: Sketch for a Theory of Collaboration', in W.R. Louis (ed.), *Imperialism*, New York, 1976, p. 129.

transmission and reception of scientific ideas gradually acquired an autonomy and momentum of their own. Everyone in the colony did not conceive of themselves as victims. Their views varied in accordance with the time and space they occupied. It would be wrong to argue that in a colony, ideas were not generated but were imported or received. Even if they are imported, ideas are not like 'finished products' and cannot be 'consumed' neat by the recipients.[61] They undergo certain amounts of refinement and sometimes distortion, and at times they change, for good or worse, even beyond recognition. The crucial questions here are: how does this happen, and what impact does it leave on colonial power-relations or on 'internal' power-equations.

Though the very nature of the colonial milieu ensured that no syncretic acculturation would take place, yet efforts were made, and the differing traditions did move together to some extent or the other. It also ensured that rejection of colonial rule would not automatically mean rejection of the culture it left behind.[62] The labours and the agonies of the cultural and scientific mediators fall within the ambit of colonial science, for it is they who finally sound its death-knell, and it is they who usher in an era of 'non-dependent' science.[63] It is perhaps patriotic to talk about 'struggle' and 'independent' science. But struggle against whom? The imperium, of course. Can they be studied separately? Are they not inseparable? The imperial and subaltern materials are not like grain and chaff to be winnowed. A comprehensive trajectory should include, highlight and analyse both.

To sum up, colonial science did help to develop what a recent work calls the 'core culture' of imperialism.[64] The colonizers explained this 'core' as an essential stage of 'natural history' and a product of their rationality, while those at the receiving end often found it 'insular, racist and arrogant', and so tried to read or reinvent rationality in their traditions. Miles away from the master-slave dialectics, the post-colonial debates discuss the subsumed contradictions, disjunctions and ambivalences.[65]

[61] Kapil Raj, 'Defusing Diffusionism', paper presented at Unesco Colloquium on *Science and Empires*, Paris, April 1990.
[62] Gauri Vishwanathan, 'Raymond Williams and British Colonialism', *The Yale Jr. of Criticism*, IV, II, 1991, pp. 47–66.
[63] R.W. Home and S.G. Kohlstedt (eds), pp. 2–3, note 55.
[64] Suhash Chakravarty, *The Raj Syndrome: A Study in Imperial Perceptions*, New Delhi, 1991, pp. 8, 44, 90.
[65] Benita Parry, 'Problems in Current Thories of Colonial Discourse', *Oxford Literary Review*, 7, 1–2, 1987, pp. 27–58.

Models, though currently in fashion, are of limited value for such a complex scenario.

Locating Pre-colonial Science: Indian Example

Studies on the state of science and technology, in certain culture-areas prior to European expansion, are vital not only for the reconstruction of pre-colonial history but also for a proper evaluation of the role played by techno-scientific developments in the process of colonization itself. No study of what is usually called 'colonial science' would be complete without reference to pre-colonial science and technology with all its 'innocence', 'guilt', limitations, appropriateness and, perhaps, usefulness.

It is not that scholars have not paid attention to this problem. But several grey areas remain and at the present level of research it is difficult to say anything with precision and authority. Three major shades of opinion, however, can be identified. To the first category belong the majority of contemporary European travellers and several subsequent British officials and scholars who found everything in India 'black and bleak'. In sharp contrast, another set of opinion is quite enthusiastic about India's scientific credentials and potentialities in the pre-colonial period. A third set of opinion treads cautiously and offers guarded comments. These opinions can be studied only in the background of certain questions—questions relating to science and technology as cultural artifacts; the nature and process of change that they undergo in different areas at different points of time; the reasons behind an innovation, a complex cultural diffusion or a simple geographical relocation; political and economic compulsions; the place and working of the 'transmitter' or interlocutor; the intricacies of the game of adoption, rejection, indifference or gradual assimilation, etc. No single model, formula or explanation can cover all these aspects. Human needs are constantly changing; a particular technology or a set of technologies appropriate to one time and culture may not necessarily be appropriate to another.

But these differences, similarities or points of interaction can certainly be compared and would provide greater insight. But in any such exercise the European experiences loom large (even in the background). I am not suggesting that modern science and technology (S&T) are to be indentified totally with the Western productive and cultural systems. Partially, yes. One can also think in terms of an Indian S&T, Chinese or Arabic S&T, each on its own terms;[66] or, still better, one can visualize a mainstream of

---

[66] Claude Alvares, *Homo Faber: Technology and Culture in India, China and the West, 1500 to the Present Day*, Delhi, 1979, p. 36.

techno-scientific activity to which different countries in different periods contributed in various ways. The three inventions that Bacon identified as the source of great changes in Rennaisance Europe—printing, gunpowder and magnetic compass—were products of Chinese, not European civilization. They revolutionized literature, warfare and navigation in Europe (not in China). Why were these three so readily adopted by the Westerners? Because, Basalla explains, Western culture was not monolithic, and the Europeans were eclectic, open to new ideas, influences and things.[67] Is this not true also for South Asia? Basalla further elaborates: technological changes are selected by certain active and productive individuals who need not represent all segments of society, nor need they necessarily be concerned with public welfare.[68] Had late medieval India no such agents?

Several leading scientists of eighteenth-century Europe were instrument-makers by profession. Science and technology became mutually dependent.[69] It was never a one-way traffic. Another significant feature was the growing economic pressure on the capacity to invent. There also came into existence scientific societies and scientific journals. All these suggest 'a scientific attitude very different from that associated with crafts or technics which change only slowly and which revere established procedures and established knowledge'.[70]

The accounts of contemporary travellers show that these symptoms of change had no echo in India. From them came the stories of the 'oriental mind' and Indian resistance to innovation and change, and these were to become the obsession of European scholarship for generations to come. However, these accounts do throw some light on the level of science and technology in pre-colonial India.[71] Astronomy, medicine, and the Indian textile and steel-making processes impressed travellers the most. Their accounts mostly begin with feelings of surprise and admiration and end on a suspicious and arrogant note. For example, Indian astronomy was

[67] G. Basalla, *The Evolution of Technology*, Cambridge, 1988, p. 176.

[68] Ibid., p. 204.

[69] The invention of the achromatic lens, for example, owed its origin to Newton's work on the compound nature of light. Similarly Fraunhofer, while working on glass technology, discovered the dark lines in the solar spectrum.

[70] D.S.L. Cardwell, 'Science and Technology in Eighteenth Century', *History of Science*, I, 1962, p. 37.

[71] Satpal Sangwan has referred to hundreds of pamphlets, travelogues and books published mostly during the eighteenth–nineteenth centuries in his *Science, Technology and Colonisation: An Indian Experience 1757–1857*, Delhi, 1991, pp. 167–87. Michael Adas uses several travelogues as evidence of 'First Encounters' in his *Machines as the Measure of Men*, Ithaca, 1989, p. 21.

lauded as 'a proof still more conspicuous of their extraordinary progress in science', and its accuracy was found to be at par with that in modern Europe.[72] The observatories were seen as 'gigantic relics of the zeal in the pursuit of science manifested in former days'. Then comes the indictment: 'It is carried on by mechanical rules, without any idea of the principles upon which they depend . . . the instruments employed are rude in the extreme.'[73] The standard criticism of Indian astronomy was:

I. 'It gives no theory, nor even any description of the celestial phenomena, but satisfied itself with the calculation of certain changes in the heavens, particularly of the eclipses of the sun and moon.'[74]

II. The Indian astronomers were satisfied with their traditional systems; they did not bother to improve upon, nor did they welcome any criticism of, the Puranic and Siddhantic systems.

Most of the travellers recorded that Indians had made remarkable progress in mathematics and astronomy in ancient times which gradually fell from grace, particularly after the establishment of the Muslim rule. Later this was uncritically accepted by several British and Indian historians. Thus grew the notion that science flourished only in ancient India and not in medieval India.[75] However, scholars like Rahman, Dharampal, Alvares and Ansari have raised strong objections to this notion. Rahman has compiled a 'Bibliograpy of Source Materials in Sanskrit, Arabic and Persian'. He argues that throughout the medieval period, scientific and technological activity, as is evident from the number of manuscripts, was both continuous and vigorous. Secondly, though the major contribu-

[72] W. Robertson, *An Historical Disquisition Cconcerning the Knowledge which the Ancients had of India*, London, 1791, pp. 302, 308, quoted in Sangwan, ibid., pp. 2–3.

[73] Murray Hugh, *Historical Account of Discoveries and Travels in Asia from the Ealiest Ages to the Present Time*, Edinburgh, 1820, pp. 310–11, quoted in ibid.

[74] John Playfair, 'Remarks on the Astronomy of the Brahmins', *Transactions of the Royal Society of Edinburgh*, II, 1790, pp. 135–92, reproduced in Dharmpal, *Indian Science and Technology in the Eigthteenth Century*, Delhi, 1971. pp. 12–13.

[75] For example, George Sarton wrote: 'Hindu culture was stifled, if not stamped out, in many places by Muslim conquerors.' George Sarton, *Introduction to History of Science*, vol. II, London, 1947, p. 107.

In a prestigious publication on the history of science in India, Subbarayappa wrote, 'India had her period of glory in the Classical Age and made remarkable progress . . . even right up to the twelfth century AD. Thereafter the creative endeavour showed signs of decay due largely to the traditional compulsion and political vicissitudes.' Bose, Sen and Subbarayappa (eds), *A Concise History of Science in India*, New Delhi, 1982, pp. 484–6.

tions lie in the field of astronomy, mathematics and medicine, they cover a wide range of scientific and technological subjects. Thirdly, as compared to contributions of a general nature, there are a large number of special treatises. The number of manuscripts listed in this Bibliography is quite large. In the sphere of astronomy alone, in Persian, 411 manuscripts are said to have been compiled from the tenth to the nineteenth centuries, of which 32 belong to the eighteenth century; in Arabic; out of 346 manuscripts, 22 were written in the eighteenth century. Sanskrit has the maximum number of manuscripts—2136—out of which 190 belong to the seventeenth century and 37 to the eighteenth century. As for the nature of these manuscripts, out of the 32 Persian manuscripts, 11 are of a general nature, 1 is of a special nature, 2 are commentaries, 2 are translations and 6 are alamanacs; out of the 22 Arabic manuscripts, 8 are of a special nature, 6 are commentaries and 8 are alamanacs; and out of the 37 Sanskrit manuscripts, 3 are of a general nature, 15 are special, 8 are commentaries, 2 are translations, 4 are anthologies and 5 are almanacs.[76] The list is impressive, but to determine whether they contain the seeds of modern science, or at least reflect the advance achieved in science, would require further study.

There is one astronomical work of the eighteenth century, that of Sawai Jai Singh, which has attracted a lot of attention. Jai Singh wanted to explore why the time of different celestial phenomena, especially the eclipses of the sun and moon, differs according to Siddhantic and Greaco-Arabic astronomy and does not very often tally with that of actual occurrence. He consulted a large number of almanacs, traditional scholars and European travellers. He was not satisfied with calculations done through astrolabs (brass instruments) and thought that stone observatories, bigger in size and fixed in one place, would give more accurate results. So he constructed observatories in Delhi, Jaipur, Mathura and Varanasi. He was also presented with a telescope by a French Jesuit. Jai Singh is credited with evolving a systematic scientific method. He sent his scholars to Central and West Asia and invited European scholars to his court. The results of his efforts were later compiled in the *Zij-i-Muhammad Shahi* (1728) which is considered to be the most important astronomical work of medieval India and several commentaries were later written on it. Jai Singh mentions in his Zij all the three traditions (Islamic, Brahmanical and European). Yet his obsession with masonary instruments (which was not the European tradition), accuracy, the calender, etc. is usually taken

[76] A. Rahman, (ed.), *Science, Technology in Medieval India: A Bibliography of Source Materials in Sanskrit, Arabic and Persian*, New Delhi, 1982, pp. xi–xvi.

to mean that Jai Singh's outlook was medieval and limited to the Ptolemaic concept of the universe. However, Rahman finds this contrary to what Jai Singh's outlook really was. 'Jai Singh does not acknowlege his debt to Copernicus and Kepler explicitly but seems to have accepted their theories.' Rahman arrived at this conclusion from Sobirov's translation of the Zij from Persian to Russian.[77] Sobirov quotes Jai Singh as saying:

The predecessors of astronomy, namely Hipparchus and Ptolemy and others, gave the principles of the movements of planets and description of the orbits of their movement but their description is far from the truth. The system of the world is in reality the movement of the planets occurring contrary to the descriptions given by the above-mentioned scientists. The orbits of the movement of the planets has a different form. Above all, it should be mentioned that the orbits have elliptical shape in one of the centres of which lies the sun.

Jai Singh's acceptance of the Copernican model, contrary to general belief and strong tradition, indicates his open-mindedness and true scientific spirit.[78] Sobirov also credits Jai Singh with the full use of the telescope. The Zij says:

As our artisans have constructed the telescope so excellent that with its aid we can see bright and luminous stars even about midday in the middle of the sky, by employing such powerful telescope, the new moon can be seen even before the time the astronomers have determined for its rays to begin emanating. And also after it has entered the prescribed limit of its invisibility, it still remains visible (through the telescope).[79]

Another important deviation that Jai Singh made from the traditional Greco-Arabic astronomy related to the so-called 'fixed stars'. The Ptolemic astronomy puts the stars into two categories, the wandering stars and the fixed stars, the latter conceived as immovable. In the seventh section of his Zij, Jai Singh refutes this theory:

Those stars that are termed Fixed Stars in the terminology of astronomers are not stationary in reality. Nor do they move with one rate of velocity, but with different velocities.[80]

[77] G. Sobirov, 'Samarkand Scientific School of Ullugh Beg', Dushanbe, 1975, quoted in A. Rahman, 'Maharaja Sawai Jai Singh II: Purposes and Contributions', paper presented at the *Seminar on Sawai Jai Singh*, New Delhi, Oct. 1989.
[78] Ibid.
[79] S.A.K. Ghori, 'The Impact of Modern European Astronomy on Raja Jai Singh', *Indian Jr of History of Science*, 15(1), May 1980, p. 55.
[80] Ibid., p. 56.

Rahman surmises that Jai Singh's aim was to bring about, through the application of science, a renaissance in India. Another scholar claims that the path to the final 'reawakening' (i.e. the scientific revolution) was blocked by the threat of colonization.[81]

These enthusiastic estimates are, however, not shared by several scholars. It is argued that Jai Singh was no theoretician and he adhered to the old Ptolemian concept. Moreover, it is said that his *Zij* is verbatim lifted from *Zij-i-Ulugh Beg*, composed three centuries earlier. There is no doubt that he at least thought that the brass astrolabs were not accurate and he was brilliant enough to devise new ways of measurement. But his obsession with finding the exact moment and with accuracy calls for some explanation. Was it motivated by astrological concern? Obsession with the exactness of time (e.g. the time of *yagna* or marriage) has been an important feature of Indian social life and Jai Singh was naturally part of it. Moreover, he was an intensely religious and ritual-minded person, and had performed difficult Vedic *yagnas* (sacrifices) such as the *Vajapeya* and *Asvamedha*. So his Zij was intended only as a means to compute accurate time and not as a treatise to show off his acumen or document his new findings.[82] The telescope had come to India even before Jai Singh's birth.[83] He was definitely aware of it but it is doubtful that he made full use of it. In the absence of a chronometer, one could see distant objects through a telescope but could not measure them, while Jai Singh's real interest lay in measurement. So, despite his enthusiasm and efforts, Jai Singh may appear as a sort of historical anachronism who intellectually belonged to the medieval astronomical tradition but chronologically lived in the modern age of astronomy.[84]

This, however, is not to undermine Jai Singh's efforts. With little more foresight and courage he could have transcended his cultural limits. It was not that the Ptolemaic system was always blindly followed. Earlier, during Shah Jahan's time, Mulla Mahmud Jaunpuri had ventured to raise doubts

---

[81] S.M.R. Ansari, 'Zij-i-Muhammadshahi: The Astronomical Tables of Jai Singh', paper presented at the *Seminar on Sawai Jai Singh*, New Delhi, Oct. 1989.

[82] K.V. Sarma, 'Astronomical Investigations of Sawai Jai Singh: Their Objective', paper presented at the *Seminar on Sawai Jai Singh*, New Delhi, Oct. 1989.

[83] One of the early followers of Kepler, Jeremiah Shakerley had emigrated to Surat and had observed with a telescope the transit of Mercury in 1651 and a comet in 1652. In 1689, Ft. Jean Richaud, a French Jesuit, used a telescope at Pondicherry.

[84] R.K. Kochhar, 'The Growth of Modern Astronomy in India', *Vistas in Astronomy*, vol. 34, 1991, p. 72.

about the system in his *Shams-e-Bazegha*.[85] Later, in a commentary on Jai Singh's Zij, Mirza Khairullah Khan argued:

Whenever we calculate the different positions of the Sun and other planets in accordance with equations of the circle, they do not conform with the actually observed ones. On the contrary, when the equations are derived, taking the orbits elliptical and calculate the positions, they generally conform with observations. Hence the orbits must be elliptical.

It is significant that this remark is based on observation only. There is no evidence to suggest that Khairullah Khan had any knowledge of Kepler.[86] In the second half of the eighteenth century, a few tracts appeared which were either translations of a European work or were composed under European supervision. For example, Abul-Khair Ghiyasuddin made a Persian translation of William Hunter's book on the Copernican system, and this was done under the supervision of Hunter himself.[87]

As in astronomy, Indian technology evoked a mixed response. Several foreign observers were awe struck by the quality of Indian steel (called wootz) as well as Indian textiles. On the basis of tributes paid to Indian textiles, metal-works, construction and ship-building techniques by several early European travellers, Sangwan argues that:

by the beginning of the eighteenth-century India had attained the distinct status of a technologically advanced country. Indian workmen excelled in their profession. They had also adopted some new techniques and technologies which appeared more useful. There was thus a continuous process of technological change in pre-industrial (pre-colonial?) India.[88]

But Sangwan refrains from giving specific examples of this 'process'. Was the 'process' 'continuous' and 'adaptive'? If this was so, Indian production methods would not have been so severely criticized by the later travellers whose accounts Sangwan cites so extensively. Wallace, Munro, Buchanan and Orlich, for example, condemned Indian agricultural technology as being of a 'very low standard, stagnated since centuries, and dominated by ill-treated and ill-instructed village mechanics'. Textile technology was considered by Martin, Fraser, Moorcraft and Dubois as 'most primitive, crude, clumsy, unscientific and stagnated

[85] W.H. Abdi, 'Interaction of West Asian and Central Asian Science with Indian Tradition', in A. Rahman (ed.), *Science and Technology in Indian Culture: A Historical Perspective*, New Delhi, 1984, p. 161.
[86] Ibid.
[87] A. Rahman, (ed.), 'Science and Technology in Medieval India', p. 285, note 76.
[88] Satpal Sangwan, p. 6, note 71.

heritage of the oriental nation'.[89] This condemnation may as well be part of the process of hegemonization. Without criticizing Indian S&T they could not have justified colonization. For European observers there may also have been some genuine difficulties in understanding an abstruse treatise or in appreciating a simple technical device which would appear 'appropriate' only when viewed against the existing socio-economic context.

The necessity to place certain scientific developments or certain tools in a comprehensive historical context has led some scholars to justify and defend (or even glorify) the indefensible. Dharampal, for example, argues:

Smallness and simplicity of construction, as of the iron and steel furnaces or of the drill-ploughs, was in fact due to social and political maturity as well as arising from understanding of the principles and processes involved. Instead of being crude, the processes and tools of eighteenth-century India appear to have developed from a great deal of sophistication in theory and an acute sense of the aesthetic. . . . In the context of the values and aptitudes of Indian culture and social norms (and the consequent political structure and institutions) the sciences and technologies of India, instead of being in a state of atrophy, were in actuality usefully performing the tasks desired by Indian society.[90]

There is no doubt that agricultural tools, irrigation methods and certain crafts were 'appropriate' and in tune with the existing capabilities and and requirements, but the 'sophistication in theory' to which Dharampal alludes, is markedly absent. In Walker's description of Indian agricultural practices (written in 1820), another scholar notices 'ecological farming'.[91] The variety of agricultural implements, the drill plough, the system of rice transplantation and rotation of different crops in the same field, do speak of the rich experience of Indian peasants. Their agricultural technology, far from being static, was constantly developing over time. A seventeenth-century text, *Dar Fann-i-Falahat*, describes the various methods of grafting, preparation of soil, harvesting techniques, water and manure requirements, etc. Interestingly it refers to male and female plants (this was two centuries before J.C. Bose proved that plants have life).[92] Later, several tracts were written in Persian on various useful plants, for example, *Nakhl-bandiya* by Ahmad Ali in 1790 and *Nuskha-i-Kukh-bad* by

[89] Ibid., pp. 7, 18.
[90] Dharampal, pp. LXIII, LXV, note 74.
[91] Claude Alvares, p. 51, note 66.
[92] Harbans Mukhia, 'Agricultural Technology in Medieval North India', paper presented in a *Seminar on Traditional Technologies in Indian Agriculture*, New Delhi, 6–8 March 1989.

Amanullah Husain.[93] Irrigation methods varied according to local conditions. In the arid zones of the Thar desert the *khadin* system was developed to store runoff water from the high-catchment area. Northwest Maharastra followed the *phad* system, while in the deep South, the *khudimarammat* system was followed. Both systems involved community management of water resources. These systems were the outcome of the experiences and collective wisdom of the 'practical' peasants. Yet they do not stand comparison with eighteenth-century Japan where row-cultivation was introduced, the number of plant varieties increased along with deliberate seed selection, while irrigation by treadmills and Dutch pumps was improved and extended.[94]

Like Dharampal, Alvares writes enthusiastically about Indian agriculture and the textile industry during the seventeenth and eighteenth centuries.[95] The Dutch and the French tried to imitate Indian dyeing techniques without much success. Letters from the Jesuit priest, Coeurdoux (1742–47), and the Roques and Beaulieu manuscripts (1734) show the importance of Indian influence on the European textile industry.[96] However, the fact remains that this technology transfer was carried out by the Europeans themselves. Similarly, Indian steel (wootz) was greatly valued. The iron-smelting centre at Konamsundram produced wootz which the traders from Isphahan purchased for making the famous Damascus blades. The superiority of Indian iron was due to the fact that it was prepared in small quantities at low heat, in an apparatus wherein the reduction of iron oxides remained incomplete and that of other oxides practically nil.[97] But its production was fairly small and localized, while Europe was fast moving towards mass production. The Indian smiths could not obtain high temperatures and opt for large furnaces because they did not know how to generate power except through the use of draught animals or charcoal. Except in one or two places, water power remained unthought of and untapped. This meant a high cost of production, and so naturally Indian peasants kept the use of iron to the

[93] Ivanow Wladimir, *Concise Descriptive Catalogue of the Persian Manuscripts in the Curzon Collection of the Asiatic Society of Bengal*, Calcutta, 1926.
[94] E.L. Jones, *Agriculture and the Industrial Revolution*, Oxford, 1974, p. 136.
[95] C. Alvares, pp. 46–64, note 66.
[96] Ibid., pp. 60–1.
[97] H.C. Bhardwaj, 'Development of Iron and Steel Technology in India During the 18th and 19th Centuries', *Indian Jr. of History of Science*, 17, 2, 1982, pp. 223–33.

bare minimum. Similarly, mining itself was done on a small scale. It involved barely more than scratching the surface of the earth, the tools used being iron crow-bars and spades. Mining below the water level and haulage were simply out of the question. Curiously enough, though gunpowder was used for armament, it was never used for mining purposes.[98] But there did exist a flourishing metallurgical industry, run almost like a cottage industry. Slags of iron and steel, and metals such as copper, zinc, lead and, to a smaller extent, silver and cobalt, in parts of Rajasthan, Bihar and Deccan bear testimony to this.[99] Zinc production in India precedes that in Europe.

In the realm of armaments, an area which concerned the State the most, one finds Indian interests most receptive and in some respects ahead of its western counterpart. Alvi and Rahman have provided an interesting sketch of a late sixteenth-century Indian technologist, Fathullah Shirazi, who crafted a multi-barrelled cannon (a precursor of the machine-gun), a machine for cleaning gun-barrels called the *yarghu* and a wagon-mill.[100] Shirazi was not an inventor but a wonderful adaptor. He was the first to use the gear-wheel (which was confined only to devices for raising water) for other purposes. But all these remained fancy devices and could not be used in serious industrial operations in the Mughal karkhana. In the late eighteenth century one finds a similar story in the use of rockets by the armies of Hyder Ali and Tipu Sultan. The Mysore rockets were much more advanced than what the British had seen or known, chiefly because of the use of iron tubes for holding the propellant which enabled higher bursting pressures in the combustion chamber and hence higher thrust and longer range for the missile.[101] The rockets consisted of a tube (about 60 mm in diameter and 200 mm in length) fastened to a 3 m bamboo pole, with a range of 1–2 km. In the battle of Pellilur (1780) the British were defeated because Colonel Baillie's ammunition tumbrils were blasted by the Mysore rockets. These caused great fear and confusion, but because of their lack of accuracy and sometimes direction, they could not

[98] A.K. Ghose, 'History of Mining in India 1400–1800 and Technology Status', *Indian Jr. of History of Science*, 15, 1 May 1980, pp. 25–9.

[99] R.D. Singh, 'Development of Mining Technology during Nineteenth-Century India', *Indian Jr. of History of Science*, 17, 2, 1982, p. 206.

[100] M.A. Alvi and A. Rahman, *Fathullah Shirazi: A Sixteenth-Century Indian Scientist*, New Delhi, 1968, pp. 4–13.

[101] Roddam Narasimha, *Rockets in Mysore and Britain, 1750–1850*, Bangalore, 1985 (memeographed), pp. 1–45.

win battles for Tipu Sultan. In the last Anglo-Mysore war, Wellesley (later the hero of Waterloo) himself was shocked by the 'rocket fire'. Several rocket cases were sent to Britain for analysis and these led to a great interest in rocketry in Europe. William Congreve took up this challenge in Britain. Scientific principles were applied, and appropriate designs were made, tested and evaluated. This whole process was completely alien to the Indians of the eighteenth century.[102] These examples serve to reveal why the Indians lost out to the colonizers. Alvi and Rahman do not consider the devices used by Indians 'entirely devoid of creative potentialities'. 'On the contrary, they are replete with suggestive ideas which if they had been taken up further could well have been the basis of a distinct technological tradition.'[103]

The catch, however, lies in the 'ifs and buts' of history, and this brings us to the third set of opinions which advocates neither an unqualified denunciation nor a naive (perhaps revivalist) appreciation of the pre-colonial S&T. Writing in the 1930s, at the peak of the national movement, B.K. Sarkar compared India and the West in terms of the following equations:[104]

   I. India in exact science (BC 600–AD 1300)
       = Europe in exact science (BC 600–AD 1300)

  II. Renaissance in India (1300–1600)
       = Renaissance in Europe (1300–1600)

 III. India in exact science (1600–1750)
       = Europe in exact science (1300–1600)

Thus it was during the seventeenth and eighteenth centuries, the post-Renaissance epoch (that of Descartes and Newton), that Europe began to outdistance India in the natural sciences. Dharampal also concedes that 'it is possible that the various sciences and technologies were on a decline in India around 1750 and perhaps had been on a similar course for several centuries previously'.[105] Irfan Habib does not accept any description of pre-colonial technology as primitive but calls for 'a wider study of the social constraints that prevented either an endogenous development of industrial technology comparable to that of modern Europe or, at least,

---

[102] Ibid., p. 19.
[103] M.A. Alvi and A. Rahman, p. 28, note 100.
[104] B.K. Sarkar, *India in Exact Science: Old and New*, Calcutta, 1947, p. 7.
[105] Dharampal, p. XXXII, note 74.

a rapid absorption of the European technology itself'.[106,107] He argues that many mechanical principles frequently employed in modern machines were in use in Mughal India, but adds that the range of their application was rather limited. Pre-colonial S&T was definitely not primitive. A better description perhaps would be 'proto-science and technology', clearly distinguishing it from the post-seventeenth-century modern scientific tradition based on the experimental method.[108]

Analyzing the Indian response to European technology during the sixteenth–seventeenth centuries, A.J. Qaiser does not hold India guilty of xenophobia, 'but as long as there was an alternative or appropriate indigenous technology which could serve the needs of Indians to a reasonable degree, the European counterpart was understandably passed over'.[109] There were several areas in which interaction between the East and the West resulted in acceptance and improvement. These were ship-building, armaments, metallurgy, cloth-printing and architecture. Several other important developments, such as mechanical clocks, the printing press, telescopes, coal, etc. remained mere curios. Since these were not found culturally compatible, they did not attract the attention of the Indian nobility. Only in war-weaponry was the need felt and, therefore, guns and gunners were imported. Indian nobility mostly patronized those devices that had an aristocratic flavour (e.g. *Itr-i-Jahangiri*, a distillation apparatus, one of which called *NoorJehan Ka Bhapka*, is still in vogue). Neither the nobility nor the merchants would invest in the upgradation of technology. Tools remained the sole concern of the poor artisans, and this poverty of tools was sought to be compensated by individual skills—skills which are manifest in Dacca muslin, brilliant dyes and wootz. Even this craft production, though superbly executed, did not stand on its own. It was heavily dependent on the agrarian system which, once under strain (as in the early eighteenth century), triggered adverse chain-reactions.

This at best is an economic explanation. Can there be a cultural or a

[106] Irfan Habib, 'Technological Changes and Society: 13th and 14th Centuries', Presidential Address, Medieval India Section, *Proc. of the Indian History Congress*, 1969, p. 139.

[107] Irfan Habib, 'Technology and Barriers to Social Change in Mughal India', *Indian Historical Review*, vols. 1–2, 1979, p. 152.

[108] R.A.L.H. Gunawardana, 'Proto-Science, and Technology in Pre-colonial South Asia', S.A.I. Tirmizi (ed.), *Cultural Interaction in South Asia in Historical Perspective*, New Delhi, 1993, pp. 178–208.

[109] A.J. Qaiser, *The Indian Response to European Technology and Culture (1498–1707)*, Delhi, 1982, p. 139.

social explanation? The latter brings to mind the adverse role of the caste system which Qaisar dismisses as 'the Cheshire Cat of the Weberians' and 'a fragment of imagination, absolutely erroneous and unwarranted'.[110] He does not come across a single example of the non-acceptance of a technique because of caste considerations. This is true, but if science and technology are treated as part of social activity, the role of the caste system cannot be wished away. It is true that castes changed and alongside professions changed, but it was this elasticity which helped perpetuate the hereditary division of labour and the segregation of skills. P.C. Ray is the first historian of science who saw in the caste structure 'something that made science a prey to creeping paralysis'.[111] Caste led to the ruinous separation of theory from practice—of mental work from manual work. He wrote:

The intellectual portion of the community being thus withdrawn from active participation in the arts, the how and why of phenomena—the co-ordination of cause and effect—were lost sight of—the spirit of enquiry gradually died out. Her (India's) soil was rendered morally unfit for the birth of a Boyle, a Descartes, or a Newton. . . .[112]

In eighteenth-century India, this was compounded by an enormous intellectual (cultural?) failure on the part of the ruling class. Jai Singh had attracted several scholars to his court but he never thought of establishing an institution which would continue and improve on his work. Science education was practically nil and lost in the maze of *nyaya* debates in Nadia and Mithila. More than three hundred years before P.C. Ray, Abul Fazl had mourned

the blowing of the heavy wind of *taqlid* (tradition) and the dimming of the lamp of wisdom . . . the door of 'how' and 'why' has been closed; and questioning and enquiry have been deemed fruitless and tantamount to paganism.[113]

Had this illustrious historian lived in the mid-eighteenth century, he would have perhaps been more harsh. Oriental learning had no 'state of the art' knowledge; yet, as an orientalist, fighting a losing battle, argued:

---

[110] A.J. Qaisar, Presidential Address, Medieval India Section, *Proc. of the Indian History Congress*, Goa, 1987.

[111] Debiprasad Chattopadhyay, *History of Science and Technology in Ancient India: The Beginnings*, Calcutta, 1986, p. 10.

[112] P.C. Ray, *History of Hindu Chemistry*, II, Calcutta, 1909, p. 195.

[113] Quoted in Irfan Habib, 'Capacity of Technological Change in Mughal India', in A. Roy, and S.K. Bagchi, *Technology in Ancient and Medieval India*, Delhi, 1986, pp. 12–13.

If we destroy it we shall degrade both ourselves and the people we undertake to improve. A history of the successive systems of science and philosophy though it may not teach the true nature of things will yet afford much valuable information of another kind. It will teach what mankind has thought and how it has reasoned about these things and the successive steps by which they have arrived at Truth. It is in short the *history of human opinions and this is at least as important as that of human actions.*[114]

[114] John Tytler to T.B. Macaulay, 26 Jan. 1835, quoted in G. Viswanathan, *Masks of Conquests: Literary Study and British Rule in India*, New York, 1989, p. 104 (emphasis added).

CHAPTER 2

Exploration and Encounter:
The Early Phase

Every accumulation of knowledge, and especially such as is obtained by social communication with people over whom we exercise a dominion founded on the right of conquest, is useful to the state: it is the gain of humanity.

Warren Hastings in 1784

It is no longer the warrior, but the man of science, who has to sigh for other worlds—not to conquer, but as a wider field for the extension of his rapidly-increasing power.

The United Service Journal in 1834

A knowledge of the local terrain, both geographical and intellectual, was as important as possessing a superior musketery and consolidation of the empire depended on this knowledge. From the colonizers' point of view, the late eighteenth century was an exciting time; a new empire was in the making, and the colonizers were interested in gathering the maximum possible information about India, its people and resources. Several tracts and travelogues appeared, the important ones being those of W. Robertson, John Capper, Hugh Murray, G.R. Wallace, J.M. Honigberger, F. Buchanan, B. Heyne, M. Martin, R. Heber, and M. Jacquemont.[1] The information was often jumbled, but the writers seldom lost sight of what they considered 'useful'. They faithfully reported what was best in India's natural resources and technological traditions, and what could be the most advantageous to their employers. The East India Company (EIC) was quick to realize that the whole physical basis of its governance depended on a geographical, geological and botanical knowledge of the area it conquered. While proposing to survey Mysore in January 1800, Mackenzie clarified that his object was

[1] S. Sangwan, 'European Impressions and Interpretations of Indian Science and Technology', *Social Science Probings*, II, 3 Sept. 1985, pp. 353–77.

to obtain as soon as possible a clearer and better defined knowledge of the Extent, Properties, Strength and Resources of a Country . . . to elucidate many objects of Natural History, connected with commercial views and therefore interesting to the Company, exclusive of the advantage in the improvement of scientific knowledge.[2]

The reference to 'improvement of scientific knowledge' is interesting and elucidates the dual mandate which a colonial explorer-cum-scientist had. Here exploration emerges more as a state enterprise. Unlike Victorian Africa where the explorers arrived much before the flag and trade, in India these three moved together, hand in hand. Notwithstanding some 'internal' conflicts and adjustments, their interests largely converged.

This, however, does not mean that 'local knowledge' and 'local power' were less significant in a colonial structure. A colonizer had to understand them and later use them to his advantage. After all, under the Baconian epistemology that he had inherited, all knowledge could be translated into technical control. The same Western sense of cultural superiority also indicated the impossibility of all men attaining knowledge and power in equal measure. So the colonizer had the added advantage of not only claiming a monopoly over knowledge but at the same time condemning the 'other' epistemologies as worthless or antiquated.[3] He negotiated from a vantage point and with a clarity, determination and aggression which his 'native' opponent could hardly match. Yet the first rumblings of 'encounter' are to be seen in this phase.

This chapter is basically introductory in nature and outlines the beginning of colonial science in India and its gradual maturation through the efforts of surveys, scientific societies, educational bodies, and indigenous and foreign interlocutors. When the Company rule ended, colonial science had entered a distinctive phase and reached its peak by the turn of the century. However, scientific enquiry began on a small scale, with halting yet firm steps. Every Asia-bound ship had on board a 'surgeon-naturalist' or a few medical men, mostly Scots and Danes, who formed a substantial body of the early botanists and zoologists. A second group of 'scientists' was drawn from the military, especially from among the army engineers, and from their ranks came the early geologists, meteorologists and astronomers. The surveyors were of course the forerunners of scientific exploration.

[2] Greville Papers, IOL, MSS. Eur. E.309/2.
[3] C. Alvares, 'Science, Colonialism and Violence. A Luddite View', in A. Nandy (ed.), *Science, Hegemony and Violence*, Delhi, 1989, pp. 90–1.

Survey Operations

Rennell stands foremost among the early surveyors. Whilst in the Navy from 1756–63, he learnt marine surveying. In 1764 he took to land surveying under the instructions of Vansittart who was then Governor of Fort William. In 1778 he returned to England to organize and publish his findings.[4] His surveys were important for revenue purposes and later figured in several law-suits. Since his surveys were made under the authority of the government and the maps were not made for any explicit purpose, in 1872 a presumption of accuracy was attached to them under Section 83 of the Indian Evidence Act.[5] In the South, Colonel Kelly surveyed the Carnatic region and his charts proved of immense value to General Eyre Coote in military operations.[6] The British could succeed against their numerically superior adversaries largely because they possessed a thorough and scientific knowledge of the country through which they marched. So survey and expansion were to move side by side. The Directors, back home, knew that the act of measurement was an act of control. In a letter dated 9 December 1784, they reiterated their interest in regularly securing all maps, charts, etc., and gave directions for the use of oil paper for tracing maps. There are interesting details in this letter as to how a surveyor should conduct his work.[7]

Towards the close of 1799, Major Lambton drew up a project for a geographical survey from the Coromandel to the Malabar coast based on geodetic principles. Colonel Wellesley, fresh from his successful exploits against Tipu Sultan, quickly grasped the significance of this proposal and probably also used his influence with his brother, Lord Wellesley, the then Governor-General.[8] Thus was created the Great Trigonometrical Survey of India (GTSI).

The idea of a trigonometrical survey was not very old; it was first conceived of by General Watson after the suppression of the Scots up-

---

[4] His early works are: *A Description of the Roads in Bengal and Bihar*, 1778; *The Atlas of Bengal and Behar*, 1779; *Memoir of a Map of Hindustan*, 1788, *Memoir of a Map of Peninsula of India*, 1793. He was interested in politics also and in 1794 wrote a pamphlet entitled *War with France, the Security of Britain*.

[5] F.D. Ascoli, 'The Legal Value of Rennell's Maps', in F.C. Hirst, *A Memoir upon the Map of Bengal*, Calcutta, 1914.

[6] B.A. Saletore (ed.), *Fort William—India House Correspondence, 1783–85*, IX, Delhi, 1959, p. XXXVIII.

[7] Ibid., p. XXXIX.

[8] *Calcutta Review*, IV, 7, 1845, p. 77.

rising in 1745. The Carnatic and Mysore wars brought home the necessity of a trigonometrical survey in India. But there were a few sceptics. The Finance Member of Madras Council, for example, felt that such a survey was utterly unnecessary. On being told that many important places were wrongly marked in all the existing maps, his answer was: 'If I wish to proceed to Seringapattam, I have only to tell the *palakeen* bearers, and they will find their way to it just as well as if it were ever so accurately placed in the maps.'[9] Fortunately this point of view went unheeded. Lambton was adequately patronized and later secured Warren, Everest and Voysey as his assistants.[10] Warren was the first to strike gold in Mysore and gradually several incidental notices on topography, climate and geology came out of these operations. Lambton had high praise for Voysey and felt that their joint investigations were beneficial for both trigonometrical and geological purposes.[11] They provided the bases of topographical, cadastral and fiscal surveys.[12]

No less important were the marine surveys, particularly for a maritime power. The Court always showed a keen interest in the improvement of charts and in the navigation of Indian seas. Rennell and Dalrymple encouraged this subject and in 1770 Ritchie was appointed the first Hydrographical Surveyor to the Company.[13] The Napoleonic wars re-emphasized the importance of hydrography and in 1809 a full-fledged Marine Survey Department was established in Bengal with Captain Wales as the first Surveyor-General.[14] Henceforth a number of hydrographers like Horsburgh, Dominicetti, Ross and Haines minutely surveyed not only the coasts of peninsular India but also the coasts and archipelagoes from Malaya to Madagascar.

These survey operations were not without administrative problems. Mr Laidlow who was deputed for a mineralogical survey of Kumaon in June 1817, was denied allowances in 1818. Although a very able man, he is said to have been treated unfairly.[15] J.A. Hodgson, the then Surveyor-General, was more appreciative of Herbert than of Laidlow and this may

[9] Ibid., p. 80.
[10] In Jan. 1810, Lord Minto himself transmitted to the Asiatic Society, Lambton's 'Account of the Measurement of an Arc on the Meridian', *Asiatic Researches*, XII, 1816, p. 1.
[11] Home, Public, no. 84, 6 July 1821.
[12] J.T. Walker, *Account of the Operations of GTSI*, I, Dehra Dun, 1870, p. xxxv.
[13] C.R. Markham, *A Memoir on the Indian Surveys*, London, 1871, p. 4.
[14] *Calcutta Review*, 67, 133, 1878, p. 576.
[15] C.R. Markham (2nd ed.), 1878, p. 207, note 13.

have caused bitterness between the two.[16] In July 1821, Captain W.S. Webb, another surveyor in Kumaon, tendered his resignation because he was superceded by Herbert.[17] In early 1828, the whole Himalayan survey project was suddenly abandoned as an economy measure, and Herbert was left high and dry.[18]

Botanical Investigations

By far the best use of science for colonial purposes can be found in botanical and geological explorations. In fact the colonies were originally known as 'plantations'. European travellers were struck by the numerous varieties of flora, and in several early European settlements in India, a few devoted themselves to the study of local plants and their medicinal usages. Notable among them are Garcia da Orta in Goa, Heinrich van Rheede in Malabar, J.G. Koenig in Tranquebar and Robert Wight in Madras. But their work was disorganized and mostly of an academic nature. Each worked according to his own way of thinking, without any official support, for, as Griffith wrote later, the Company was not interested in the means but in the results and dried plants counted as means.[19]

The commercial and military importance of botanical investigations came to be realized only after the Company secured a firm grip over Bengal and Madras. In 1778 James Anderson, a surgeon with the Madras Army, obtained a large piece of wasteland near Ft St George, and there he experimented with the cultivation of sugar-cane, coffee, American cotton and also European apples.[20] In Bengal, Robert Kyd conceived of the idea of supplying the Company's navy with teak timber grown near the port where it could be used in ship-building. So in June 1786 he submitted a scheme for the establishment of what he appropriately called a 'Garden of Acclimatization' near Calcutta which was promptly accepted.[21] This scheme included proposals for introducing the cultivation of cotton, tobacco, coffee, tea and other commercial products. Later, in 1791, he produced another report in which he wrote about the use of coconut coir for navigation purposes and about how teak plantations were

[16] Home, Public, no. 52, 5 Oct. 1821.
[17] Home, Public, no. 35, 13 July 1821.
[18] Home, Public, no. 20, 14 Feb. 1828.
[19] Quoted in I.H. Burkill, *Chapters on the History of Botany in India*, Delhi, 1965, p. 14.
[20] Anderson Papers, IOL, Photo Eur. 85.
[21] Home, Public, nos. 13–14, 16 June 1786.

necessary to meet naval requirements, particularly in times of war.[22] Kyd was equally interested in finding the best possible rope-yarn for navigation purposes and successfully acclimatized a Malayan tree, *Gomatoo*, the substance of which produced the toughest and most durable rope.[23] The Company was quick to realize these advantages and henceforth its patronage was seldom found to be wanting. While approving of Kyd's proposals, the Court acknowledged: 'So sensible are we of the vast importance of the objects in view that it is by no means our intention to restrict you, in point of expense, in the pursuit of it.'[24] What had hitherto been a private and unofficial enterprise became a part of government policy which later led to what Sangwan terms 'plant colonialism'.[25]

Roxburgh succeeded Kyd in 1793. Earlier, at Samalcotta near Madras, he had experimented with pepper, cardamum and indigo. He was moved by the poverty of the people there and called for the introduction of plants (like jack-bread fruit) that would furnish sustenance to the poor in times of scarcity.[26] The Court approved of such efforts but appreciated more that part of his activity that had a 'commercial' bearing, and on this basis chose him for Calcutta.[27] Roxburgh added to the Calcutta botanic garden about 2,2000 species of plants, besides more than 800 species of trees. His works on economic botany (fibres, etc.) earned him three gold medals from the Society of Arts. As for the scientific value of his labours, G. King later wrote, 'I have worked a good deal with Roxburgh's Flora and among Indian plants and it takes a good deal to convince me of a Roxburghian blunder'.[28]

Perhaps the most notable find of Roxburgh was Nathaniel Wallich, a surgeon attached to the Danish settlement at Serampur, who was taken a prisoner of war in early 1809 when several French, Dutch and Danish settlements in and around India were attacked by the British during the Napoleonic wars.[29] Even though Wallich was earning three to four

[22] Robert Kyd Papers, IOL, Mss. Eur. F. 95. Roxburgh also showed great interest in timber for ship-building and teak plantations came up on a large scale at Sylhet and Bankura. *Selections from Records of Bengal Government*, xxv, 1857, pp. 2–3.

[23] Major-Gen. Hardwick Collection, B.M. Add. Mss. 12615.

[24] Quoted by Buchanan, Home, Public, no. 94, 15 April 1816.

[25] S. Sangwan, 'Plant Colonialism', *Proc. of the XLIV Session of the Indian History Congress*, Burdwan, 1983, pp. 414–24.

[26] For scarcities he blamed the system and the administration. Home, Public, no. 10, 5 Dec. 1799.

[27] I.H. Burkill, p. 22, note 19.

[28] *Annals of the Royal Botanic Garden of Calcutta*, v, 1895, p. 6.

[29] Home, Public, no. 26, 10 Feb. 1809.

hundred rupees a month in private practice, he offered to serve as a botanist on 'whatever allowance' the government would grant and declared that his object was to gain knowledge and not make money.[30] But when the question of finding a successor to Roxburgh arose, Wallich did not find favour. Later, James Hare, the new superintendent, and Wallich did not get on well together and Wallich had to leave the garden.[31] Carey came to his rescue and lobbyed on his behalf with several officers in high posts.[32] This proved beneficial and after some time Wallich got charge of the botanical garden, but this incidentally gives an insight into how things moved in science-administration.

Wallich soon earned the reputation of being an avid plant-collector. He surveyed Bengal, Bihar, Assam and Nepal and even went to Penang, Singapur, China and Burma.[33] He not only collected but also tried to distribute his specimens as widely as possible. In the mid-1830s he took to England thirty barrels of dried plants containing almost everything he had collected in India, and there he drew up a list of forty scientists and institutions to whom he would give samples for further study. This caused some heart-burning among British botanists. The following extract from the notes of Alphonse de Candolle, an Italian botanist who worked with Wallich in London, is illustrative of the problem faced by Wallich, a colonial scientist *vis a vis* the metropolis:

It must be said that the English botanists disapprove of Wallich's exotic liberalities; as soon as he is back in India everything he is not able to distribute will be placed under lock and key and buried for ever . . . , I see clearly that Brown is annoyed by Wallich's generosity. He said to Kunth (who had been sent by Prussian Government to obtain species from Wallich for Berlin Museum) that he had at least succeeded in putting an end to the scandal of Wallich's distributions . . . Wallich's health has been much impaired and he needs to stay in Europe to restore it; however his request to stay an extra year, while granted by the Board of the Company, has been denied by the Government Board of Control. This has driven Wallich into a state of frantic anger and despair because without him to sort out, classify and supervise the facts, his collections will be lost to science. Sinister motives seem to be involved, the Admiralty especially fearing that Wallich's generosity would set an example and force it to make available its own enormous herbariums.[34]

[30] Ibid., no. 52.
[31] Hardwick to Banks, 15 May 1817, Hardwick Papers, B.M. Add. Mss. 9869.
[32] W. Carey to H.T. Colebrooke, 14 June 1817, Carey Papers, IOL, Eur. Mss. B230.
[33] *The East India and Colonial Magazine*, XI, 67, June 1836. pp. 499–512.
[34] Roger de Candolle, 'N. Wallich: How the Largest Botanical Collection was Brought to Europe by a Single Man', B.M. (Natural History), BMSS, CAN.

The matter apparently caused a sensation. The Helvetic Society of Natural Sciences passed a resolution and Augustin Pyramus wrote a letter on behalf of scientists abroad that Wallich be allowed to complete his essential work. This sort of pressure worked and Wallich got the extension he was seeking.

Differences of opinion and mutual suspicions were manifested in Calcutta also. McLelland and Griffith were less than satisfied with Wallich and felt that his frequent tours had resulted in the neglect of the garden itself. Roxburgh had laid out a Linnaean garden (i.e. a border of selected plants spaced to exhibit the Linnaean system of classification), but Griffith wanted just the opposite (i.e. a garden exposing the natural system flanked by a garden of medicinal plants).[35] So, when in late 1842, Wallich left for England on leave, Griffith and McLelland got the opportunity to introduce their plans and with official approval destroyed the Roxburghian layout. But before the new system could be thoroughly implemented, Wallich returned in August 1844 and stopped the revamping. For a long time the Calcutta Botanic Garden, the biggest in the country, was to remain a hotch-potch of several thousand species. In 1856, T. Thomson, then the Superintendent, reported: 'The library, herbarium, museum have been starved from want of funds, and in consequence, the scientific character of the establishment has been so entirely lost, that its existence is scarcely known in Europe.'[36] Even as a gigantic nursery garden, it was going down. As compared to about 110 cases of specimens annually despatched to Europe during 1836–40, the average came down to 50 in the 1850s.[37]

But such a state of affairs did not mean that botanical investigations had reached a dead end. Gardens were, rather, proliferating. In 1854 a botanical garden at Madras was asked for.[38] Dedicated workers like Forbes Royle subjected the Himalayan flora to closer scrutiny and better utilization. He experimented on the rhea fibres of Assam and the hemp of the Himalayas and corresponded with the Royal Society of Arts and the Commercial Association of Manchester for their possible utilization. During 1848–50, J.D. Hooker travelled in Bengal, Sikkim, Nepal and the Khasia Hills, and in 1855 produced two excellent works: *Flora Indica* and the *Himalayan Journal.* He was the first to explore the passes into Tibet and his work was found indispensable when Colonel Younghusband

[35] I.H. Burkill, p. 91, note 19.
[36] *Selection from the Records of Bengal Government,* xxv, 1857, p. 63.
[37] Ibid., p. 57.
[38] Home, Public, no. 13, 21 Dec. 1855.

entered the Forbidden City in 1903.[39] Botanical investigations were thus of commercial, military as well as scientific importance.

The Cash-Crops Boom

The problem of how to make Indian agriculture most remunerative was also very important. The colonizers fully recognized India's potentialities as an agricultural country and called for sinking more and more capital in agriculture. An influential journal thus exhorted the British to 'embark some portion of your redundant riches in speculations of Indian Agriculture: grow there sugar, rice, cotton, tea, indigo, even wheat, encourage the silk worm, and culture of tobacco, opium etc'.[40]

Cash crops were naturally favoured. In 1793 the net profit from a *bigha* of wheat yielded only 7 annas, while hemp cultivation could give 3 to 7 rupees.[41] Hemp had a clear edge over wheat. Its economic as well as military significance lay in its use for the rigging of ships. Russia had to stop its supply during the Napoleonic wars, and British shipping greatly suffered as the price of hemp, which in 1792 was only 25 rupees per ton, rose to 118 in 1808.[42] Therefore the colonizers turned to India for their supply of hemp. The Agricultural and Horticultural Society of India (AHSI) took up the challenge, offered inducements and rewards, and in 1841 produced a very exhaustive report on flax-culture.[43] Soon the Court felt that no special official effort was needed as the cultivators, aware of its high profitability, were themselves shifting to it.[44]

Experiments in both cotton and tea were carried on simultaneously. The government established experimental farms for cotton as well as gardens for tea plants. At the outset cotton farms received the most careful attention. Pecuniary advances were made to individuals; seeds were procured from Egypt, Brazil and North America; saw gins were imported; and large prizes were offered for the best samples of cotton. Metropolitan business houses realized the stakes involved, and in 1838, several chambers of commerce at home petitioned the Court.[45] The Court itself was

[39] Thiselton-Dyer to Haldane, 1.6.1907, Campbell Bannerman Papers, B.M. Add. Mss. 41218, f. 188.
[40] *The East India and Colonial Magazine*, VIII, 48, Nov. 1834, p. 430.
[41] R. Wisset, *On the Cultivation and Preparation of Hemp*, London, 1804, pp. 113–18.
[42] Home, Revenue, nos. 1–10, 13 April 1840.
[43] Home, Revenue, nos. 14–17, 22 Nov. 1841.
[44] IOR, E/4/771, *Bengal Despatches*, XXXII, 1842, p. 288.
[45] Home, Revenue, Letters from Court, no. 4, 1839.

worried and called upon 'the Government of India to promote and extend throughout India *any* plan for this purpose'.[46] American planters were brought to nurse experimental cotton farms.[47] But the project was doomed from the very beginning. The Superintendent of the American Cotton Project, J.H. Pearly, frankly confessed: 'Their object in coming to India was to make money which they found they could more easily accomplish in their own country.'[48] Despite repeated failures, the Court was very keen on the mechanization of cotton-cleaning, and in 1840, even announced a reward of £100 'for the invention of the best instrument for cleaning cotton, adapted to the use of the natives of India'.[49]

Like cotton, tea generated a lot of enthusiasm. Kyd had grown China tea in his botanical garden in 1780. Banks wanted to induce the tea-experts of Canton and Hunan to migrate with their tea-shrubs and tools to Calcutta 'where they will find the Botanic Garden ready to receive them'.[50] But the authorities probably discouraged further experiments as they presented a rival to the China tea trade, which was a source of much wealth to the Comapny.[51] Bentinck showed keen interest in the possibility of extensive tea cultivation, as he had done in steam transport and coal explorations. A Tea Committee was formed in 1835 and Gordon was sent to China to procure seeds, plants and Chinamen, and for this purpose was given a credit of 25,000 dollars.[52] In 1839, the government asked for a rough map of the tea-bearing areas in Assam.[53] In 1847, W. Jameson, Superintendent of the Saharanpur Botanical Garden, asked for government tea-plantations in Pauri, Dehra Dun and Bheem Tal at a cost of Rs 22,980 per month, involving 1000 acres of land and the import of six Chinese tea-cultivators. This was sanctioned in deference to the Court's desire that the 'experiments should be conducted on the most liberal scale'.[54]

Till famine struck grievously in the mid-1860s, requiring it to formulate an agricultural policy, the Government of India had tried to act by proxy, i.e. through the agency of the Agricultural and Horticultural

---

[46] IOR, E/4/758, *Bengal Despatches*, XIX, 1839, p. 1099 (emphasis added).
[47] For interesting instances, see American Planters Papers, IOR, Mss. Eur. C. 157.
[48] Home, Revenue, nos. 2–5, 28 Feb. 1842.
[49] Revenue, Agri., Fibres and Silk, nos. 18–27, 22 June 1840.
[50] J. Banks, 'Memorial on Tea', 27 Dec. 1788, IOL, Mss. Eur. D. 993.
[51] *The Tea Cyclopaedia*, Calcutta, 1881.
[52] W.N. Lees, *Tea Cultivation and other Agricultural Experiments in India*, Calcutta, 1863, p. 11.
[53] Home, Revenue, nos. 18–20, Oct. 1839.
[54] *Selection from the Records of NWP Government*, I, Calcutta, 1856, pp. 250–3.

Society of India. The Court had clearly laid down that 'excepting in special cases where it may be considered desirable to carry on important experiments under the immediate control of government, we are of opinion that those objects will be best attained through the medium of public societies in aid of whose funds a subscription on the part of government would be properly applied'.[55] Government officials and influential zamindars were members of this society and its funds came through private subscription, the sale of plants and government aid. In 1830 the government gave the AHSI Rs 20,000 for distribution among the most successful cultivators of cotton, tobacco, sugar, silk and other articles of 'raw produce'.[56] The Society had ambitious plans. Basically engrossed in the acclimatization of several foreign varieties of fruits and vegetables,[57] it often assumed the role of an adviser to planters and advised them as to where and how to invest capital in new agricultural ventures.[58] It experimented with flax-culture and tea.[59] The AHSI, however, worked under severe limitations and famines completely exposed its inadequacies to pull Indian agriculture through difficult times.

Geological Explorations

The science of mineralogy did not lag far behind in getting the attention given to plants right from the beginning. As early as in 1804, one Abraham Hume, a ship owner in the Company, argued that 'mineralogy requires a distinct establishment in which the whole grounds of mineralogy and geology illustrated by specimens and authenticated by chemical analysis can be investigated'.[60] The Royal Institution drew up a plan saying that in India 'are to be found the most valuable mineral treasures that are known in this globe; and from the wisdom and liberality of the East India Company, great and effectual assistance may be hoped for in aid of the execution of a plan by the adoption of which the intrinsic value of those

---

[55] IOR, E/4/765,*Bengal Despatches*, XXVI, 21 April 1841.
[56] *Calcutta Gazette*, 20 May 1830.
[57] Records of the AHSI, Feb. 1827, p. 115.
[58] Ibid., 13 May 1835, n.p.
[59] Dwarka Nath Tagore, an effective member of the AHSI, offered his extensive premises at Manicktolla, rent free, for a school where a limited number of local workmen could be trained in flax-culture. Home, Revenue, nos. 14–17, 22 Nov. 1841. The Court appreciated this gesture. IOR/E/4/771, *Bengal Despatches*, XXXII, 1842, p. 291.
[60] *Dictionary of National Biography*, X, London, 1908, pp. 208–9. Hume later helped found the Geographical Society of London.

treasures may be ascertained and brought into use'. It went on to request Lord Wellesley, as well as the Governors of Madras and Bombay, to extend government patronage to mineralogy.[61]

The economic value of geological investigations was of immediate concern to the Company, with the coal-fields of India looming large. The year 1807 saw the birth of the Geological Society of London and it was no coincidence that in the very next year Lord Minto ordered an investigation into the Raniganj coal-field.[62,63] Within a few years Colebrooke brought out a report on Sylhet coal.[64] Franklin looked for coalmines in Palamu,[65] while Herbert wrote about the occurrence of coal within the Indo-Gangetic tract of mountains.[66] In 1840–45 Lieutenant Newbold published a series of papers on the minerals of South India with particular reference to manganese ore.[67]

The economic as well as military value of these early explorations is obvious in what Herbert wrote while listing the important Himalayan minerals. 'Sulphur appears to deserve the first notice, if it be only for its value as an ingredient in the manufacture of gun-powder. During the late war (Napoleonic) its price rose to £30 per ton in Europe and it would seem a subject not unworthy of attention, to ascertain in what quality and at what price we could draw it from our own provinces.'[68]

The only idea the government then had of the duties of a geological surveyor was that he should move from place to place, and report upon real or fancied discoveries of minerals.[69] Making these reports could be very annoying. Herbert once complained: 'In scientific enquiries nothing is more prejudicial than to have the attention continually called from the proper subject to attendant official details which while they occupy much time in their preparation yet fail to afford any just idea of the nature of the researches.'[70] The discovery of minerals had to be followed by their

[61] Quoted from the Royal Institutions' Archives, in M. Berman, *Social Change and Scientific Organization: The Royal Institution, 1799–1810*, Ph.D. thesis, Johns Hopkins University, 1971, p. 231.

[62] For details, see H.B. Woodward, *The History of the Geological Society of London*, London, 1907.

[63] *Transactions of the Mining and Geological Institute of India*, I, 1907, p. 64.

[64] Home, Public, no. 22, 20 Sept. 1814.

[65] Home, Public, no. 90, 19 May 1829.

[66] Home, Public, no. 42, 1 Feb. 1827.

[67] *JASB*, vii, 1843, pp. 212–14; xiii, 1844, pp. 992–5.

[68] *Asiatic Researches*, xviii, 1, 1829, pp. 228–9.

[69] C.R. Markham, p. 216, note 13.

[70] Home, Public, no. 42, 1 Feb. 1827.

proper commercial utilization. So in 1827 the Government of India requested the Court to send from the mining districts of Engand or Germany a few 'practical mineralogists'.[71]

From all the minerals, coal got the maximum attention and was indeed the most important. Coal had been known in India from time immemorial as is evident from outcrop fires in the Damodar valley. Yet there is no recorded history of coal in India until John Sumner and S.G. Heatly discovered it in 1774 near Sitarampur in Bengal, and applied to Warren Hastings for permission to work coal-mines.[72] As wood was abundant, the Indians had naturally ignored coal. After 1774 this industry languished for nearly fifty years because of lack of transport facilities and appreciative consumers. Official interest was aroused with the possibility of establishing iron-works, steam-ships, etc., and between 1826–51, the search for coal was encouraged and intensified.[73] In 1836 the Company appointed a 'Committee for the investigations of the Coal and Mineral Resources of India' with the chief object of determining 'the existence, extent and relative accessibility of the beds of mineral coal in different parts in India and their immediate practicability to the increasing demands of the steam navigation'.[74] This Committee stands as a milestone in the evolution of colonial science in India because here for the first time various types of coal and minerals were listed along with map illustrations of the sites as well. And also for the first time was raised the question of employing trained geologists in India to investigate coal formation in the country. In October 1843, the Government of India urged upon the Court 'the expediency of adopting a system of scientific enquiry both with the view of securing success to such attempts as are made by the Government and of promoting confidence among private speculators to embark upon undertakings for obtaining coal'.[75]

In addition, some sort of private pressure was also mobilized. Urging follow-up action on the Coal Committee Report, the editor of an influential Calcutta journal wrote to Charles Lyell, the noted British geologist. Lyell gave that letter to R.I. Murchison, then President of the Geological Section of the BAAS, who replied, lamenting that 'whilst very

[71] Ibid., no. 55.
[72] *JASB*, XI, 1, 1842, p. 811.
[73] *Memoirs of the GSI*, LVII, 1931, pp. 1–3.
[74] *Report of the Committee for Investigating the Coal and Mineral Resources of India*, Calcutta, 1838, p. 1.
[75] C. Fox, 'Coal Committee and Geological Works', *Nature*, 160, 4078, 1947, p. 889.

large sums have been spent upon enquiries into the botanical productions of Hindustan, and eminent botanists have been liberally employed, no geologist has been engaged, regularly and systematically, to work out the relations of the rock masses, containing various mineral substances useful to man'. Such pressure tactics did have an effect.[76]

The result was that, in 1846, D.H. Williams was sent out as Geological Surveyor to the Company. The Court's direction was 'that you will direct your principal attention to those localities which promise to afford supplies of coal, and which are so situated with respect to water-carriage, as to give a real commercial value to the coal which they may produce'.[77] He prepared two reports, one on the Damodar Valley and the other on the Ramgarh coal-fields. In his second report, Williams, interestingly enough, discussed the unfortunate results of the Permanent Settlement on mineral deposits if zamindars were allowed to assume mineral rights.[78] His death in 1848 was followed by a brief interregnum during which McLelland discovered the Giridih coal-field but could not succeed in establishing his credentials as a full-fledged in-charge of the survey operations. During this period Dr Fleming was conducting a parallel survey work in the Salt Range without any authority from his Bengal counterpart.[79] Things were thus in a confused state during this prelimi-nary period of geological investigations in India. This state of affairs improved in November 1850 with the appointment of Thomas Oldham who had already distinguished himself as the Director of the Geological Survey of Ireland. His arrival in March 1851 marked the establishment of a 'continuous' GSI. Inspite of the fact that his work was intended to be more economic in nature, he was able, in 1856, to classify his observations and lay the foundations of stratigraphical classification in Indian geology.[80]

Mechanical Innovations: The Telegraph

No government, much less a commercial one, could have ignored the importance of transport and communications. Steamboats, locomotives and electric telegraph were brought to India as soon as their value had been demonstrated in England. The 1830s and '40s were clearly ripe for such

---

[76] *The Calcutta Jr. of Natural History*, III, 12 Jan. 1843, pp. 615–16.
[77] Home, Revenue, no. 3, 28 Nov. 1846.
[78] L.L. Fermor, *First Twenty-Five Years of the GSI*, New Delhi, 1976, p. 33.
[79] *JASB*, XVII, 2, 1848, pp. 500–26; XXII, 2, 1853, pp. 229–79, 333–68, 444–62.
[80] *Memoirs of the GSI*, 51, 1, 1926, p. 7.

technological ventures but these were motivated by separate considerations. Steamboats and railways were largely initiated and financed by private merchants for the expansion of trade while the electric telegraph was entirely financed and managed by the government. In England, the telegraph grew as a commercial adjunct to railways, but in India it came ahead of railways. The reasons were basically political.[81] Externally, overland and sea routes had to be made invulnerable to any hostile, putative Russian advance.[82] Internally, the successive annexations of local states had considerably expanded the range of imperial activities. No wonder the authorities felt the pressing need for an electric telegraph system in India.

Much earlier, in 1802, Dr Dinwiddie, who taught botany at Ft. William, had made certain galvanic experiments.[83] The Calcutta officials were quite aware of the uses of signals in war. In early 1825, for example, a set of Connolly's telegraphic signals was presented to the Asiatic Society under the impression 'that they will be found useful in the hostilities at present carrying on in Burmese territories'.[84] Semaphoric communication was introduced in the late 1820s and semaphore stations were carried as far as Kedgeree and then Sagar island at the cost of Rs 25,000.[85]

Adolphe Bazin was the first to submit, in 1839, a plan for electric telegraphs in India and wanted it to be examined early in view of his 'political engagements', and 'the unsettled state of European affairs'.[86] Meanwhile, a surgeon, W.B. O'Shaughnessy, had gradually shifted his interest from medico-chemical research to the field of electromagnetism. His experiments on telegraphs signalled his switch over from theoretical science to 'utilitarian' (or, perhaps more appropriately, 'imperial') technology. He erected a line of wires 21 miles in length, the first long line of telegraph ever constructed in any country.[87]

The whole system, however, remained at an experimental stage and it

[81] S.K. Ghose, *The Introduction and Development of the Electric Telegraph in India*, Ph.D. thesis, Jadavpur University, 1974, p. 71.

[82] Russophobia had been an important feature of the post-Waterloo British foreign policy.

[83] W.H. Carey, *The Good Old Days of John Company*, II, Simla, 1882, p. 164.

[84] Captain Marryat had introduced these signals with success in the division of ships under his command at Rangoon, and later ordered for an additional number of these sets. *Calcutta Monthly Journal*, XLVII, 363, Jan. 1825, p. 51.

[85] W.H. Carey, p. 166, note 83; *The Asiatic Journal*, XIX, 1825, p. 684.

[86] *JASB*, VII, 1839, p. 436.

[87] G.W. Macgeorge, *Ways and Works in India*, Westminster, 1894, p. 500.

was only in 1849, when Dalhousie was engaged in continuous conflict with the Indian states, that the electric telegraph was officially proposed.[88] The railway contracts were signed on 17 August 1849 but the need for a telegraph system was so urgent that the Court decided not to wait for the construction of telegraph lines along with the railways.[89] The very next year, O'Shaughnessy was asked to set up an experimental line between Calcutta and Chinsura.[90] Henceforth there was to be no obstacle. Even though Dalhousie had recommended the construction of lines in different phases, the Court sanctioned the entire length of 3150 miles.[91] So keen was the Court, that in November 1853, construction began, and within fifteen months O'Shaughnessy was able to connect Calcutta, Agra and Attock, and also Agra, Bombay and Madras.

That the telegraph could cause politically sensitive issues to explode was well realized by O'Shaughnessy. In July 1855 Captain Franshaw, the Calcutta correspondent of the *Lahore Chronicle*, sent a telegram to his editor which read: 'The Sontal insurrection is spreading in Bancoorah and Midnapore. A foreigner is reported to be among them, and Russia is suspected.' Intercepting this, O'Shaughnessy wrote to Cecil Beadon, the Home Secretary, 'that in time of open insurrection it is not proper to telegraph through a government line, messages for publication announcing the spread of rebellian, the defeat of government troops, etc'.[92] The telegraph played a crucial role when a large-scale revolt broke out in 1857. This was perhaps the first 'telegraph war',[93] and as Beadon acknowledged, 'the electric telegraph has been worth incalculable sums to government in this frightful event'.[94] Through a series of telegrams exchanged between the provincial governments and the Supreme Council during 16–29 May 1857, a huge troop movement could be accomplished in a very short time. The result was one more victory for the British.

[88] S.K. Ghose, p. 88, note 81.
[89] Court to GG, 26 Sept. 1849, *Despatches to India*, IOR, E/R/801, pp. 1081–5.
[90] Home, Public, no. 47, 4 April 1850.
[91] Court to GG, 20 Oct. 1852, *Despatches to India*, IOR, E/4/817.
[92] Home, Public, 116, 10 Aug. 1855.
[93] Ghose has contributed an excellent chapter on 'Telegraph in Indian Mutiny'. He regrets that Kay, Malleson, Holmes, Forrest, Sen, Mazumdar, Collier and others who wrote on 1857, are generally silent over the role of the telegraph. S.K. Ghose, pp. 168–94, note 81.
[94] Beadon to O'Shaughnessy, 1 June 1857, *Board of Control Letters on Telegraph*, IOR, F/2/26.

Science Education

One of the intentions mentioned in Section 43 of the Charter of 1813 for the grant of Rs 1 lakh to be spent on education was the 'introduction and promotion of knowledge of the sciences among the inhabitants of the British India'.[95] But the Court gave no directive as to which system of science, indigenous or European, was to be preferred. The Court perhaps tried to avoid taking sides and took refuge in the neutrality of the engraftment principle, calling for a fusion of the scientific and medical techniques of the East and West. As a result, the whole issue got bogged down into what is known as the Anglicist–Orientalist controversy, which the former finally won.[96] To quote Trevelyan:

> Our object ought not to be by any means of translations, to make at the best an imperfect graft of the tree of knowledge on a trunk, the heterogenity of which will not admit of its flourishing upon it, but by the introduction of our own literature and the instruction of the natives in it from their earliest youth we ought to plant a young and flourishing tree which with the encouragement it is in the power of government to afford it, will shoot out and spread its branches far and wide while the trunk of the old system will be left to a natural and neglected decay.[97]

Nothing illustrates better the prevailing attitude of the colonizers. They thought that Western ideas would seep into the Indian mind in 'a benignly osmotic fashion'.[98] It is not that some of the officials were not aware of the local techno-scientific traditions. In 1824, the Committee of General Instruction, Bengal, reported that 'the arithmetic and algebra of the Hindus lead to the same result and are grounded on the same principles as those of Europe'.[99] Cultural pluralists like Colebrooke, Prinsep, Adam, Campbell and Tytler repeatedly advocated the full use of indigenous systems and institutions. Colebrooke wanted 'to investigate the sciences of Asia . . . with the hope of facilitating ameliorations'.[100] But

[95] Arthur Howell, *Education in British India*, Calcutta, 1872, p. 5.

[96] For meticulous details, see H.M. Griffin, *T.B. Macaulay and the Anglicist-Orientalist Controversy in Indian Education*, Ph.D. thesis, Pennsylvania University, 1972.

[97] C.E. Trevelyan, Kota note, 21 May 1830, American Philosophical Society Papers, Philadelphia, Mss. 954, T725.

[98] G. Viswanathan, *Masks of Conquest: Literary Study and British Rule in India*, New York, 1989, p. 16.

[99] Quoted in B.D. Basu, *Education in India under East India Company*, Calcutta, n.d., p. 60.

[100] H.T. Colebrooke, *Miscellaneous Essays*, I, London, 1837, p. 2.

such arguments fell on deaf ears. One by one, Prinsep, Tytler and Adam had to resign in disgust. Prinsep regretted that 'the publication of Bhaskaracharya's work with commentary was proposed but rejected. An edition of Euclid in Sanskrit was prepared, it too was rejected. Hutton is now suspended, and a vade-mocum of anatomy half through the press is also laid aside; all that might be done is thwarted and frustrated . . . .'.[101]

The imperial ethos had already assured Macaulay an easy victory. Here was a man who, within just a few weeks of his arrival, found the indigenous medical system such that would 'disgrace an English farrier' and a knowledge of astronomy that would 'move laughter in girls in an English boarding school'.[102] Such a contemptuous tenor is obviously the result of pre-conceived notions. He had no first-hand experience of Indian scholarship, nor any translation of the Indian medical and astronomical treatises existing at the time. In the 1830s Sanskrit-Persian literature was virtually untapped. Jones, Colebrooke and Wilson had mostly translated works of poetry and drama. What is more, Macaulay vehemently denied that the government owed its subjects anything at all, least of all a share in the reconstruction of their educational systems. His personal distaste for science led to a curriculum which was purely literary.[103] The introduction of science as a subject was thus delayed. In July 1835, the General Committee of Public Instruction even recommended the abolition of the existing science professorship at the Hindu college and discontinued the instruction of chemistry there.[104] A contemporary journal wrote that 'more useful knowledge is to be gained from the study of one page of Bacon's prose, or Shakespeare's poetry, than from a hundred pages of Euclid'.[105] Later in 1854, F.J. Mouat, then Secretary to the Council of Education, Bengal, also felt that the study of English literature was 'more conducive to the improvement in the strength and tone of mind' than 'a particularly exclusive mathematical education'.[106]

In 1844 the idea of having a Professor of Natural and Experimental Philosophy was revived. But a controversy arose as to whether the emphasis was to be on pure science or applied science. One section of the Council

---

[101] Home, Public, no. 12, 1 July 1835.
[102] Home, Public, no. 15, 7 March 1835.
[103] H.M. Griffin, p. 488, note 96.
[104] Home, Public, no. 4, 8 July 1835.
[105] *The Hurkaru*, 28 April 1838; also quoted in the *Calcutta Monthly Journal*, XLVIII, Oct. 1838, p. 206.
[106] F.J. Mouat to C. Beadon, 10 March 1854, *Selections from Records of Bengal Government*, XIV, Calcutta, 1854, p. 55.

of Education felt that 'the course of Natural Philosophy would be only such as is considered necessary as a branch of general education without being specially adapted to or intended for professional men'. The other section thought that 'the course should be a combined experimental and mathematical course . . . without which it is quite impossible for the engineer, architect, surveyor, navigator or mechanist to become proficient as practical men in their several departments'.[107] A decade later, Mouat regretted the extreme emphasis on mathematics alone and pleaded for 'a demonstrative or experimental course of study'.[108]

As a 'demonstrative' as well as a 'remunerative' science, geology caught the attention of officials and publicists alike.[109] As coal-explorations progressed, demands were raised 'for proper instruction in economic geology and appointment of geological professors in the colleges'.[110] Even the Military Board intervened and wrote to Dalhousie that 'considering the extent and nature of the resources of British India . . . geology must be regarded as one of the most essential and necessary branches of instruction'.[111] There was some confusion as to where the geological classes were to be located. The Council of Education wanted them to be at the Calcutta Medical College, while the Court wanted them to be attached to the civil engineering classses at the Presidency College.[112] Another difficulty was that geology was lumped with natural history and this meant a course on practically every natural object on the earth. No individual could have done justice to such an encyclopaedic course. Initially Oldham volunteered to teach, but later discovered that in the absence of facilities concerning practical and field work, it would be 'utterly impossible' to teach geology in Calcutta.[113,114] Still, lectures on geology were introduced during the 1856–57 session, but as they could attract only a very small number of students (because geology was not made a subject for the B.A. degree), the lectures were discontinued and the professorship abolished.[115] The Court, however, expressed dissatisfac-

[107] Home, Public, nos. 19–30, 13 Sept. 1845.

[108] F.J. Mouat to C Beadon, 10 March 1854, p. 55, note 106.

[109] As revenue from opium became discredited at home, a need was felt to look for less objectionable resources like minerals.

[110] *Calcutta Review*, XII, 23, 1849, p. 233.

[111] Home, Public, no. 29, 2 Sept. 1848.

[112] F.J. Mouat to C. Beadon, 10 March 1854, p. 63, note 106.

[113] Home, Public, no. 33, 27 July 1855.

[114] *Selections from the Records of Government of India*, LXXVI, Calcutta, 1870, p. 59.

[115] Home, Education, nos. 3–7, 2 Oct. 1857.

tion over the propriety of this decision, with the result that the chair of geology was re-established in July 1858.[116,117] But this did not make geology as a subject more viable. The Syndicate of Calcutta University resolved in August 1858 to oppose the introduction of geology into the academic curriculum.[118]

Geology received at least some attention, but agriculture, from which the company extracted maximum revenue, was left in the cold. Wellesley was perhaps the first Governor-General to have talked of model farms as forming a branch of agricultural instruction.[119] In 1829, Prinsep proposed the establishment of a model agricultural college, complete with an experimental station and stock farm for Europeans and Indians to run jointly.[120] Bentinck liked the idea but could not implement it. In 1839, a contemporary complained that 'in Germany, in servile Russia, in bigotted Spain, in distracted Italy, we find schools and professorships of agriculture, but in British India *depending wholly on the soil for its revenue*, there is nothing of kind'.[121] The Court kept quiet about it; perhaps it was too obsessed with the acclimatization of foreign seeds to think of agricultural education and research.

Science education thus did not fit into the exigencies of the Company Raj. Even the much-publicized Wood's Despatch of 1854 did not pay the required attention to it. Primary education was preferred, for higher learning would have created more awareness among the Indians and thereby fuelled discontent. A journal enquired as to 'whether in our generous enthusiasm in the cause of Indian education, are we not rapidly annihilating the superiority which constitutes the tenure of our empire, and preparing, by our own efforts, our own political extinction in Asia'.[122] But a need was certainly felt to have a class of apothecaries, hospital assistants, surveyors and mechanics to serve the fast-growing medical, survey and public works departments. Training local youths was obviously much cheaper than getting technical prsonnel from abroad. So there opened, in 1822, a medical school, and in 1843, an engineering class at the Hindu College.

[116] IOR, E/4/851, *Bengal Despatches*, CXII, pp. 601–8.
[117] Home, Education, nos. 1–2, 23 July 1858.
[118] Home, Education, no. 4, 27 Aug. 1858.
[119] Home, Public, no. 2, 6 June 1805.
[120] *Asiatic Journal*, 1 Feb. 1830, p. 67.
[121] H. Piddington, *On the Scientific Principles of Agriculture*, Calcutta, 1839, p. 8 (emphasis as in original).
[122] *The Bombay Quarterly Magazine*, 1, 2, Jan. 1851, p. 10.

Pre-British India had a vigorous medical system, well suited to satisfy local needs.[123] But, as in other spheres, the new rulers dubbed this system 'unscientific' and 'superstitious'. Only a few, like Adam, wanted the government to attempt a fusion of 'both exotic principles and local practices, European theory and Indian experience', and thereby 'revive, invigorate, enlighten and liberalize the native medical profession in the *mofussil*.[124]

As the demands of subordinate health workers grew, a medical school was proposed in 1822 with the twin purpose of teaching both the Western and Indian systems of medicine. Medical classes were also started at the Calcutta Sanskrit College and the Calcutta Madarsa. Similar experiments were conducted in Bombay and Madras as well. In early 1826 a medical school was founded in Bombay by Elphinstone, the governor, with the loftier objective of 'general diffusion of medical science among the natives by educating native youths to a knowledge of the European system and then sending them into the districts to practice'.[125] But since the government's needs were primarily of a military character, this utilitarian object became secondary, and the education of hospital assistants, an object not mooted in the first instance, was taken up. But even this did not progress well due to the government's lack of interest and paucity of funds. After a brief existence of six years, this school was abolished.

The progress of medical education through the only surviving Calcutta Medical School (experiments in Bombay and Madras having failed) took a curious turn in the year 1833, when the 'language controversy' arose. In this controversy, Dr John Tytler, the Principal of the school, sided with the orientalists. He admitted that the indigenous systems were medieval, but they did contain a grain of truth. For him, the only solution was to allow the students to draw comparisons, sort out errors, and then work towards the improvement of their own system.[126] The Anglicists, as expected, found no merit in Tytler's views. Tytler found himself in trouble when he started preparing Arabic translations of a few European textbooks. The problem of vocabulary was most serious, for in order to translate one word of English, he spent hours in searching through Arabic

[123] A.L. Basham, 'The Practice of Medicine in Ancient and Medieval India', in Charles Leslie (ed.), *Asian Medical Systems*, California, 1977, p. 40; and S.H. Askari, 'Medicines and Hospitals in Muslim India', *Journal of Bihar Research Society*, XLIII, 1–2, 1957, pp. 7–21.

[124] W. Adam, *Report on Vernacular Education*, Calcutta, 1868, pp. 322–3.

[125] Home, Public, no. 18, K.W., 18 July 1838.

[126] *Centenary Volume of Calcutta Medical College*, Calcutta 1935, pp. 7–9.

lexicons, only to find that its counterpart did not exist. He concluded that translations were unprofitable, and that many years would elapse before the Indians rejected the 'crude fallacies' which their medical system upheld. He thus provided the Anglicists a lever with which to drive their point home.[127] The result was that in early 1835 the medical classes at Sanskrit College and Madarsa were abolished along with the Native Medical Institution itself, and a new college was founded wherein all pupils were required 'to learn the principles and practice of medical science in strict accordance with the mode adopted in Europe'.[128] This was an important event, for henceforth, through syllabi and language, was to be fostered a 'dependent science', and Indians were made to look for Western models in every field of medical science.

It was easy to dangle Western models and flaunt the superiority of Western systems, but when some financial investment was required to realize its proclaimed objectives, the government would develop cold feet. To quote a despatch from the Court: 'The plan of establishing a laboratory at your presidency (Bengal) similar to that at Apothecaries Hall in this country (England), with an establishment of chemists, aided by a steam engine, and other expensive apparatus, will, we apprehend, be found an inexpedient and unnecessary measure and we desire accordingly that it be not carried into effect.'[129] This veto was given in the very year in which Madhusudan Gupta had become the first Indian to dissect a human corpse—an event to commemorate which Fort William had even boomed a 21 gun salute.

Pumping resources into the country was thus no easy matter. The local rich came forward. A magnificent galvanic battery was presented to the College by public subscription. D.N. Tagore offered annual prizes to the tune of Rs 1000, and Mutty Lal Seal later gave a large piece of land. In 1845 four Indian medicos were sent to England for higher studies; two were sponsored by D.N. Tagore and one each by Dr Goodeve and the Nawab of Murshidabad.

The rate of progress was certainly steady and well-geared to meet the immediate requirements of the government. By 1838 the demand for 'native doctors' in the army became so pressing that a Hindustani class had to be opened in which anatomy, medicine and surgery were taught in Urdu, the original scientific nomenclatures, however, being retained. Later, in 1851, a Bengali class was also opened. Despite the Macaulayan

[127] H.M. Griffin, p. 135, note 96.
[128] Home, Public, no. 20, 7 March 1835.
[129] Court to GG, 26 Sept. 1836, IOR, E/4/752 *Bengal Despatches*, XIII, 1837.

verdict, subdued voices were still heard in favour of the vernacular. The Hindustani and Bengali classes were extremely popular, and it was felt that only through them could European science be popularized. Academically also the students of Hindustani (Military) classes were often found better. In 1848 a teacher noted that 'the dissections of the English class were for the most part decidedly inferior to those of the Military class. Whereas the dissecting rooms of the Military class were found generally full of diligent dissectors, and the subjects were never thrown away only partly dissected, the reverse was the case with the English class'.[130]

The 1830s were important, but not only for medical and general education purposes. Talks about steamers, telegraphs, drains and railways and the expansion of survey and revenue operations had brought to the fore the necessity of raising a subordinate class of surveyors, mechanics and overseers. During the last Anglo-Maratha war, Maitland had noticed how difficult it was to secure the services of local artificers, and later recalled that 'there was a terrible dearth of practical men for the public service, and may account for the very great expenditure of artillery carriages, carts and machines in the ordinance departments'.[131] So in 1840, he on his own, without government assistance, set up in Madras a school for ordinance artificers. But in Bengal, the government took the initiative and an engineering class was instituted at the Hindu College in 1843. Two years later, Baird Smith started private engineering classes at Saharanpur.

The stage was thus set for a bigger experiment. The need for a vigorous prosecution of the Ganges canal provided the pretext, and in October 1847, a full-fledged engineering college at Roorkee was started. Three courses were offered. The first was an advanced one, the second was exclusively for European soldiers, while the third course was in Urdu for local youths. Unlike the medical courses, these classes were an instant success, and within five years the college acquired a workshop for scientific instruments, an observatory and a geological museum.[132] The success at Roorkee inspired the Court to initiate a degree course in civil engineering in the proposed university curricula. It had found the Roorkee course 'far more useful than other lectures'.[133] In 1855 a separate engineering college at Calcutta was readily sanctioned.[134] Bombay and Madras did not lag

[130] *The Indian Register of Medical Science*, I, Calcutta, 1848, pp. 329–32.

[131] Home, Public, nos. 3–4, 19 April 1850.

[132] Thomason's Despatches, *Selections from the Records of the Government of NWP*, II, Calcutta, 1858, p. 318.

[133] IOR, E/4/826, *Bengal Despatches*, LXXXVI, 19 July 1854.

[134] IOR, E/4/830, ibid., XCI, 2 May 1855.

behind and here also the Roorkee model was followed. Maitland's Artificers' school was left untouched and a separate engineering college was added.[135] Preparations were thus in full swing for the expansion of what was then termed 'public works'.

Scientific Societies

Pre-British India did not have anything like a scientific society, let alone a journal, which could provide a platform for scientific workers. As a result, research remained esoteric and tended to get lost. Jones was the first to realize this and founded the Asiatic Society in 1784. This society soon became the focal point of all scientific activities in India. On 7 September 1808, it resolved to form a physical committee 'to propose such plans, and carry on such correspondence as might seem best suited to promote the knowledge of Natural History'.[136] But the idea languished in the absence of adequate support till January 1828 when the committee was revived and a series of lectures called the 'physical class' was started.[137]

In the 1830s, however, the Asiatic Society came under criticism.[138] Corbyn found its meetings lacklustre; he regretted that 'there has been a great falling off and other societies in consequence have been founded'.[139] Another contemporary narrates an interesting instance of how the government bought off vocal members:

The Society's Agricultural Committee had the advantage of possessing a very active secretary, intent on progress, who procured sundry excellent papers for his Committee. . . . This Committee which was begining to know so much, and what was still a greater offence, to tell the public so much about India, alarmed the jealousy of a bureaucractic despotism, which determined to silence it at any price. The first thing was to find or make good place for the secretary and stop his mouth; and accordingly this votary of progress one morning took his colleagues by surprise (of whom one or two never spoke to him afterwards) by announcing his promotion to a Government appointment. After this blow the Agriculture Committee withered away under the frowns of government, and the Asiatic Society found it expedient to confine itself to the most harmless antiquarian researches for the future.[140]

[135] IOR, E/4/987, *Madras Despatches*, 127, 11 Feb. 1857.
[136] Asiatic Researches, XVIII, 1, 1829, p. I.
[137] Ibid., p. II.
[138] Incidentally, this was the time when Charles Babbage was attacking the Royal Society in London.
[139] *The Indian Review and Journal of Foreign Science and Arts*, V, 1841, p. 167.
[140] John Dickinson, 'Government of India under Bureaucracy', *India Reform Tract*, VI, London, 1853, pp. 78–9.

Official interference came in other forms as well. In July 1856, for example, the government decided to take away the Museum of Economic Geology from the Society and give it to the GSI. This time the president of the Society was T. Oldham who, as Superintendent of the GSI, naturally had an eye on the Society's precious geological collections. He not only supported the government's move, but may have even pulled strings to influence the decision. The majority of the Society's Council opposed the move. One of its members, Rajendralal Mitra, felt that 'by giving up the geological department altogether, I very much fear we shall soon be under the necessity of asking the Medical College authorities to take away the zoological collection, and the curators of the Public Library to relieve us of the books'.[141] The government had to withdraw, and the issue developed into the bigger question on how to have a *real* imperial museum.

There were several other societies that are worth mentioning. The activities of AHSI have already been noted. The Calcutta Medical and Physical Society, established in March 1823, broke the social and professional isolation of the doctors, and without any government aid, was able to publish its *Monthly Circular* and *Selections* regularly.[142] In 1818 the Literary and Scientific Society was established in Madras and in 1804, the Literary Society was founded in Bombay. Both these societies became branches of the Royal Asiatic Society which was established in 1825 in London. The Bombay society became a branch of the London society in 1829.[143]

These societies started as exclusive European clubs. No Indian was accepted by the Asiatic Society of Bengal till 1829. Manekji Cursetji was the first Indian who sought membership of the Bombay Branch. His nomination was proposed in 1833 but was squashed during the elections. Afterwards he was elected a member of its parent society in London. This was a very piquant situation—a person accepted as a member in London remained non-grata in Bombay. Later, in the early 1840s, the Bombay Branch had to accept him as a member.[144] Bal Shastri Jambhekar also experienced a similar situation. He had to send his research papers to the Secretary of the Society who then read them out at the meetings.[145]

Notwithstanding discrimination against Indians, these societies ren-

[141] Home, Public, no. 49, 7 Oct. 1859.

[142] *Medical Selections*, 1, Calcutta, 1833, pp. III–IV.

[143] *Journal of the Bombay Branch of Royal Asiatic Society*, Centenary volume, 1905, p. 20.

[144] Ibid., Index volume, 1886, pp. 12–13.

[145] G.G. Jambhekar (ed.), *Memoirs and Writings of Bal Gangadhar Shastri Jambhekar*, I, Poona, 1950, p. XXXIV.

dered invaluable service, particularly through their journals whose standard compared very favourably with the European journals. It was no mean achievement that Calcutta, with a reading public of little more than two thousand, could produce and support scientific journals like *Gleanings in Science*, and *Calcutta Journal of Natural History*. The latter even attempted to establish an Indian Association for the Advancement of Natural Science on the pattern of the British Association for Advancement of Science (BAAS).[146]

'Native' Response[147]

An important concomitant of the evolution of colonial science is the repeated condemnation of colonial society as non-scientific. The empire rested upon prestige which in turn depended upon racial myths. The British came to India with a definite sense of superiority. To quote Lushington: 'The English are really a superior race compared with the Hindus; they know better than the inferior race itself what is suitable to it.'[148] Prinsep struck a different note:

While we endeavour to push our own systems of instruction and science in this country, we are too apt to spurn and decry the literature, the science, and even the language of the East, as if they were not only incapable of imparting the smallest particle of knowledge, virtue or truth, but incapable also of improvement by engrafting upon them the new growth of western knowledge, which has sprung ahead of the Asiatic and elder stock only within the last century or two.[149]

But the real dilemma was most succinctly stated by Spilsbury who, writing about the Narmada region in a personal letter to his brother, pointed out that 'they (the natives) are rather barbarious in this part, but

---

[146] The Prospectus of this Association was published in *Calcutta Journal of Natural History*, 1841, pp. 8–14.

[147] The use of the term 'native' in contemporary documents seems to be an offshoot of racialism. It was chiefly used for 'non-Caucasian people of any region spoken as if strange'. *Webster's New International Dictionary of English Language*, Massachussets, 1955, p. 1630.

From the order appointing Asst. Surgeon J.G. Vos in 1832, it appears that the definition of a 'native' then employed was the 'son of parents of whom either one or both were of pure unmixed native extraction'. D.G. Crawford, *A History of the IMS*, I, London, 1914, p. 502.

In this work, the term has been used only to retain its original flavour.

[148] C. Lushington, 'Progress of Education in British India', *Asiatic Journal*, XXI, March 1826, p. 318.

[149] *JASB*, III, 1834, p. 519.

I dare say as soon as they get civilized they will become as great scoundrels and villains as they are in our own provinces (Bengal)'.[150] In the eyes of the British, before exposure to the West, the native was an ignorant 'barbarian', but afterwards he was bound to become a 'scoundrel'.

There is no doubt that most Indians were (and perhaps still are) grossly superstitious. When steamers first appeared on the Ganges, thousands flocked on the banks to worship them and implore mercy, believing them to be engines of a supernatural creation.[151] However, this does not mean that society as a whole was irrational, averse to or incapable of any technological improvement. It may be true that 'fire-carriages' and 'lightning posts' left the people bewildered, but bewilderment does not necessarily generate antipathy. The people in Europe and America were also bewildered with the amazing performance of the telegraph.[152] From among the multitude of Indians who ignorantly worshipped the steamers, came Mutty Lal Seal who immediately recognized its importance, invested in shipping and ran a very profitable business. Here Bernstein finds 'an indication that the essence of modern science was as native to a Bengali as to a European mind, when properly developed'.[153] Earlier in Bombay, Jamsetjee had earned fame as a master-builder at the Company's dockyard. Jamsetjee could not bear the contempt of being called a 'black fellow' and lodged his protest in a unique way. When the first frigate built of teak was about to be launched, he went down to the ship's hold and carved—'This ship was built by a black-fellow, AD 1800'.[154]

Indian peasants were almost rhetorically dubbed as conservative and resistant to innovation. Addressing the AHSI, Bentinck observed: 'It is impossible not to deplore the same defective state in the agricultural, as in every other science in this country. Look where you will, and you find the same results—poverty, inferiority, degradation in every shape.'[155] An average Indian cultivator had certain reservations, but perhaps they were valid ones. Piddington gives an interesting instance. Once he offered the seeds of Bourbon cotton to some ryots who became convinced of its advantage but said that they dared not cultivate any new crops. The reason

[150] G.G. Spilsbury to E.A. Spilsbury, 25 Nov. 1820. G.G. Spilsbury Papers, IOL, Mss. Eur. D. 909.
[151] Thomas Bacon, *First Impressions and Studies from Nature in Hindustan*, II, London, 1837, p. 434.
[152] S.K. Ghose, p. 82, note 81.
[153] H.T. Bernstein, *Steamboats on the Ganges*, Calcutta 1960, p. 194.
[154] W.H. Carey, II, pp. 184–5, note 83.
[155] Quoted on the cover of H. Piddington, note 121.

assigned was that their 'zamindar would make demands far beyond the profits'.[156] Similarly, they would not easily adopt the Italian filature system for silk winding or gins for cotton-cleaning. These machines were owned by those who were not the actual users but had the money to invest in them. This meant a further strengthening of the control of brokers over peasants and the subordination of producers to gin and filature owners. The peasants naturally resisted and, as Bhattacharya puts it, 'it was not always rejection of change itself, but rejection of a change for the worse'.[157]

As regards the natives' powers of exploration and sophistication in techniques, an argument was assiduously built up that the Indians lacked these. There was certainly no overwhelming enthusiasm for technological change, but there are examples to prove that the Indians did not shut their eyes to innovation and invention. Moshin Hussain of Arcot with his theodolite had impressed George Everest, the Surveyor-General of India (1830–43), as a remarkable mechanic with a talent for invention.[158] Of Radhanath Sikdar, Everest observed: 'In his mathematical attainments there are few in India—European or native—that can at all compete with him . . . even in Europe these attainments would rank very high.'[159] Sikdar was one of the first to compute and find out in 1852 that a peak designated XV which had been observed from six different stations, was the highest point on earth. When Colonel Waugh, the then Surveyor-General, got this information from Radhanath, he promptly named it after his predecessor and benefactor, George Everest, even though this peak had local names. The Nepalese called it Devadhunga and Gauri-shankar; the Tibetans, Jamokangkar. Hodgson, the British Agent in Nepal, brought this to the notice of the Royal Geographical and Royal Asiatic Societies. While the Royal Geographical Society approved of Waugh's action, the Royal Asiatic Society supported Hodgson and repudiated the name of Everest. Later, S.G. Burrard, another Surveyor-General, admitted that 'when the Gaurishankar controversy opened, the name of Everest was an interloper on the map of Asia; but its trespass has long since been condoned'. Nowhere did he mention the contribution of

---

[156] Ibid., p. 18. A similar dilemma was also noticed by Roxburgh while at Samalcotta in Jan. 1793. Home, Public, no. 10, 5 Dec. 1799.

[157] S. Bhattacharya, 'Cultural and Social Constraints on Technological Innovation and Economic Development', *IESHR*, III, 3 Sept. 1966, p. 261.

[158] R.H. Phillimore, *Historical Records of the Survey of India*, III, Dehra Dun, 1954, p. 458.

[159] Ibid., IV, pp. 340–1.

Sikdar; only a passing reference was made to the fact that this peak was discovered by the computers at the Calcutta office.[160]
.The natives were receptive to scientific knowledge. Rammohun's petition to Amherst asking for a proper science education is well known. In 1833 the young Derozians brought out a bilingual monthly called *Bignan Sar Sangraha* to spread scientific knowledge. It is no wonder that in 1842, J. Long found Calcutta to resemble Cambridge or Oxford in atleast some respects.[161] There were several Indians who responded very favourably. Radhakant Deb wrote papers on the use of manures and the chemical elements in Indian soils and was a founding member of the AHSI.[162] The Medical and Physical Society of Calcutta elected four Indians, all of whom were orthodox Hindus—Radhakant Deb, Ramcomul Sen, Madhusudhan Gupta and Raja Kalikrishna Bahadur—as corresponding members of the Society in 1827 and they produced a few papers on indigenous drugs for the Society.[163] The Mechanics Institution, which was founded in 1839 to serve Eurasians, avoided Indians, but others like Calcutta Lyceum (1844) and the Society for the Promotion of Industrial Arts (1854), sought the active support of D.N. Tagore, Pearychand Mitra, Rustomji Cowasjee and Rajendralal Mitra.[164]

The first to declare that India needed a Bacon was Akshay Kumar Dutta. He wrote *Bahya Vasterer Sahit Manab Prakitir Sambandha* (Relationship between External Objects and Nature of Man) in 1852 and later *Padartho Vidya* (Physics) in 1856.[165] He was sacked from the editorship of the *Tattvabodhini Patrika* because of his rationalist views. Rajendralal Mitra edited *Vividhartha Sangraha* and *Rahasya Sandarbha* and structured them on Baconian principles. The fourth issue of *Vividhartha* carried a long article on the 'Baconian System of Philosophy'.[166]

In north India, the spirit of scientific rationality was epitomized in the teachings of Master Ramchandra who taught mathematics at Delhi College in the 1840s.[167] Through his journals, *Fawaid-ul-Nazrin* and

[160] S.G. Burrard, 'Mount Everėst: The Story of a Long Controversy', *Nature*, 71, 10 Nov. 1904, pp. 42–64.

[161] Gautam Chattopadhyaya (ed.), *Awakening in Bengal*, I, Calcutta, 1965, pp. XII–XXV.

[162] *Transactions of the AHSI*, XXV, March 1828, p. 352.

[163] *Transactions of the Medical and Physical Society of Calcutta*, III–V, 1827–31.

[164] Rajat Sanyal, *Voluntary Associations and the Urban Public Life in Bengal (1815–1876)*, Calcutta, 1980, pp. 269–73.

[165] B.·Bhattacharya, *Banga Sahitya Vijnan*, Calcutta 1960, pp. 61–4.

[166] Siddharth Ghosh, *Rajendralal Mitra: Baconian Axis in a Colony*, memeographed.

[167] S. Irfan Habib and Dhruv Raina have brought forth this historic figure from

Mohabb-i-Hind, he relentlessly attacked superstition, and wrote numerous articles on science and technology, scientists and the scientific
temper. Like his peers in eastern India, he was also an avid translation-
enthusiast, and wanted the benefits of modern civilization to reach the
masses through the medium of an Indian language. It is difficult to
gauge to what extent he was successful. The significant point is that he
was very sincere in his efforts. Like Rammohun, he too could not
comprehend, much less articulate, colonialism in political and economic
terms.

Master Ramchandra was slightly different because he also showed an
interest in original research. His *Treatise on the Problems of Maxima and
Minima* (1850) and *A Specimen of a New Method of the Differential
Calculus* won him laurels and were reviewed by Professor Reynolds of
Cambridge and Professor Kellaud of Edinburgh. The DPI of the North
West Provinces considered the book to be 'of great scientific merit', but
added that the work 'is one of little practical utility'. The Lieutenant-
Governor of NWP wanted to honour Ramchandra by making him a
Fellow of Calcutta University; the Governor-General refused and instead
offered him a khillut (a pecuniary reward).[168]

The scientific talent of Shubhaji Babu and Onkar Bhatt Joshi was
evident in central India; their respective works, *Siddhanta Shiromani
Prakasa* and *Bhugolsar,* were rated very high.[169] The latter, published in
1841 by the Agra School Book Society, is an extremely interesting tract.
It exposes the Pauranic myths and makes a comparative study of the Siddhantic and Copernican systems of astronomy. It is written in a 'teacher-
taught' dialogue style. The questioning pupil enquires: 'The Great Vyasa
has written that the oceans are made of nectar and butter-milk. How
could he be wrong?' The teacher replies: 'What Vyasa Muni was doing was
in praise of God, he was not doing *Vigyan* (Science)'. Thereafter he says
that the British had navigated several oceans and had found them to be
made of only saline water. Subhaji Bapu's works on the Euclid, algebra
and trigonometry (all in Hindi) were hailed as 'models of what educational works ought to be'.[170]

In Bombay, Bal Gangadhar Shastri Jambhekar (the first Indian

---

virtual oblivion. For details, see 'The Introduction of Scientific Rationality into
India: A Study of Master Ramachandra—Urdu Journalist, Mathematician and
Educationist', *Annals of Science,* 46, 1989, pp. 597–610.

[168] Home, Education, nos. 11–13, 8 May 1857.

[169] *Calcutta Monthly Journal,* XXXI, June 1837, p. 393.

[170] J.R. Ballantyne, *A Discourse on Translation,* Mirzapur, 1855, p. 13.

assistant professor of mathematics at Elphinstone College) was also, like Ramchandra, a translation-enthusiast and wrote on theories of equation and calculus in Marathi. He was convinced that higher education could be obtained only through the medium of English, but he valued the role of regional languages for mass education.[171] He was perhaps the first Indian to start a journal (*Bombay Durpan* in 1831) for popularizing science and established the Native Education Society which later did a commendable job in translating European works. He also worked for medical education and wanted it to be in the vernacular so that the native vaidyas and hakims could study. He was a member of the Bombay Geographical Society but was denied entry into the Bombay Branch of the Royal Asiatic Society. His contemporary, Hari Keshavji Pathare (1804–58) wrote *Siddhapadartha Vijnyan Sastra* (natural philosophy), *Rasayana Sastra* (chemistry) and *Sastriya Jnanadarshana* (a work on general science). During 1818–57, a dozen tracts, influenced by Western thought, were written in Marathi.[172] Indian students themselves came forward; boys from Elphinstone College established a students' Literary and Scientific Society. Around five to six hundred local students heard lectures on chemistry by Ardaseer Framjee and later attended a course on mechanics, hydrostatics, pneumatics, optics, heat, electricity and magnetism by Dada Bhai Naoroji, then president of the Vernacular Society.[173] That the students imbibed this new knowledge well can be ascertained from what J. Prinsep wrote to O'Shaughnessy after examining the papers of the chemistry students of Calcutta Medical College: 'All the essays are extremely creditable; indeed the extent and accuracy of the information, on the single subject selected to test the abilities of the pupils, has far surpassed my expectations, and I do not think that in Europe any class of chemical pupils would be found capable of passing a better examination.'[174]

Better employment opportunities would have helped popularize science education. With this view, J. Thomson (Lieutenant-Governor, NWP) drew up a scheme for the examination and employment of natives as sub-assistant executive engineers, but the Governor-General allowed

[171] Home, Public, no. 18, K.W., 18 July 1838.

[172] E.M. Gumperz, *English Education and Social Change in late 19th-century Bombay (1858–1898),* Ph.D. thesis, University of California, 1965, p. 446.

[173] *The Bombay Quarterly Magazine,* I, 2 Jan. 1851, pp. 121–2. Dada Bhai was a student of Bal Shastri Jambhekar and was to later become the ablest advocate of Indian cause in British forums.

[174] *Calcutta Monthly Journal,* III, 1837, p. 826.

him to train only two or three natives on an experimental basis and postponed the adoption of the scheme.[175] For the same reason, engineering classes at Elphinstone college did not draw students, and the engineering school at Lahore also had to be shut down.[176,177]

The Indian response to Western medical education was lukewarm at the beginning. The abolition of medical classes at the Madarsa led to a signed petition by 8312 Muslims of Calcutta accusing the government of 'causing the science of Arabia to cease'.[178] The students of the Sanskrit College also reacted in the same way when their stipends were discontinued.[179] Gradually the situation changed for the better. In 1849, the Hindustani class had 100 students, out of which 75 were Muslims.[180] These classes were meant to train students for the job of dressers and compounders. The English class, which obviously had in-built advantages, continued to be dominated by the Hindus, particularly the Kayasthas.

The establishment of the Calcutta Medical College was welcomed by an influential section of the natives. Ramgopal Ghosh presented nineteen volumes of new medical works to the College, Rustomji offered a gold medal and Dwarkanath instituted scholarships. The monetary value of these offers was not much. But it clearly showed that the newly emerging educated group in Calcutta was prepared to overcome a deep-seated prejudice to master Western science.[181]

However, the traditional Kavirajas and Vaidyas, who were confident of getting no encouragement from a foreign government, kept themselves aloof from the modern scientific world.[182] And even those who went to medical college, found the course extremely lengthy and expensive. In 1845 the course was extended from four to five years,[183] and the result was that the very next year the number of students fell from 57 to 32. Since judicial and commercial jobs offered higher salaries, naturally the brighter students preferred courses in these fields. The question was not of

[175] Home, Public, nos. 19–30, 13 Sept. 1845.
[176] IOR, E/4/803, *Bengal Despatches*, LXIV, 1850, p. 362.
[177] Home, Public, no. 105, 20 Nov. 1847.
[178] Home, Public, no. 9, 13 March 1835.
[179] Home, Public, nos. 44–5, 8 April 1835.
[180] *Report of Medical College of Bengal,* 1849–50, p. 3.
[181] Udayan Mitra, *Social Factors in the Rise and Development of the Bengali Middle Class*, Ph.D. thesis, Rabindra Bharati University, 1982, p. 40.
[182] G. Mukhopadhyay, History of Indian Medicine, II, Calcutta, 1923, p. 18.
[183] This was an academically sound decision—a reform that was carried out in England itself forty-five years later. D.G. Crawford, II, p. 442, note 147.

unwillingness on the part of students but of inducement given by the government. In the army the commander-in-chief refused to accept natives as surgeons; only sub-assistantship was offered.[184] Hence several medical graduates took to private practice.

In 1848, two Indian doctors (B.N. Bose and G.C. Seal) returned from England with a recommendation from the Court to be absorbed in suitable positions. The Council of Education found them qualified for a professorship at the Calcutta Medical College. But the Government of Bengal raised the objection that to give natives such good assignments would inflame the ambitions of others. So one was sent to the medical college hospital and the other to a dispensary.[185] While in England, they had won several medals which obviously caused some heart-burning in Calcutta, for a contemporary disdainfully noted that 'parading and laying stress on these things (medals) takes the bloom off them', and even warned that 'nothing can be more dangerous or injurious to the future professional progress of these native surgeons than the over-laudation bestowed on them'.[186] One of them, S.C. Chuckerbutty, later got into the IMS when its examinations were thrown open to the Indians. Henceforth there was to be no dearth of eager candidates whenever opportunities arose in any field, and, as we shall see later, a major obsession with British officials in the subsequent decades was how to curtail such opportunities.

Pattern Discerned

The numerous facts cited above clearly show that the British developed and employed skilfully what Morris Berman calls 'the ideology of science, i.e. that part of the complex of ideas which deals with a certain class or group's attitude towards science, especially as a factor in promoting its own interests, economic or otherwise'.[187] The colonizers, in both their administrative as well as individual capacities, certainly subscribed to a particular ideology of science. Their concept of science was closely related to the needs of the empire. They were not oblivious to the great strides science was making in their home country, but were also quick to recognize the role and importance of science in empire-building.

The most interesting feature of this early phase of colonial science lies in its highly individualistic character. During this phase science owed

[184] Home, Public, no. 41, 6 Jan. 1841.
[185] Home, Public, no. 10, 29 April 1848.
[186] The Indian Register of Medical Science, I, Calcutta, 1848, p. 333.
[187] M. Berman, p. VII, note 61.

much to individual enthusiasm and some of these individuals merit separate study. Not without reason did R.J. Bingle plead 'for less emphasis on government policy and more on the individuals who pushed the Government of India, to sponsor their work, form policies and establish scientific institutions'.[188] These were certain individuals on the spot who largely determined what was advantageous to both trade as well as the country. Thus Rennell, Kyd, Roxburgh, Carey, Lambton, Williams, O'Shaughnessy and others emerge as pioneers. These colonial scientists tried their hand in several fields simultaneously and were in fact botanists, geologists, zoologists, physicists, chemists, geographers and educators— all rolled into one.[189] This had its positive as well as negative points. As data-gatherers they had no peers; but for analysis and recognition, they had to depend on the metropolitan scientific culture whose offshoots they were and from which they drew sustenance.

The scientists in the metropolis always showed a keen interest in the works of colonial scientists. Banks was a patriarch—perhaps the most formidable pillar on which numerous scientific workers scattered all over the globe tried to lean at some point or the other. In this regard, the Roxburgh–Banks correspondence makes for a very fascinating study. Roxburgh wanted the Directors of the EIC to publish his botanical notes and drawings and for this he looked to Banks for support—'my great distance from every sort of help and with only a limited library renders your support doubly necessary'.[190] A decade later, he wrote: 'It would only require a hint from Sir J. Banks to the Directors to induce them to send me all new publications on Natural History.'[191] He even got his eldest son, William, appointed as his assistant with the help of Banks. It was not without some difficulty that this was done, for to quote Banks: 'I can do nothing for him (William) yet because the present chairman (of the Company) does not like me, but he will soon pass away and if he is succeeded by some friend of mine, be assured . . . .'[192] Banks' name awed

[188] *Proceedings of the Sixth European Conference on Modern South Asian Studies*, Sevres, July 1978, p. 32.

[189] For example, while seeking lecturership at the Calcutta Medical College, O'Shaughnessy offered not only to teach chemistry and experiment with medical plants but also to 'give practical instructions to non-medical as well as Indian and European students, on the chemical arts of dyeing, bleaching, calico-printing, distilling, sugar-refining, melting of ores and manufacture of drugs'. Home, Public, no. 15, 5 Aug. 1835.

[190] Roxburgh to Banks, 30 Dec. 1790, B.M. Add. Mss. 33979, f. 64.

[191] Roxburgh to A.R. Lambert, 21 Feb. 1801, B.M. Add. Mss. 28545, ff. 173–4.

[192] Banks to Roxburgh, 7 Jan. 1799, B.M. Add. Mss. 33980, f. 138.

and inspired even those who had no contact with him. One L Davis sent to Banks a tract on agriculture written by a zamindar, Raja Mit Jeet Singh, and in his letter wrote: 'The Raja will be encouraged if he is informed by Lady Loudoun or Lord Moira that his treatise has come under the notice of Banks.'[193]

In July 1838 the presidents and fellows of three distinguished societies, the Royal Society, the Geological Society and the Royal Geographical Society, collectively presented a memorandum to the Company in which, along with tidal and magnetic experiments, they pleaded for 'an improved topographic map of India, engraved in the style of the British and other European surveys, accompanied by geological, statistical and other informations similar to that preparing for the Ordnance Survey of Ireland'.[194] This proposal was not accepted. Nothing more was heard of it until Humboldt urged the then King of Prussia to induce the Company to reconsider the subject. This resulted in the scientific mission of the Schlagintweit brothers to India during 1854–55.[195]

Both the workers in India and their mentors at home were thus equally enthusiastic, and their interests no less diversified. Anderson's interests, for example, were wide, not only on the botanical side, but also in the development of the industry of the countryside. The Court had to once scold him for interfering in what was not considered his business.[196] The Court wanted its officials to give the maximum attention to their officially assigned scientific works and many of them often complained of the paucity of time or funds. Roxburgh, for example, wrote in 1790:

The greatest, if not the only obstacle to the progress of knowledge in these provinces is our want of leisure for general researches; and as Archimedes, who was happily master of his time, had no space enough to move the greatest weight with the smallest force, thus, we, who have ample space for our enquiries, really want time for the pursuit of them. 'Give me a place to stand on', said the great mathematician, 'and I will move the whole earth'. Give us time, we may say, for our investigations, and we will transfer to Europe all the sciences, arts and literature of Asia.[197]

But the Court remained firm in its resolution. In 1837, commenting upon MacLelland's and Griffith's visit to Burma, the Court made its position very clear: 'The practice of employing medical officers apart from their official duties to carry on agricultural or horticultural experiments,

[193] L. Davis to Banks, n.d., B.M. Add. Mss. 8968, ff. 60–1.
[194] Major T.B. Jervis Papers, Archives of the Royal Geographical Society, London.
[195] *The Athenaeum*, 1764, 17 Aug. 1861, p. 215.
[196] Anderson Papers, IOL, Photo, Eur. 85.
[197] Roxburgh Papers, IOL, Eur. Mss. D. 809.

or to prosecute enquiries into matters connected with natural history, is one of which we generally disapprove.' It agreed that such visits would be 'useful to the cause of science' but added that they would be 'only remotely connected with the particular question of the practicability of cultivating the tea-plant with a view to its manufacture as an article of commerce'.[198]

The Court maintained a tight administrative control, no doubt, but always felt gratified when an employee did something 'useful' in his leisure time without any extra cost to the Company. Captain Campbell, for instance, was officially a surveyor at Salem. But he set up a small laboratory there on his own, and experimented on the best mode of smelting iron, on the manufacture of carbonate of soda from the soda soil of Baramahal, and also on reeling silk and in constructing an improved winding reel. He made meteorological observations for four years and wrote separate manuals on both meteorology and chemistry. His *Manual of Chemistry* contained information on how muriate of morphia could be easily made for 1/100th part of the price in Europe. The government was duly impressed with these works, but wanted them to be done in his leisure time and without government money. Campbell was informed that the Governor-General could not 'authorize the employment of a military officer on scientific pursuit which are not professional or incumbent on him as part of his proper duties'.[199] The government was thus willing to reap the benefits but not pay the price.

The spirit of the age permitted a colonial scientist to conveniently shift from one branch of science to another (e.g. from botany to geology) in accordance with local conditions and circumstances. W.B. O'Shaughnessy, for example, underwent a remarkable metamorphosis. He began his career as a chemistry teacher at the Calcutta Medical College in 1835, but within a decade abandoned it in favour of the more practical work of telegraph installation. The chemist in O'Shaughnessy gradually got interested in electrical problems. He experimented with an induction coil on dead as well as living bodies and observed the effect of electric shocks in healing some diseases. He did not restrict his ideas to theoretical concepts or laboratory experiments. While experimenting on an electric motor, his ultimate goal was to make 'a light, economic and harmless engine' which could replace 'the preponderous and expensive steam engines'.[200] Such engines could not be made but the experiments helped him to gain knowledge of telegraph engineering. He worked without

[198] IOR, E/4/752, *Bengal Despatches*, XII, 23 Aug. 1837, pp. 408–9.

[199] Home, Public, no. 4, 5 Jan. 1842.

[200] W.B. O'Shaughnessy, 'On the Employment of Electromagnet', *Journal of the Calcutta Medical and Physical Society*, I, 1847, p. 40.

yielding to the 'absentee landlordship' of European scientists, and displayed an uncommon adaptivity to colonial conditions and requirements. But this does not make him 'a scientist belonging to the period of transition from Phase II to Phase III in Basalla's model', as Ghose would have us believe.[201] O'Shaughnessy remains a colonial scientist, conscious of, and dedicated to, the practical problems of his government.

Personal relationships often proved vital. Wilkinson narrates an interesting instance. Rana Madhav Singh of Rewa got suspicious when Captain Paton entered his territories for survey purposes and the latter was asked to withdraw. Wilkinson had a good rapport with the Rana; he called the Rana's astronomer (Joshi) and with the help of the 33rd verse of the *Siddhanta Shiromani* (wherein Bhaskar gives succinct general directions for the measurement of an arc of the meridian) he satisfied them of Captain Paton's real objectives. Henceforth the Rana extended his full co-operation.[202] Later Colonel Waugh recollected how his personal friendship with Maharaja Gulab Singh helped Captain Montgomerie to complete his survey of Kashmir.[203] Without such tact and conciliation, it would have been impossible to carry out survey operations which were not only of scientific but of immense strategic value.

This sort of dependence upon individual whims was not confined to the Indian situation. The Court itself was no better. A change at the directors level did not necessarily mean a sharp turn in policy but it did often bring a change in *modus operandi*. Banks himself complained that the directors were always changing and that he was tired of 'urging schemes on such an unstable body'.[204] Calder later accused the Court of 'supineness and indifference respecting the Natural History of her eastern dominions'.[205] The Court's role was all the more important because it always maintained a very close and tight control over its Indian administration. Such a policy of remote but absolute control invariably led to long procedural delays. In 1834, for example, the Bombay Geographical Society applied to the local government for permission to inspect the public records of the marine department. This had to be referred to the Court, and a minor thing like this took three years to materialize.[206]

Another important feature of this early phase is the relative neglect of

[201] S.K. Ghose, pp. 59–67, note 81.
[202] *JASB*, III, 1834, p. 512.
[203] Waugh Papers, IOL, Mss. Eur. F. 181/5.
[204] Banks to Roxburgh, 9 Aug. 1798, B.M. Add. Mss. 33980/159–60.
[205] *Asiatic Researches*, XVIII, 1, 1829, p. 1.
[206] *The East India and Colonial Magazine*, XI, 65, April 1836, p. 303.

medical and zoological sciences and this is in sharp contrast to larger investments in botanical, geological and geographical surveys from which the British hoped to get direct and substantial economic and military advantages, while medical or zoological sciences did not hold such promises. Some of the early naturalists did not conceal their disinterest in zoological studies. William Jones, for example, gave a very strange explanation; he would not patronize such studies because they involved pain to living objects. 'Why deprive the butterfly its natural enjoyment because it has the misfortune to be rare or beautiful', he observed.[207]

In 1804, an 'Institution for promoting the Natural History of India' was set up at Barrackpur near Calcutta, with a menagerie and an aviary. But the Court showed little enthusiasm for this brain-child of Wellesley,[208] who had envisaged the institution as a scientific annexe of a college at Ft. William where the Company's servants were to be trained. Although the Company refused to finance the larger scheme, the menagerie came into existence and was put under the charge of Buchanan. Fleming and Gibbons helped him on a voluntary basis. Specimens began to pour in from all parts of India; paintings were made of these specimens and meticulous notes prepared and sent to London. After 1808, this work was allowed to lapse, and the menagerie lost its scientific purpose and degenerated into a place of amusement.[209] Later, Blyth, Sykes and Hodgson tried to win recognition for the subject. Still writing in 1836, Hodgson had reasons to lament that Indian zoology was carried on either by men who knew about animals but were 'inexpert in science', or by those 'rapid passengers' who had no time to observe creatures and thought up their accounts afterwards.[210]

Research in physics or chemistry was simply out of the question. John Warren did experiment on the refraction of light in air and also on the oscillation of a pendulum at Madras, while Goldingham made some measurements on the velocity of sound. Prinsep made a few chemical analyses of natural waters and rocks. Nothing more was possible, for there were no laboratories, equipment or specialized training facilities.

The reigning spirit remained that of exploration. Systematization or analysis of results had to wait for some time. Western medical classes, for

---

[207] *Centenary Review of the Asiatic Society of Bengal*, III, Calcutta, 1885, p. 57.
[208] Home, Public, no. 23, 26 July 1804.
[209] M. Archer, 'India and Natural History: The Role of the East India Company', *History Today*, IX, 1 Jan. 1959, p. 738.
[210] Quoted in H.J.C. Larwood, 'Western Science in India before 1850s', *Journal of Royal Asiatic Society*, 1962, p. 71.

instance, were started in 1822, but it took another thirty years to produce the first exhaustive compilation of information on tropical diseases in India.[211] The treatment and study of tropical diseases was undertaken by individuals who were separated both geographically and professionally and so, naturally, a consistent body of knowledge failed to develop.[212] This was true for every branch of knowledge.

However, even disjointed and often haphazard studies served some purpose. The positive achievements were the establishment of scientific bodies and museums. Archer found a utilitarian spirit in 'that zealous band of surgeons and engineers, sweating in the torrid plains or struggling through malaria-ridden forests, who gave the world its basic knowledge of India's natural history'.[213] This was part of a once fashionable tendency to attribute all works of improvement, all scientific approaches to Indian problems to the utilitarians.

The Company itself, notwithstanding the stinging criticism referred to above, did earn a few encomiums from its scientific workers. Rennell declared forcefully that 'whatever charges may be imputable to the managers of the Company, the neglect of useful science, however, is not among the number. The employing of geographers, and surveying pilots in India; and the providing of astronomical instruments, and the holding out of encouragement to such as should use them, indicate, at least, a spirit somewhat above the mere consideration of gain'.[214] Goldingham found the company desirous 'to have everything done for the benefit of science', and thus 'truly deserving the thanks of the scientific world in Europe'.[215]

Several governor-generals also shared the lofty views of these pioneers. As early as in 1804, Wellesley wrote that 'to facilitate and promote all enquiries which may be calculated to enlarge the boundaries of general science is a duty imposed on the British Government in India by its present exalted situation'.[216] Addressing the College of Ft. William on 27 July 1825, Lord Amherst assured that 'while the means of introducing European science are diligently sought, there is no desire hastily to

---

[211] This was by Charles Morehead, *Clinical Researches on Disease in India*, 2 vols, London, 1856.

[212] M. Worboys, *Science and British Colonial Imperialism*, D. Phil. thesis, University of Sussex, 1979, p. 88.

[213] M. Archer, p. 743, note 209.

[214] J. Renell, *Memoir or a Map of Hindustan*, London, 1788, p. VIII.

[215] J. Goldingham, *Astronomical Observations at Madras*, III, Madras, 1825, p. 1.

[216] Home, Public, no. 23, 26 July 1804.

supercede what exists; no attempt abruptly to introduce improvements, before the way is paved for their reception'.[217] Ten years later Bentinck declared that 'the great object of the British Government ought to be the promotion of European literature and science among the natives of India'.[218]

These were good intentions indeed, which, to be realized, required far more liberal educational and job opportunities than the government was willing to concede to the natives. Metcalfe found it inexpedient 'to force them unnaturally into new employment for the sake of a theory. Let them be employed whenever it is deemed desirable'.[219] Macnaghten voiced the same opinion.[220] The catch, however, lay in the word 'desirable'. These pious expressions of goodwill, filtering through bureaucractic shelves, ended only in empty verbiage.

The Company officials spoke the idiom of rationality but only to the extent that matched the rhetoric of administration. Macaulay, the cultural essayist, and Macaulay, the administrator, were not different or incongruent. The roles were interchangeable and were meant to instil the same effect—domination. The Orientalists were equally involved in this project, albeit in a slightly different way. The Anglicists and the Orientalists look like polar opposites but in reality they are not. A recent critique treats them 'as points along a continuum of attitudes'.[221] The continnum can be marked but not divided. These attitudes also influenced the nature of local response and set the limits to which the 'natives' would go along. The response was largely defensive, but it was neither xenophobic nor slavish. There are several instances of a rational assessment on the part of the colonized. Much, however, depended on the opportunities offered or created. The colonial state was obviously not supposed to work like a welfare state.

It would be wrong to accuse the East India Company of ignorance, innocence or deliberate neglect of scientific explorations. Here was a trading company par excellence; it could easily discern where the advantages lay. Administratively, it enjoyed a great deal of autonomy. It was virtually a 'state' within the state, and so could effectively opt for state-

---

[217] *Calcutta Monthly Journal,* Aug. 1825, p. 96.
[218] Home, Public, no. 19, 7 March 1835.
[219] Minute by Charles Metcalfe, 11 Oct. 1829, in C.H. Phillips (ed.), *The Correspondence of Lord William Bentinck,* I, Oxford. 1977, p. 178.
[220] Home, Public, no. 30, 17 March 1835.
[221] G Viswanathan, p. 30, note 98.

supported scientific and technical activities. The fact that such activities lacked coherence does not mean that a policy did not exist. Banks found it inherent in the very law of nature. For him and his people:

A colony like this (India) blessed with advantages of soil, climate and population so eminently above its mother country, seems by nature intended for the purposes of supplying her with all raw materials and it must be allowed that a colony yielding that kind of tribute, binds itself to the mother country by the strongest and most indissoluble of human ties, that of common interest and mutual advantage.[222]

[222] J. Banks, Tea Memorial, 27 Dec. 1788, IOL, Mss. Eur. D. 993.

CHAPTER 3

Administering Science:
Organizational Imperatives

The other day the Under Secretary had the cheek to propose that all my writings, reports, etc. should pass through his hands for revision before they are issued. Secretariat interference is the curse of every Indian branch of special or scientific work. . . . I wish to be free. I object to admitting that any purely administrative officer is capable of correcting my writings or of doing more than obtaining the annual grant for scientific investigation.

George Watt, Economic Reporter to the Government of India,
November 1884

Council is an administrative body. The place of the expert is to advice, not to administer, and his place is accordingly outside it, not on it.

Lord Curzon on the possibility of a scientist in his Council,
January 1904

When the Crown formally took over the Indian administration in 1858. exploratory activities in the country were already beyond their formative stage. The problem now was more of consolidating and institutionalizing the gains which individuals had indefatigably laboured for. In some areas like survey operations, medical education, etc. the Company had taken the initiative and made some headway, but the results were not up to the expectations of the science-enthusiasts, both in India and at home. Hence, there was no remorse at the transfer of power from the Company to the Crown. A botanist, stationed at Karachi, wrote: 'I fancy no scientific man in England is sorry at the change of masters we have got, for I am certain the present Government will encourage science much more than the Directors ever did, especially among their own servants.'[1]

[1] N.A. Dalzell to J.D. Hooker, 12 June 1859, Kew Papers, Indian Letters, 1853–1900, vol. 154, f. 26.

This hope was largely misplaced, and thirty-seven years later, a publicist in Madras regretted that 'the State—the great Government of India—has not yet been able to quite get rid of the mercantile monopolist policy of the East India Company'.[2] The basic tenets of administration, the objectives and goals remained the same. But the area of operations was extended and penetration was deeper. The need for sound institutional experiments was felt to consolidate the imperial gains.

Survey Organizations

The Company had realized the importance of survey works for military and revenue considerations. As late as in 1901, when a survey party in Kashmir was asked to make a survey of the glaciers for purely scientific purposes, the then Surveyor-General of India privately wrote to Curzon: 'The main reason of the survey is to improve our maps for military purposes, and where so much country remains to be surveyed, it is a pity to delay the work in order to make accurate surveys of the glaciers generally.'[3]

In 1818 the GTSI was created to supervize all types of survey operations, whether topographical, trigonometrical or related to revenue matters. Lambton, Everest and Waugh enjoyed this centralization of power. In 1862, after Waugh's retirement, the office of the Surveyor-General was separated from that of the superintendency of the Trigonometrical Survey; Colonel Thuillier was appointed to the former post, and Colonel Walker to the latter office. But the division of labour was not absolutely clear, and the sphere of operations overlapped quite often. In 1864 Colonel Dickens submitted an elaborate report on the Survey Department. He envisaged the position of the Surveyor-General as that of 'a consulting officer to the Government of India', while the trigonometrical and topographical surveys were to be under separate superintendents attached to each local administration.[4] But this scheme of total decentralization did not find favour.

Between 1861 and 1878, under Thuillier's command, revenue surveys were extensively carried out and were given a more professional name— the Cadastral Survey. Henceforth this was to occupy the maximum time and money of the entire survey establishment. In 1891–92, for example,

[2] *Indian Engineering*, XIX, 23 May 1896, p. 324.
[3] G.C. Gore to Curzon, 9 May 1901, Curzon Papers, IOL, Mss. Eur. F. 111/203.
[4] *Report of the Indian Survey Committee*, Calcutta, 1905, p. 8.

out of 21 survey parties, 12 were engaged in remunerative operations, i.e. on work connected with the enhancement of land or forest revenue. The 21 parties were:[5]

Specific work	No. of parties
Trigonometrical	1
Topographical	3
Forest	4
Cadastral	7
Traverse	1
Scientific (tidal, electro-telegraphic, astronomical, etc.)	3
Geographical	2

Revenue enhancement thus engaged more than 50 per cent of the total workforce of the survey, and the expenditure was even greater. The average cost of revenue surveys was more than double that spent on topographical surveys. The following chart illustrates the cost and average rate of surveys during 1870–71:[6]

	Square miles covered	Total cost in rupees	Average rate of survey per square mile
Topographical Survey	14,592	3,24,225	Rs 22 (about £2.50)
Revenue Survey	16,938	7,64,745	Rs 45 (about £5)

These survey operations came under a severe financial strain during the 1870s. Several survey parties, both topographical and revenue related, were broken up; while the native subordinates were retrenched, several European assistants were asked either to retire on pensions or to go to other departments.[7] The budget was almost static during 1869–76, and savings remained a major concern. The following table bears this out:[8]

---

[5] *General Report on the Operations of the Survey of India*, 1891–92, p. 1.
[6] Taken from *General Report of the Topographical Survey of India*, 1870–71, p. 7.
[7] Thuillier, the Surveyor-General, remonstrated that 'the diminution of the trained machinery as recently insisted by the Government . . . appears to be as much a mistake financially, as it must be injurious to the crying wants and necessities of the Administrations, which must have the survey they ask for, sooner or later', ibid., 1876–77, p. 7.
[8] Ibid., p. 8.

Financial year	Sanctioned budget grant in Rs	Actual expenditure in Rs	Savings	Deficit
1869–70	27,08,282	24,58,938	2,49,340	...
1870–71	23,67,273	20,87,103	2,80,170	...
1871–72	24,04,710	22,18,409	1,86,301	...
1872–73	24,56,190	24,10,796	45,394	...
1873–74	24,15,570	24,55,198	...	39,628
1874–75	24,15,930	24,35,124	...	19,194
1875–76	23,81,000	23,38,614	42,386	...

Apart from this administrative and fiscal pruning, a need was felt to give the survey works a more compact and cohesive look. So in 1878 the three branches—trigonometrical, topographical and revenue—which till then had been virtually separate departments, each with its own cadre and establishment, were amalgamated. The post of the Superintendent of the GTSI was abolished, and his duties were undertaken by the Surveyor-General in addition to his topographical work, while a Deputy Surveyor-General was placed in charge of revenue surveys. In 1882 the original programme of the GTSI was completed; and as topographical work was regarded more a luxury than a necessity, it was decided to transfer parties, as far as possible, to cadastral work or other operations, the cost of which fell to local administrations. The result was that the energies of the surveyors were totally diverted to forest and cadastral work. Whereas, in 1883–84, there were eight topographical parties, in 1890 only four survived. The Indian Survey Committee, constituted by Curzon in 1904 to enquire into the working of the Survey of India, expressed grave concern at this shift in priorities.[9]

Geological explorations had a far more direct economic bearing. The East India Company was aware of its importance but was not willing to sanction a separate full-fledged geological department and had lumped it with topographical surveys. The Company's policy was to derive the maximum benefit at minimum cost. The arrival of Thomas Oldham in March 1851 marked the establishment of the Geological Survey of India (GSI) on a regular basis.[10] He started, as far as material conditions and

[9] Report of the Indian Survey Committee, pp. 11, 38, note 4.
[10] C.S. Fox, 'The Geological Survey of India', Nature, 160, Dec. 1947, pp. 889–91.

opportunities were concerned, literally from scratch—a room, a box and a messenger.[11] His job was a contractual one (for five years) and was always renewed until he himself sought retirement in 1876 on account of failing health. During the first phase of his contract, he chose seven geologists with great care (of whom three, H.B. Medlicott, W.T. Blanford and H.F. Blanford, later became Fellows of the Royal Society) and laid a solid foundation for the GSI. At this time the nomenclature of the officers was a simple one. Oldham was Geological Surveyor and the others were all assistants. The department was treated as a personal affair of its head.[12] The Government of India was satisfied with his work, but not the Court, which passed adverse remarks even while granting him a second term in 1856.[13] Perhaps the court wanted him to concentrate more on the survey of mineral resources than on the study of geological structures. The Government of India, however, defended Oldham and even authorized him, in anticipation of the concurrence of the Court, to recruit three or four more assistants. In 1856, a permanent office-building and a museum was also sanctioned, and publications such as memoirs and annual reports were started in the same year. Senior assistants like J.G. Medlicott and the two Blanfords were placed in charge of parties, a development in the organization that was to lead to the introduction of the term 'Deputy Superintendent' in 1862. By 1860 twelve more assistants had been appointed and in 1862 the number rose to sixteen.[14] But the staff members could never stabilize. Many of them died due to climatic exposure and a few left because of stagnation in service. In 1862 J.G. Medlicott chose to become an Inspector of Schools and was lost to science, while H.F. Blanford turned to meteorology. In almost every annual report, Oldham alluded to difficulties in obtaining properly qualified assistants. To make the service more attractive, in May 1866, the assistants were divided into three grades. The first grade was to receive a salary of up to Rs 1000 per month, the second up to Rs 700 and the third up to Rs 500. A new entrant was to get a minimum of Rs 300 per month and all the assistants were entitled to an annual increment of Rs 50. Promotion was to depend on merit and on the existence of a vacancy, but in the absence of a vacancy, the candidate's claim on annual increment was to be honoured.[15] This was a different practice from that followed by

[11] C.R. Markham, *A Memoir on the Indian Surveys*, London, 1871, p. 154.
[12] L.L. Fermor, *First Twenty-Five Years of the GSI*, Calcutta, 1976, p. 42.
[13] Ibid., p. 157.
[14] This was to be the maximum strength for the next forty years.
[15] Home, Public, nos. 82–4, Jan. 1867.

other departments. Oldham even wanted to induce a few of the good students of Calcutta University to join the Survey as apprentices by offering them scholarships but this was vetoed by the Finance Department.[16] Finance officials offered an absurd solution—since scholarships involved money, they suggested that inducements be given in the form of extra marks at the University examinations.[17] Oldham refused to accept this offer. In 1871 he pleaded for a substantial hike in salary (involving an extra expenditure of Rs 39,600) and also for a special relaxation of leave rules so that his assistants could be sent out of India during summer recess on geological missions, again of course at the government's expense.[18] The government refused to grant GSI these special favours because acceptance of Oldham's proposals would have encouraged other surveys to raise similar demands. A comparison of all the officers drawing Rs 400 and more in each survey shows that the GSI was already in a favourable position:[19]

	No. of officers	Average salary per month (Rs)	Average years of service
Geological Survey	16	770	11
Great Trigonometrical Survey	27	717	16
Topographical Survey	17	583	15

Still, Oldham did a lot to enrich the organization. When he retired in 1874, he left behind an excellent collection of specimens and books, an impressive array of publications and a dedicated cadre. He was not averse to the 'natives' and had pleaded for science education (especially geology) at the university level. When this did not materialize, he even attempted to remedy the situation by introducing a system of apprenticeship.[20] Ram Singh was the first Indian to join the GSI as an apprentice in March 1873 and was made to attend the elementary science classes at the Shibpur Engineering School. Two more apprentices, Kishan Singh and Hira Lal, who joined in January 1874, were asked to attend Physical Science lectures at the Presidency College, Calcutta. Oldham's successor, H.B.

[16] Home, Public, Surveys, no. 65, 10 July 1869.
[17] Home, Public, nos. 116–17, Jan. 1870.
[18] Revenue, Agriculture, Surveys, nos. 1–10, Jan. 1872.
[19] Ibid.
[20] Revenue, Agriculture, Surveys, no. 38, June 1874, pt. B; *Records of GSI*, VII, 1874, p. 8.

Medlicott, however, reversed the trend. He was disillusioned with persons such as P.N. Bose and P.N. Datta who had been educated in England and who owed their appointment in the GSI directly to the Secretary of State. Geology had attracted the attention of Indian students in the 1890s. Still later, Holland, who superceded Bose as the director in 1903, wanted geologists to be trained in Britain itself on a regular basis for employment in the colonies. He preferred to allow the work to suffer but would not settle for anything less.[21] Professional jealousy and discrimination on grounds of race harmed not only the Indians but in some cases the Europeans also. When Waagen was being tipped for the post of palae-ontologist in the GSI after Stoliczka's death in 1874, A.O. Hume, then Secretary to the GOI, remarked that 'though I have the greatest liking for Germans, I do not think that our whole Survey should consist of foreigners'.[22]

The botanical survey could never acquire the structural homogeneity and cohesiveness which characterized the GSI or the Survey of India. It was not that the colonizers were oblivious of its importance, rather they preferred to keep it administratively flexible so that it could be con-veniently put to fairly diversified uses. The botanists could not get an imperial cadre. Some of them were attached to the revenue department, some to agriculture, and a few to forestry and medicinal works. The nature of assignments was equally varied. After Wallich, scientific surveys and classification had ceased to be the major occupation of botanists. Accli-matization of imported exotic plants and seeds was given top priority. Concern for vegetable technology (namely dyes, oils, fibres and other miscellaneous products of vegetable origin) superceded everything else and Royle (1823–31) and Jameson (1842–75) gave practically their whole attention to it. Brandis and Gamble, who began as botanists, turned to the management of forests. Pharmacography also owed some-thing to the botanists. The botanical portion of Ainslie's *Materia Medica* (1813) was written by Roxburgh, that of O'Shaughnessy's *Bengal Pharmacopoeia* (1844) by Wallich, and that of U.C. Dutt's *Materia Medica of the Hindus*(1877) and K.L. Dey's *The Indigenous Drugs of India* (1896) by G. King. From the importation of the cinchona plant to the manufacture of quinine, their association was required. The Superinten-dent of the Botanic Garden in Calcutta was also the Superintendent of Cinchona Cultivation in Bengal. Such a multitude of objectives could not

---

[21] Mss. GSM 2/284, Institute of Geological Sciences, London.
[22] Revenue, Agriculture, Surveys, nos. 6–11, April 1875.

have been efficiently pursued under a strictly centralized system. In fact the botanists themselves did not want such a system. To quote Gamble: 'No good will be done by making the (Botanical) Department Imperial— the officer in each Province must be under the Local Government in its Revenue Secretariat.'[23] Local demands and exigencies in this area were much more pressing than in geological or topographical surveys. This led to certain administrative anamolies. The Superintendent of the Saharanpur Botanic Garden worked under the Director of Agriculture, NWP, while those of Bengal and Madras were under the Revenue Department. When the forest department was established, some of the botanists were asked to lend their services; and the Superintendent of the Royal Botanic Garden in Calcutta had to assume for a while the duties of the Conservator of Forests in Bengal. When agricultural departments were created in the provinces and officers had to be found to man the new departments, the botanists, already endowed with some experience of Indian conditions and plants, fitted in ideally. This, however, meant a loss to scientific botanical investigations. Apart from full-scale transfers, services were also sought on a temporary basis for a specific purpose. In 1872 the curator of the Calcutta Botanic Garden, John Scott, was deputed to the poppy districts of Bihar. He was asked to make two sets of reports—one to the Board of Revenue regarding practical points relating to the poppy plant and opium, and the other to the Superintendent of the Botanic Garden regarding the scientific points in physiological botany.[24] This dual reporting, both in administration and research, remained the chief characteristic of the botanical survey throughout the century. In contrast, geologists and topographers had the advantage of belonging to distinct administrative units though as investigators they also often had to perform double roles.

Botanic gardens suffered the most in the post-Wallichian years. Lack of direction and administrative apathy eroded their credibility. King gives the most vivid description:

Jameson, who for about 40 years was superintendent of the Saharanpur garden, was a botanist only in name; and under him the botanical part of the work of Saharanpur had been utterly lost sight of. When I was at Saharanpur (1868–69), there was hardly any material for a herbarium, there was no botanical library; in fact there were *absolutely no appliances for botanical work*. I asked for sanction for

[23] J.S. Gamble to Thiselton-Dyer (Director, Royal Botanical Garden, Kew), 28 March 1888, Kew Papers, Miscellaneous Reports, Botanical Survey, 1884–1920, ff. 28–30.

[24] *Bengal Administration Report*, 1875–76, p. 173.

the purchase of a few of the more necessary botanical books, but to my application only a partial and grudging assent was given. The entire function of the garden were horticultural, and the superintendent's energies were confined to the cultivation of vegetable and other seeds for the soldiers' gardens, and to the care of two tea gardens which the Local Government then owned in Kumaon. The impression which was foced upon me was that the Local Govt. took neither interest nor pride in the garden, and that it would gladly be rid of it. . . . When I took charge of the Calcutta garden in 1871, it was under the Govt. of India, and I must say it was in a very poor state . . . *more than half its area was occupied by tall thatching-grass.*[25]

The cyclones of 1864 and 1867 destroyed whatever little the Calcutta garden had to offer. A private planter who visited this garden a day after the storm, described it as 'a perfect wreck' and advised Anderson (the then Superintendent) to start afresh on 'a new site somewhere near Darjeeling out of the range of our awful monsoon gales'.[26] It was proposed to convert the botanic garden into a large wet dock to meet trade requirements in Calcutta. Luckily these suggestions fell to the ground, but they are illustrative of the state of affairs existing at the time. The Government of India was reluctant to make sufficient grants. So in 1875–76 a plan to provincialize the garden was put forward. Initially King viewed the proposal with some alarm, as possibly implying a further reduction of funds and a curtailment of the scientific functions of the institution. General Strachey (a member of the Viceroy's Council and a personal friend of J.D. Hooker), however, assured him that the provincial government would do more for the garden than the Government of India could or ever would do.[27] The Calcutta garden did benefit from the shift to the Bengal Government. Funds were increased, its functions were not curtailed and its all-India character was retained. But the fiscal and administrative independence of Roxburgh or Wallich could not be brought back. The Bengal Government laid a firmer grip on administrative matters and secretariat interference increased. This was pointed out by G. Watt (the Economic Reporter) to T. Dyer (Director, Kew Garden) in a private letter:

Witness Dr King. If he wishes to go to Nepal or Afghanistan, as Wallich and Griffith did in turns, can he do so? He is fixed down to being an officer in charge of a certain section of the Bengal Secretariat and can no more take up a position

[25] Note by G. King, 17 June 1887, Kew Papers, Misc. Reports, f. 38, note 23 (emphasis as in original).
[26] Kew Papers, Indian Letters (Bengal), 1853–1900, vol. 153, f. 343.
[27] Note by G. King, Kew Papers, Misc. Reports, f. 39, note 23.

of independence like what Wallich did. He has allowed himself to be localized and had taken over so much office work that he has no time for scientific work. . . . The other day the Under Secretary had the cheek to propose that all my writings, reports, etc. should pass through his hands for revision before they were issued. Secretariat interference is the curse of every Indian branch of special or scientific work. . . . A Mr Leotard has left Mr Buck's office (Rev. Agricultural Department) because everything he wrote was marred by the alterations effected in the office. He felt that life was not worth living under such humiliating circumstances. Mr Buck (the Secretary) tried to have a say in the Dictionary (of Economic Products of which Watt was the editor) which I threatened to resign if I was not left to carry out the Government's requisitions without having to submit to any supervision. He took objection. I suspect, he had written you on the matter. Do me the favour to say as little as you possibly can on the subject for if once Mr Buck thinks that you are likely to back him instead of me, I shall have a poor chance of establishing a scientific department which would be able to do any good. . . . I wish to be free. I object to admitting that any purely administrative officer is capable of correcting my writings or of doing more than obtaining the annual grant for scientific investigation.[28]

E.C. Buck (Secretary, Department of Revenue and Agriculture, GOI) was dissatisfied with his own department. Convinced of the usages of science in the service of the Empire, in 1884, he wanted to bring all scientific bodies under one umbrella and create an Imperial Department of Science manned by an imperial scientific cadre. That he would not succeed was a foregone conclusion because the provinces were not likely to agree. But he did try. Buck's diagonosis was:

The provinces—at any rate the larger ones are ambitious of setting up their own scientific departments—each one for itself. But in the first place they cannot afford a first class scientific staff, and in the second place they have to do a great deal of work over again. And again the small provinces which can afford nothing, are left out in the cold altogether. Madras has set up a temporary scientific staff which it does not pretend that it can afford to keep for more than two years. The botanical institutions at Calcutta and Saharanpur, tied as they are to their own provinces, can do no national work. All this is very unsatisfactory. I find the Finance Department obdurate. . . . My central idea of what might be done is this. The great obstacle is want of money. Let us then get attached to a money procuring Department. Here is one, the Forest department which ought to belong to our hands. It has a net profit of 75 lakhs increasing within a few year to 100 lakhs. It is an Imperial Department and wants a scientific corps and has made already a step towards establishing one in the Dehra Dun Forest School.

[28] G. Watt to Thiselton-Dyer, 9 Nov. 1884, Kew Papers, Indian Letters (Bengal), ff. 468–9, note 26.

All that would be wanted if the forest came under Rev. Agri. Department would be to expand this school into the imperial College of Forests and Agriculture. The establishment of a strong scientific staff would at once be justified. We could have a botanical section, a microscopist, an entemologist, a chemical examiner, etc. All these officials would be useful to Government in the fields of agriculture and forests. All scientific work done in the provinces under Local Government would be referred to it. The College would also take the lead in scientific education. These are my castles in the air. Whether they will be built or not on a more solid foundation I cannot tell. . . .[29]

Meanwhile the Madras Government sought the advice of Dyer (the Director at Kew) over the constitution of a botanical department for its area. Dyer took the opportunity to suggest a close-knit federation of the various botanical departments of different provinces. This was in February 1885, and it gave Buck the opportunity to translate some of his ideas into practice. Buck, as noted earlier, was against the provincialization of science. 'The country cannot afford to allow each province to pay separately for high class science. They must either club together in groups for the maintenance of scientific progress, or they must look to Imperial Government to supply once for all the scientific establishment required. . . .'[30] The Secretary of State agreed with Dyer but added the proviso that this should be done 'without interfering with the control exercised by the Provincial Governments'.[31] This was a delicate task: how were provincial claims and funds to be reconciled with imperial needs? A fairly long battle of wits and correspondence ensued over the next five to six years,[32] and it was only in February 1891 that the Government of India could announce publicly the birth of what it called the Botanical Survey of India (BSI).[33]

It was a strange arrangement. The botanical gardens were to remain provincial but botanical investigations were made imperial. To the Superintendent of the Calcutta Botanic Garden was given one more title, that of 'Director of the Botanic Survey of India'. This directorship was in addition to the existing superintendentship of the garden and chinchona plantation works of G. King. The Calcutta garden was to be the nucleus

[29] E.C. Buck to J.F. Duthie (Supdt. Botanic Garden, Saharanpur), 20 July 1884, Kew Papers, Misc. Reports, 5, Indian Botanic Survey, ff. 13–15.

[30] Note by E.C. Buck, 1 Jan. 1886, Revenue, Agriculture, nos. 36–41, Feb. 1886, K.W.

[31] Ibid.

[32] Such wrangles did not take place when the Survey of India and GSI were established.

[33] Revenue, Agriculture, nos. 3–6, May 1891.

(Dyer had called it the Kew of India), but in practice, it had no 'directing' functions whatever except in Bengal, Burma and Assam (the areas accredited to it). King was not even recognized by the other provincial botanists as a referee, much less as a central authority.[34] Duthie was separated from the Saharanpur garden and given the designation of Director of the Botanical Department, Northern India, and under him was placed a vast area—the NWP, Punjab, Central Provinces and Rajputana. The principal of the Poona College of Science (who belonged to the provincial education department) was given charge of the whole presidency of Bombay, including Sind. The Government Botanist of Madras was to also look after the states of Hyderabad and Mysore in addition to his own presidency.

This was how division of work took place with King as the titular head. He was the Director only in so far as to have the power to determine whether the programme of work which the other botanists proposed for given seasons or the reports of work accomplished during the given years, were satisfactory. He was not empowered to lay down for them the work they should undertake nor was he able to direct two officers to combine their forces on some particular piece of work.[35] Another weak point was that Watt's credentials were not taken into account; as the Reporter on Economic Products he wanted to have the power of investigations in agricultural botany in his hands. He was prepared even for a transfer to the BSI on a designation such as the Economic Officer of the Survey, but did not want to relinquish his claim as the authority on economic botany. This was denied, and instead a sop was offered in early 1894 in the form of a notification that 'he should be prominently associated with any scheme which the Director may think proper to arrange'.[36]

The result of these half-hearted measures was that the BSI remained a non-starter for a long time. The Agriculture Department always loomed large in the background. The Agricultural Conference of 1893 wanted the BSI to concentrate more upon agricultural botany, and later in 1902, the post of the cryptogamic botanist (then held by Dr Butler) was detached from the BSI and given to the newly-created office of the IG of Agriculture.[37] J.B. Fuller, Buck's successor, treated botanists as minor

[34] Ibid., K.W.
[35] Prain to Gage, 8 May 1907, Kew Papers, Misc. Report, F. 127, note 23.
[36] Revenue, Agriculture, nos. 18–19, Feb. 1894, file 37.
[37] Kew Papers, Misc. Reports no. 5, India Agriculture, 1869–1921, ff. 227–9.

partners,[38] and the Botanical Survey remained under constant threat of being devoured.[39]

The botanists had at least some structure, however weak it might be, but the unfortunate zoologists had none. A few zoologists were attached to the Indian Museum, no doubt, but its Board of Trustees was dominated by the senior officials of the GSI.[40] In 1895, A.W. Alcock, a zoologist and the Superintendent of the Museum, submitted to his trustees a scheme for a Zoological Survey, but no one took notice of it. Rather, in early 1903, Lefroy, an entomologist attached to the Indian Museum, was shifted to the office of the IG of Agriculture (as Butler had been taken from the BSI). On leaving the country for good in mid-1906, Alcock raked up the issue again. In a letter (27 May 1906) to the Secretary, Department of Commerce and Industry, GOI, he wrote:

Zoology here is overshadowed by its sister science in a way that is unparalleled in the world. . . . We zoologists feel hurt at being treated like the musty old museum-mongers of a century ago, whose little lives were surrounded with stuffed skins and cabinets of butterflies and shells, and who in other affairs were as helpless as owls at midday. . . . An easy remedy is to establish a Zoological Survey, on similar inexpensive lines to the Botanical Survey; to let the Director of Zoological Survey be the Superintendent of the Zoological Museum, as the

[38] Burkill (of Calcutta Garden) wrote to Dyer: 'Fuller our new Secretary in the Revenue and Agriculture Department is full of bustle. Did you know how Sir Edward Buck and he went under the names of the Buck and the Buck stick. Now one of Fuller's chief idea is to run counter to all Sir Edward Buck did.' I.G. Burkill to Thiselton-Dyer, 21 Dec. 1901. Kew Papers, Indian Letters (Madras, Bombay, Bengal) 1901–14, vol. 160, f. 222. Later D. Prain (then Supdt. of Calcutta Garden) informed Dyer that Fuller had 'a keen dislike for you and for myself and a determination to do just the reverse of his predecessor'. D. Prain to Thiselton-Dyer (marked private), 1 Jan. 1903, Kew Papers, India Letters, Calcutta Botanic Garden, 1901–14, vol. 158, ff. 17–19.

[39] The amount of insecurity under which the botanists in India worked can be gauged from what Gage wrote to Prain: 'I fully agree with Sir J.D. Hooker that a stand must be made against the encroachments of Agriculture upon purely scientific botany and I think the proper line to take is to insist that the Botanical Survey is primarily and essentially for the purely scientific work of exploration and the writing of floras and monographs. As far as economic, including agricultural-botanical work is concerned the Agriculture Department should as far as possible be self-sufficient.'

A.T. Gage (Director, BSI) to D. Prain (Director, Royal Bot. Garden), 26 June 1907, Kew Papers, Misc. Reports no. 5, ff. 136–7, note 29.

[40] Commerce and Industry, Practical Arts and Museums, no. 1, Feb. 1903, file 10.

Director of the Botanical Survey is Superintendent of the Royal Botanic Gardens; and to give him the same position, both officially and on the Board of Trustees, as the Directors of other natural science Surveys.[41]

Denzil Ibbetson, a Member of the Viceroy's Council, did not 'feel greatly interested in Alcock's scheme'. 'I doubt', he added, 'its producing any practical results; I doubt whether in view of the enormous needs of the country in the way of applied science, we are justified at present in spending much more than we now do upon mere systematic or pure science'.[42] Another Member, C.L. Tupper, felt differently. He thought it 'quite legitimate for such a Government as ours to spend some money on pure science as distinguished from applied science'. But he was under the impression that the botanical Survey would be abolished some day and then 'it might be a reasonable thing to transfer the grant made for that to zoology'.[43] It was a rather naive assumption. Obviously one hand of the Government of India did not know what the other was doing. The Agriculture Department had opposed the proposal right from the begining. The IG of Agriculture wanted the Indian Museum to 'give up all connection with entomology rather than expand that branch'. Even Tupper changed his opinion in ten days' time and dropped the proposal as 'unnecessary' on the ground that Blanford's *Fauna of British India* series had accomplished all that was then feasible in this field.[44]

In sharp contrast to the botanical and zoological surveys comes the meteorological survey. Meteorology perhaps received the most attention after the GTSI and the GSI. The Company was well aware of the importance of 'Astronomy, Geography and Navigation' in India, and so observatories were established in Madras, Bombay, Calcutta, Trivandrum, Simla, Ootacamund and Karachi in the years 1792, 1823, 1829, 1836, 1841, 1847 and 1852, respectively. But meteorological observations in these places suffered from several types of inaccuracies, divergence, faulty instruments, lack of trained personnel and the absence of any standard method of work.[45] The Asiatic Society tried unsuccessfully to salvage it by appointing a Meteorological Commitee in April 1857. In 1862 the Society asked the Government to constitute a committee on the model

[41] Ibid., nos. 1–2, July 1906.
[42] Note 3 Sept., ibid., nos. 1–2, Oct. 1906.
[43] Note 6 Sept. 1906, ibid.
[44] Note 18 Sept. 1906, ibid. It was only in 1916 that the issue was raised again and a Zoological Survey of India created. *JRSA*, LXIV, 18 Aug. 1916, p. 960.
[45] Podgson Papers, IOL, Mss. Eur. B. 251.

of the Meteorological Committee of the Board of Trade in London. A storm-warning system was very vital for an empire whose life-line passed through the sea. In October 1864, one of the most destructive cyclones on record struck Calcutta in which more than 80,000 persons lost their lives. Weather-forecasting now became a matter of the most immediate concern. The next year the Sanitary Commission probed into how far climatic and weather conditions were linked with diseases in India and called for a systematic record of the meteorological phenomena.[46] On its recommendations meteorological reporters were appointed in the provinces. As many as seventy-seven observatories were in operation by the year 1874. But this system of independent reporters proved unsatisfactory. It generated confusion about methods of observations; there was no co-ordination and no supervision. So, in early 1875, was set up a meteorological department on an all-India basis with H F Blanford as the first Imperial Meteorological Reporter to the Government of India. The existing provincial and local reporters were retained; this was significant because they could draw upon the provincial funds. Blanford himself was placed under the Revenue and Agriculture Department of the GOI, but he enjoyed (unlike his botanical counterpart) a fairly large amount of autonomy and funds. A central observatory was created for him in 1877 in Calcutta. The government agreed to accept his plans without any modifications, even if they involved a big increase in expenditure. Henceforth he was able to introduce a uniform method of observation, train observers and standardize instruments.[47] Blanford succeeded in laying a solid foundation for the department, and his successor, John Eliot (1889–1903), added to it the prestigious observatories at Madras and Bombay. The Colaba Observatory in Bombay was known for its magnetic survey works while the one in Madras was famous for its minute solar observations. Eliot was not content with works of a routine nature such as local time services, tide tables, weather forecasts, etc. Norman Lockyer had for long been pressing for an observatory devoted to solar-physics. Lockyer was a British astronomer of great repute and had made a special study of sunspots. A tropical country like India offered plenty of 'raw-materials' (sun-rays) for solar research. Apart from scientific benefits, he

---

[46] The data of Dr Murray Thomson (who in 1868 had produced a 'Report on Meteorological Observations in NWP') showed an excess of cholera on a sudden fall of temperature. *The Athanaeum*, no. 2143, 21 Nov. 1868, p. 683.

[47] A good account of the early years is given in *Hundred Years of Weather Service*, 1875–1975, Poona, 1976, pp. 8–28.

claimed that his studies had important economic bearings also. Citing data from India's meteorological reports, he showed that rainfall occurred either at the time of maximum (solar) heat pulse or minimum heat pulse and thus famines took place only during the intervals between these two pulses. He petitioned the Indian Famine Commission of 1881 on this relationship between sun-spots and famines, and tried to impress upon the government the need for an exclusive observatory in India to serve as his data-bank.[48] Eliot took advantage of the opportunity. In August 1893 a solar-physics observatory was sanctioned under the meteorological budget, and Kodaikanal was chosen as the site. Similarly, the Colaba Observatory was to be the base for magnetic surveys. But both these proposals ran into rough weather because of opposition from the Survey of India which wanted to keep them under its own supervision. In 1898 Norman Lockyer and WHM Christie (the Astronomer Royal) visited India to observe the total solar eclipse, and the Government of India took the opportunity to ask them to make a report on the state of Indian observatories.[49] Christie and the Surveyor-General (Strahan) were not very enthusiastic about an exclusive observatory on solar-physics, preferring instead a general observatory at Kodaikanal devoted to astronomy, magnetic survey as well as solar observation. Lockyer and Eliot dissented. As for magnetic survey, Christie wanted it to be given to the Survey of India, while Lockyer suggested that it should be conducted by a specialist under the general control of the Meteorological Reporter. Eliot's contention was that 'the results of investigations of solar-physics and of terrestrial magnetism belong more to the science and sphere of meteorology than to those of the Survey of India', and the Government of India concurred with this view.[50] He was given a new designation—the Director-General of Observatories—and was even acknowledged by the Governor-General-in-Council as 'our principal scientific adviser'.[51]

[48] For details, see W.L. Lockyer, *Life and Work of Sir Norman Lockyer*, London, 1972.

[49] Revenue, Agriculture, Meteorology, nos. 1–12, March 1899.

[50] He almost begged: 'The inclusion of the scientific observatories under the control of the Meteorological Reporter will improve his status and increase his prestige. The Survey of India has the prestige natural to its connection with the army, numbers, to its important works, and needs no extension of its field work to increase its value and prestige.'

J. Eliot to the Secretary to GOI, Nov. 29, 1898, ibid.

[51] Curzon to Hamilton, 2 Feb. 1899, ibid.

Other Scientific Institutions

Apart from survey organizations, there were a few other institutions, the establishment and smooth functioning of which were no less vital for both the imperial economy and the cause of science itself.[52] A museum, for example, was indispensable. The Asiatic Society had one since 1815, but there had gradually accumulated such a vast collection that the Society often found it difficult to manage.[53] So, in May 1857, it called upon the Government of India to establish in Calcutta an Imperial Museum where its whole treasure (except the library) could be housed, classified and preserved for the benefit of everybody. The government's first reaction was to refuse, perhaps due to the costs involved.[54] But the pressure for an Imperial Museum continued. Requests were made even for a Museum of Economic Botany where specimens of vegetable products could be exhibited.[55] Finally, in 1866, the GOI agreed to establish a public museum (to be called the Indian Museum) incorporating all the branches of natural history and archaeology. It was put under a Board of Trustees but overall command was retained by the government through legislation passed in 1866, 1876, 1887 and 1910. Interestingly, at no point did these laws refer to research or to the needs of science students.[56] The provincial museums were no better. They had little academic orientation even though colleges existed beside them. In 1899–1900 George Watt made a tour of the local museums in Delhi, Lahore, Karachi, Jaipur, Surat, Bombay, Bangalore and Rangoon with a view to suggest some system of co-operation and exchange between them and the Indian Museum. He found them 'all starved to a ruinous extent'.[57]

What museums were to natural history, experimental farms were to agriculture. A year before the Imperial Museum was sanctioned in Calcutta, the Madras Government had established an experimental farm at Saidapet. The object was 'to afford facilities for testing the merits of

[52] In this section only those institutions have been taken which were under the direct control of the government; those under Indian management have been discussed in Chapter VI (i.e. Response and Resistance).

[53] Home, Public, no. 49, 7 Oct. 1859.

[54] Ibid.

[55] Home, Public, no. 76, March 1864, p. 387; WBSA, Revenue, Forests, no. 10, Aug. 1869, p. 13.

[56] It was only in 1912 that a set of by-laws took care of this omission. *The Indian Museum, 1814–1914*, Calcutta, 1914, App. II, p. XXXI.

[57] Watt suggested three remedies:

little known machines, implements, manures, crops, systems of culture, livestock, etc. believed to be suited more or less to the requirements of Southern India'.[58] In 1876 an agricultural college was opened in Madras. Three years later an agricultural unit was opened at Poona which subsequently made important experiments in cotton and plough-technology.[59,60] In 1897–99 agricultural classes were opened at Shibpur and Kanpur. These however, were half-hearted attempts and had neither an all-India perspective nor sufficient research orientation. Curzon sought to fill this gap. In January 1903 an American philanthropist, Henry Phipps, gifted Curzon $100,000 for the establishment of 'a laboratory to determine the economic value, and the medicinal qualities of the plants of India—or to be used in any other way that promises enduring good to India'.[61] Ibbetson (Member, Viceroy's Council) favoured a laboratory for indigenous drugs while the Home Secretary, H H Risley, wanted the Pasteur Institute to be assisted with the money given by Phipps.[62] It was finally decided to establish a laboratory for agricultural research designed to form what Curzon called 'a centre of economic science'.[63] Phipps made a further offer of $50,000 and the proposal for the agricultural laboratory was then expanded to include an agriculture college and experimental farm also.[64] The spirit shown at Pusa led Curzon to make more ambitious plans. In October 1905 he proposed 'to establish in each important province an agricultural college and research station, adequately equipped with laboratories, and the staff consisting of an expert agriculturist,

---

I. Periodic inspection by the scientific staff of the GOI.

II. Administrative control be shifted from the trustees to the Directorate of Agriculture which granted funds.

III. The local college staff be asked to help in classifying the exhibits.

G. Watt, *Memorandum on the Organization of Indian Museum*, Calcutta, 1900, pp. 1–2.

[58] Revenue, Agriculture, nos. 13–14, Jan. 1884.

[59] Home, Education, nos. 27–30, March 1879.

[60] Revenue, Agriculture, no. 20, March 1883.

[61] Jay Phipps to Curzon, 30 Jan. 1903, Curzon Papers, note 3.

[62] Home, Medical, no. 1, March 1903, Deposit.

[63] Ibid.

[64] Curzon to Phipps, 20 Feb. 20, 1903, Curzon Papers, note 3.

Welcoming the establishment of this Institute, Thiselton-Dyer prophesied that it would some day become the 'Rothamstead of the East', but added that 'the characteristic irony, I might almost say cynicism, of the British race is that it should own its foundation to the large-minded munificence of an Americal Gentleman'. *Nature*, LXXIII, 19 April 1906, p. 587.

economic botanist, agricultural chemist, entomologist and mycologist, etc'.[65] But the idea was scuttled at the India Office. Morley vetoed it and instead advocated 'a policy of cautious advance'.[66] Minto, Curzon's successor, however, persisted; he argued that 'we are greatly behind other countries in the application of science to agriculture, and now that funds are available we consider it one of our first duties to remove this reproach. The total expenditure proposed by us is insignificant compared with the interests involved and work to be done. We strongly urge for sanction to full 30 lakhs'. But Morley refused to budge and sanctioned only Rs 4 lakhs. Minto approached Morley again, citing the examples of colossal losses from wheat-rust, sugar cane disease and cotton pests, for the appointment of mycologists and entomologists in the provinces. Morley refused, calling the proposal 'premature'.[67]

Pusa nevertheless, served as a model for a similar exercise in forestry. Dehra Dun had a school for forestry since 1881 but it was confined only to elementary forest education for the subordinate staff. The higher level staff came from Cooper's Hill which itself was dependent upon the French and German forest schools. The sanction given to Pusa raised hopes of converting the Dehra school into a Forest Research Institute (FRI) and college. A scheme was accordingly formulated and approved of.[68]

Research in bacteriology was needed to fight against epidemics. There was no centralized or co-ordinated effort, though chemical examiners attached to some provincial governments were making some effort. E.H. Hankin, for example, with a small laboratory at Agra was supposed to look after practically the whole of north India. In early 1891 a bacteriological department was opened at the Poona College of Science and its professor, Dr Lingard, was given the designation of 'Imperial Bacteriologist'. His job was 'to investigate diseases of domesticated animals in all provinces in India and to ascertain as far as possible by biological research, both in the laboratory and when necessary at the place of the outbreak, the means for preventing and curing such diseases'.[69] Since he was attached to the Poona College of Science and was supposed to also teach the principal of the

[65] Revenue, Agriculture, no. 44, Nov. 1905, file 81.
[66] Revenue, Agriculture, no. 107, Aug. 1906, file 66.
[67] Revenue, Agriculture no. 38, Jan. 1907, file 14.
[68] Revenue, Agriculture, Forests, nos. 18–20, Jan. 1906; *Indian Forester*, XXXII, June 1906, pp. 283–7.
[69] Revenue, Agriculture, Horse Breeding and Agricultural Stock, nos. 27–34, Aug. 1892.

College and students of the Education Department of Bombay, government exercised control over him. It was an interesting situation but not uncommon in those days. Here was an official, 'imperial' in the scope and area of his work but under 'provincial' control.

Two important events took place in 1893. Haffkine was sent by Pasteur to India to try out his remedy for cholera by vaccination.[70] He was an instant success. The municipalities of Calcutta and Madras invited him to visit their cities and introduce his system.[71] In 1896 he set up a Plague Research Laboratory in Bombay under the control of the General Department of the Bombay government. Apart from the preparation of anti-plague serums and anti-cholera vaccinations, his institute also undertook the diagnosis of obscure diseases in men and animals, the examination of pathological specimens, instruction in bacteriological work, research work, etc.[72] The year Haffkine reached India, a proposal came from Lahore for the establishment of a Pasteur Institute in Punjab. A committee was also formed for this purpose. The Government of India promised to lend the services of a medical officer to the Institute.[73] Since bacteriological research and preparation of serum involved considerable pain to certain animals, the government feared that an open patronage might hurt the feelings of the local people. The anti-vivisectionists were no less active. The government opted for a neutral stance. Privately it would encourage research because it appreciated the utility of the project but officially it would not.[74] Meanwhile the Raja of Dholpur took up the cause. Elgin welcomed it in the hope that if native chiefs were involved, there would be less opposition.[75] But later Curzon found that the important chiefs of Punjab, Mysore and Hyderabad were reluctant and so the Dholpur move was quietly dropped.[76] In 1900, after seven years

[70] About Haffkine, the then Secretary of State wrote to the Viceroy: 'One is naturally sceptical about remedies but Pasteur's eminence is such that any suggestion of his is worthy of attention.' Kimberley to Lansdowne, 18 Nov.1892, Lansdowne Papers, NAI, Micro. 1050–1661.

[71] *Nature*, vol. L, 21 June 1894, p. 177; 5 July 1894, p. 227.

[72] See *Summarized Report of the Work Done in the Bombay Plague Research Laboratory, 1896–1902*, Bombay, 1903, pp. 1–37.

[73] Home, Medical, nos. 103–6, March 1902.

[74] At the House of Commons, the Secretary of State gave evasive replies when asked about his government's commitments. *Parliamentary Debates on Indian Affairs*, 1896, p. 351, and 1898, p. 82, IOR, V/3/1599, 1601.

[75] Elgin to W.M. Young (Lieutenant Governor of Panjab), 19 May 1897, Elgin Papers, IOL, Mss. Eur. F.84/70.

[76] Hamilton to Curzon, 16 Feb. 1899, Curzon Papers, IOL, Mss. Eur. F.111/158.

of dithering, the government lent a medical officer to the Lahore Committee and a begining was made at Kasauli. To pacify the anti-vivisectionists, Curzon took the Bishop of Calcutta into confidence with an assurance that the provisions of the English Cruelty to Animals Act would be fully observed.[77] Protests were not as vociferous as originally feared. Soon a Pasteur Institute for southern India was being contemplated.[78]

Scientific Cadres

The establishment and development of various scientific departments and institutions called for a distinct cadre. The Survey of India and the GSI constantly faced a shortage of trained personnel but they somehow managed to develop a compact, though small, cadre. The biggest and the oldest was the Indian Medical Service (IMS). A board had been governing the IMS since 1773. It had served the Company well through its military operations and turbulations, but gradually rigidity had crept in. Dalhousie tried to revitalize the IMS. He enunciated two principles: (1) 'To provide a substitute for the collective control of the Military Board by creating unity of authority coupled with direct responsibility' and (2) 'all appointments to be governed by the principle of selection, not succession'.[79] In November 1857, the Medical Board was abolished; in its place every Presidency got a Director-General (DG) and between the DG and the senior surgeons were appointed Inspector-Generals (IG), two in Bengal and one each in Madras and Bombay. In early 1859 the DG was redesignated as the Principal IG. This new title was abolished in September 1866 and in March 1869 only one IG was left as the head of the Bengal Medical Service. In 1873 this title was changed to that of Surgeon-General. Finally, in 1895, the title of Director-General was reintroduced with the major difference that the DG was now the head of the entire IMS throughout the country.[80] These frequent changes, however, showed that all was not well with the IMS. It was closed to competition from October 1860. For the next four years the government toyed with the idea of amalgamating the IMS with the Army Medical Department (AMD of the UK). But this scheme fell through. There were two hurdles: (1) Admission to the AMD was restricted only to men of unmixed European

[77] Curzon to Lord Bishop of Calcutta, 26 June 1900, ibid., F 111/201.
[78] Ampttil (Governor of Madras) to Curzon, 5 July 1903, ibid., F 111/208.
[79] D.G. Crawford, *A History of the IMS*, vol. II, London, 1914, p. 271.
[80] Ibid., p. 41.

extraction, and this would have debarred the natives from joining the service and (2) the superior attractions of the IMS would have drawn the best of the AMD cadets and this would have adversely affected the interests of the British army.[81] The idea was raked up again in 1879 and in early 1880 certain changes were made which amounted to practically a partial amalgamation of the military medical administration of the two services. Henceforth the AMD and the IMS were to form one department for the medical administration of the army in the three presidencies, and the Surgeon-General was to be always taken from the AMD. The IMS had to acquiesce in this, but it emerged on top again in early 1896 when the medical services of Bengal, Madras and Bombay were put together under the direct administrative control of the GOI. The IMS was never short of able men and seldom lost its charm. But administrative fluctuations did sometimes affect its strength. In 1861 it had 819 members, but the cancellation of its examination over more than four years brought the membership down to 570. It gradually picked up, and stood above 700 by the turn of the century. There was a qualitative improvement also for better and more qualified persons began to join the IMS as compared to those in the 1860s and 1870s.[82]

The IMS was raised and maintained basically for army purposes. But since the army did not require their services all the time, the government appointed the officers as professors, sanitary commissioners, chemical analysts or even superintendents of lunatic asylums. Some IMS officers often held appointments totally unconnected with the medical profession. One became the Postmaster-General in Bombay, another Conservator of Forests, while many others worked as Mint and Assay Masters. To quote E. Hart, editor of the British Medical Journal, who toured India in 1894–95:

It is a bizarre and stereotyped system with an anomalous state of things in which the IMS man seems to be expected by Government to be fit for any post that may be vacant. . . . The system is radically wrong in which men worked their way up by seniority to a position in which they fill the dual capacity of Principal Medical officer of the Army, and Sanitary Commissioner with the Government'.[83]

Meanwhile a fairly large class of native medicos had come up from the

[81] Ibid., p. 284.
[82] *The Indian Medical Gazette*, April 1906, p. 142.
[83] 'Papers Relating to the Improvement of the Position and Prospects of Civil Assistant Surgeons in India', 1891–9, *Selections from the Records of GOI*, no. 377, 1900, p. 28.

medical colleges of Calcutta, Bombay and Madras. They had almost the same qualifications and expertise as those in the IMS but were placed much below in rank as sub-assistant surgeons, with hardly any scope for promotions. Even the apothecaries who had to pass only one medical examination drew more salary than the Indian assistant surgeons. The reason was that apothecaries came exclusively from European and Eurasian stock.[84] Indian surgeons began with Rs 100 a month and ultimately drew Rs 200. (In contrast a new IMS got Rs 400 a month which was double the market value at home.) In 1866 the sub-assistant surgeons of Bengal, NWP and Punjab submitted a memorial, airing their grievances, but the Government of India rejected it. Again, in 1882, a somewhat similar appeal was made by Dr Simpson (the Principal of the Temple School of Medicine, Patna). In reply, Dr Payne, the Surgeon-General for Bengal, remarked that assistant surgeons as a class were overpaid for the work they performed, that the cost of educating them fell mainly on the State and that they rarely failed to make a large addition to their income by private practice. The discontent among the native doctors continued to simmer. Dr K.N. Bahadurji and Dr A.R. Pandurang took the lead in Bombay. In April 1896 they submitted a memo which called for:

I.   A separation, complete and absolute, of the civil from military medical service, with the formation of one military medical service for India—and

II. the reorganization of the civil medical service, constituting it a purely civil service, recruited from the open profession of medicine in India and Europe, with a liberal admixture of indigenous talent in the event of proved merit and ability.[85]

This demand was repeated several times at the Indian National Congress and other forums. But the government would not budge. Rather it regarded these representations as merely 'another indication of the need for placing on a better footing that portion of the medical service which is recruited from the medical students trained in this country'.[86] Only two concessions were announced: (1) The creation of a senior grade of civil assistant surgeons on a monthly pay of Rs 300 for a select few and (2) the posting of some of them to the independent charge of a few selected civil stations in each province. This was the maximum concession that Curzon would allow. He found the existing system useful, but he did not

[84] Ibid., p. 3.
[85] Ibid., p. 28.
[86] Elgin to Hamilton, 27 Jan. 1898, ibid., p. 68.

like the IMS men charging exhorbitant fees from gullible chiefs, and framed rules limiting the fees to a modest sum. The British Medical Association denounced this restriction as 'an insult to the Service', but it had to withdraw quickly when Curzon threatened to publish such cases and called them 'a disgrace to the integrity, and reputation of the British name'.[87]

While the medical services were well organized, the most disorganized sector was that of agriculture. The government derived maximum revenue from agriculture, but the upgrading and improvement of the sector was conveniently left in the hands of private agricultural societies. The government responded with aid only when goaded by the Famine Commissions of 1866, 1880 and 1901. Commercial interests perhaps also had a decisive influence. For example, the Cotton Supply Association of Manchester wanted the creation of a Department of Agriculture so that concentrated attention could be given to cotton cultivation.[88] Lord Mayo accepted the suggestion, but his own council members, notably Richard Temple and Henry Durand, opposed it. For Durand, the creation of separate secretariats for the railways and legislative departments was of greater necessity than the creation of a department for agriculture.[89] Even so, the new department was opened in 1871,[90] only to be closed in 1879 for reasons of financial stringency. A second attempt was made at the behest of the Famine Commission of 1880, resulting in the establishment of such departments in each province.[91] But the provincial departments were unable to make much progress mainly because they were not given a staff of specialists skilled in agriculture and its allied sciences.[92] They could not go beyond the collection of revenue data and famine-relief operations. The government preferred to depend on traditional agencies like the AHSI rather than introduce innovations. The Finance Department had always been unhappy with the performance of these societies and had even asked for their grants to be stopped in 1873, but the government had not obliged the former.[93] The provincial agriculture departments remained repositories of good intentions and revenue statis-

[87] Curzon to Brodrick, 14 Jan. 1904, Curzon Papers, IOL, Mss. Eur. F. 111/163.
[88] Home,Public, no. 92, 2 April 1870; and no. 83, July 1871.
[89] Minute by Durand, 31 March 1870. Home, Public, no. 100, 9 April 1870.
[90] Home, Public, no. 134, 6 June 1871.
[91] Home, Public, no. 97, July 1881.
[92] F.G. Sly, 'The Departments of Agriculture in India', *Agricultural Journal of India*, vol. I, 1906.
[93] Revenue, Agriculture and Horticulture, nos. 21–3, May 1873.

tics. Dr Voelcker, who was sent to India in 1890 to report on Indian agriculture, found that technical knowledge of agriculture was 'the missing element' in what was then called the Department of Land Records and Agriculture. Voelcker sincerely felt that agriculture should form a department quite separate from that of land records and that its director and assistants should be experts, not civilians, on the pattern of geological, botanical and meteorological departments. But he doubted its feasibility and, therefore, confined his job only 'to suggest what can be done; to graft improvements upon the existing systems, rather than to suggest the subversion of the latter'.[94] In 1892 an agricultural chemist was appointed and a cadre began to evolve.[95] No permanent posts could be given but liberal contracts were offered. Dr Leather, an agricultural chemist, was engaged for a period of five years. On the expiry of that period the GOI recommended the abolition of the post of agricultural chemist and asked for the appointment of an IG of agriculture. This idea was first mooted in 1897 but no suitable man could be found, and the vacancy was filled only in 1901 when J. Mollison, a deputy director of agriculture in Bombay, accepted the post.[96] Simultaneously, a mycologist and a cryptogamic botanist, and a little later, an entomologist were appointed. They formed an imperial cadre. Soon they got a platform at Pusa and also a Board of Agriculture in 1904. But Curzon realized that a central institution like Pusa under the direct control of the GOI could only be the apex of a scheme and that such an institution would be worthless unless there was, at the same time, a real development of agriculture in the provinces. The Secretary of State was not very enthusiastic about further expansion, but the GOI prevailed upon him to sanction an additional fifteen experts to the already existing agricultural staff of twenty-four.[97] Curzon worked for this expansion in terms of an organized cadre with attractive and clear-cut commitments on pay, promotions and other conditions of service.[98] This led to the formal constitution of the Indian Agricultural Service in 1906.

[94] J.A. Voelcker, *Report on the Improvement of Indian Agriculture* (2nd edn), Calcutta, 1897, pp. 304–5.

[95] Revenue, Agriculture, no. 7, Dec. 1892.

[96] J. Mackenna, *Agriculture in India*, Calcutta, 1915, p. 5.

[97] Revenue, Agriculture, nos. 33–6, April 1906.

[98] Curzon wanted to organize an agricultural service on the pattern of the Indian Educational Service with the same grade plus an added advantage of local allowance. In a despatch of 14 Sept. 1905, Curzon informed Brodrick (the Secretary of State) that 'we are fully prepared to pay at the market rate for the valuable requirements which are desired'. Revenue, Agriculture, nos. 14–18, Sept. 1905. The plea for local allowance was not acceptable to Brodrick. Even the scales of pay he sanctioned 'with

Fairly close to agriculture were the problems of cattle diseases and their prevention, breeding, etc. and these were very sympathetically considered by the Famine Commission of 1880. There already existed a department devoted to horse-breeding but it was totally under military control. The famines had, however, brought to the fore the necessity of a civil veterinary department which could think and work beyond mere horse-breeding. The Revenue and Agriculture Department was entrusted with this task.[99] But it was not easy to make inroads into the domain of the Military Department. A meeting of the senior officials of these two departments was held at Simla on 2 May 1885. The over-all imperial control was conceded to the military but every province was to employ in its agriculture department a European veterinary surgeon and a number of native assistants.[100] Horse-breeding was to remain the most important work and to make it more economical the responsibility devolved upon the civil district officers.[101] This was a new responsibility, and to supervise the work, a cadre of veterinarians was created in 1891 with an IG (Veterinary) assisted by four senior veterinary professors, four scientific investigators and nine other officers appointed to oversee secondary education and carry out executive duties in this connection. But the emphasis was clear—military work at civil cost. This approach was not calculated to help veterinary investigations, much less serve the cause of Indian agriculture. A contemporary found in it yet another sign 'of the rampant militarism India has always suffered and under which she still groans, while the soil cries out for improvement'.[102] The result was that it could never muster enough strength and even its scientific work invited adverse remarks, particularly from the Indian Advisory Committee (IAC) of the Royal Society.[103] Due to lower salaries and difficult working conditions this service did not attract competent men, either foreign or native. Not even one out of 26 vacancies for the post of deputy superintendentship all over the country could be filled till 1907, and for

some hesitation'. Despatch from the Secretary of State, 24 Nov. 1905, Revenue, Agriculture, nos. 33–6, April 1906.

[99] Revenue, Agriculture, Cattle Breeding and Cattle Development, nos. 1–3, Sept. 1885.

[100] Ibid.

[101] Despatch from the Secretary of State, 17 Nov. 1887, Revenue Agriculture, Horse-Breeding and Animal Stock, nos. 16–72, Nov. 1891.

[102] *The Indian Agriculturist*, XIII, 1 Sept. 1888, p. 480.

[103] Revenue Agriculture, Civil Veterinary Administration, nos. 47–52, Nov. 1908.

141 posts of inspectorship, only 54 suitable men could be found. It is clear from this chart how unpopular this service was:[104]

Province	Sanctioned posts			No. employed in 1906–7		
	Dy. Vet. Suptd.	Vet. Inspector	Vet. Assistant	Dy. Vet. Suptd.	Vet. Inspector	Vet. Assistant
Punjab	...	24	121	...	14	112
NWPF	...	2	20	...	2	14
United Provinces	2	20	210	...	8	81
Bengal	9	30	229	...	15	40
Bombay	4	24	215	...	3	27
Madras	3	15	132	...	...	15
CP & Berar	6	126	...	...	4	60
Burma	5	12	120	...	7	72
East Bengal & Assam	2	8	120	...	1	11
Sind,Baluchistan & Rajputana	...	...	...	...	...	8
Total	31	261	1,167	...	54	440

An attempt to create a Board of Veterinary Science on the lines of the Board of Agriculture also failed.[105] The interests of veterinary science each time were sacrificed at the altar of administrative expediency.

Perhaps a better organized service was that of the foresters. It was, after all, a money-spinning department. Dalhousie commissioned it in 1856, and within thirty years it generated a net annual revenue larger than that of France and equal to that of Prussia.[106] Dr Brandis in Burma and Cleghorn in Madras laid the foundations of this service. In 1864 the Indian Forest Service was first organized on a permanent and an all-India basis.[107] At the top of the hierarchy was of course the IG of Forests and each province had at least one conservator. Each conservatorship, or forest circle, was divided under deputies or assistants and each of these divisions sub-divided into ranges. The superior officers were trained at Cooper's

[104] Ibid.

[105] Agriculture, Civil Veterinary Administration, no. 16, Aug. 1906, and nos. 19–31, June 1907.

[106] In 1889–90 the gross receipts from the Indian forests were £1,86,363 and the net revenue £706,268. *The Indian Forester*, XVIII, Feb. 1892, p. 73.

[107] Earlier Forestry was under the PWD. WBSA, Revenue, Forest, no. 104, Dec. 1862.

Hill or in Germany or France, while Indian assistants and rangers were trained at the Dehra Forest School. But this service could scarcely be called scientific. Revenue was the only obsession. The officers were not competent to undertake a study of injurious pests or of the chemistry of widely varying soils. Nor were service conditions such as to attract qualified and able persons.[108] Presiding over the botany section of the BAAS in 1899, G. King attacked the Indian Forest Department for its 'insufficient knowledge of systematic botany'. In defence, a former IG of Forests, W. Schlich argued:

The Government of India does not wish every Indian forest officer to be a botanist. . . . Forestry is perhaps not a science, but an industry based upon various branches of science, amongst which botany, geology and entomology are the most important. The forest officer cannot be an expert in each of these. To demand such a thing would be just as unreasonable as to demand that a medical man should be an expert in chemistry. . . . The Indian forest officer is an estate manager on a large scale; he must manage his estate in such a manner that they yield the largest possible amount of useful produce with the least possible outlay. . . .[109]

This led to a lively debate, and later, to recognition of the value of research in the establishment of the FRI.[110]

Forest officials were basically estate-managers; veterinarians were there for horse-breeding; the IMS was for the army; and agriculture was for revenue—then where did a scientific cadre exist? Was it only in survey organizations? The Survey of India and the GSI, of course, could develop a homogenous cadre. The IMS was a well-knit unit, and had made important contributions in various fields. Though the botanists had acquired a distinct identity, they could not maintain it and switched to agriculture. The rest had a sprinkling of individual talent, but co-ordinated or sustained effort was sadly missing.

Control from Above

Administering scientific organizations and institutions, no doubt, was the direct responsibility of the local governments and on-the-spot decisions were often taken as and when required, yet the final authority rested in

---

[108] The IG of Forests drew less pay than the Collector of a single district. *The Indian Forester*, XII, Dec. 1886, pp. 531–2.

[109] *Nature*, LXI, 2 Nov. 1899, pp. 6–7.

[110] The Indian Forester, XXVI, Feb. 1900, pp. 56–7.

London, and metropolitan interests, pressures and pulls weighed heavily on colonial administrators. The different 'trade museums' (e.g. the Indian Museum in London), certain societies (particularly the Royal Society of Arts) and merchant associations did influence several decisions regarding India. We have seen how the colonial botanists from Roxburgh's days looked to Kew for both personal and official gains. Higher education, recruitment of superior scientific staff, its training, etc. had to be centred in London. For example, in 1871, an engineering college was opened at Cooper's Hill (in England) to train 'royal' engineers for the growing railway, PWD, forestry and army works in India. They were supposed to guide those churned out by the Indian engineering schools and colleges. But the cost of the upkeep of Cooper's Hill was met from Indian revenues.[111] With the gradual strengthening of engineering institutions within India, the value of Cooper's Hill diminished. The Government of India recognized this fact, yet in a despatch to the Secretary of State (11 March 1890) it declared that 'we have no hesitation in saying that having been created, it should be maintained as long as possible, if only for the purpose of avoiding radical and frequent changes'.[112] The best students of this college preferred services in England itself. In 1903 a committee enquired into its affairs and recommended its closure.[113] This imperial institution ultimately failed though it was established with great hopes.

Perhaps more illustrative of the vicissitudes of imperial control is the story of the Imperial Institute, founded in 1887 as a monument to Queen Victoria's Golden Jubilee. Here was a metropolitan scientific institution which aimed at providing a spine to a hitherto cordless gamut of imperial science. The Institute's Charter stated eight objectives, ranging from the organization of an industrial museum, commercial intelligence and exhibitions to 'the furtherance of systematic colonization'. The success of the Great Exhibition of 1851 and the Colonial and Indian Exhibition of 1886 had already created a favourable climate; jingoism was at its peak. No wonder, science-enthusiasts like T H Huxley fell for the idea. Huxley saw in its establishment a 'public and ceremonial marriage of science and industry'. In a letter to *The Times.* (20 January 1887) he fancied:

[111] Between 1870 to 1905 the GOI had contributed £170,000 for the upkeep of Cooper's Hill. *Parliamentary Debates on India*, 1905, IOR V/3/1608, p. 501.
[112] Finance and Commerce, Salaries and Establishment, no. 152, April 1890.
[113] *Report of the Royal Indian Engineering College Committee*, Simla, 1903, p. 13.

I pictured the Imperial Institute to myself as a house of call for all those who are concerned in the advancement of industry . . . as a sort of neutral ground in which the capitalist and the artisan would be equally welcome. . . . I imagined it a place in which the higher questions of commerce and industry would be systematically studied and elucidated; and where, as in an industrial university, the whole technical education of the country, might find its centre and crown.[114]

But the merchants in London, being more practical, were a little chary and initially refused to subscribe.[115] Even the Royal Society refused to issue a formal appeal for subscriptions.[116] But the Government of India came forward with a handsome contribution of £114,528 and a few Indians donated about £35,000.[117] This was more than one third of the total contribution from the whole of the British empire (all other colonies could pool in only £64,132 but within the UK £247,439 was collected). The fund-raising operations and the construction of the building took almost six years and delayed the Institute's formal opening to May 1893. Meanwhile enthusiasm petered down and was replaced by some amount of professional jealousy. Thiselton-Dyer, the Director at Kew, was not happy. He feared that the new Institute might overshadow Kew.[118] Finally (in December 1893) he chose to resign his post as Indian Governor at the Imperial Institute.[119] There was discontent even within the Institute. The Curator of the Institute, J.R. Royle, was very sore at the behaviour of his Director, F.A. Abel. In a private letter to Dyer he complained:

I left the India Office nominated by Sir E. Buck, on behalf of the Government of India, to organize the Indian Section. . . . But I have found that Sir F. Abel has

[114] T.H. Huxley Papers, Mss. 42, 158, Imperial College of Science, London.
[115] T. Christy to T.H. Huxley, 12 Feb. 1887, ibid., Mss. 12. 190.
[116] Proc. of a Meeting of the Imperial Institute Committee of the Royal Society, 10 March 1887, Royal Society, Mss. 446, CMB. 19
[117] The Maharaja of Bhavnagar gave £17,000; Cowasjee Jahangir gave £12,500, and M.M. Bhownagaree gave £3000. J.R. Royle and B.J. Rose, *Report on the Indian Section of the Imperial Institute*, London, 1902, Morley Papers, IOL, Mss. Eur. D 573/43.
[118] In a letter to Thiselton-Dyer (30 June 1891) the India Office tried to allay his fears: 'The assistance to Imperial Institute is intended to enable it to provide for the manufactures and merchants of the United Kindom universally accessible information regarding the general commercial life of India; whereas the grant to Kew is made in return for scientific services rendered directly to the Secretary of State for India and the GOI.' Kew Papers, Misc. Reports, 1–2, Imperial Institute, p. 15.
[119] Ibid., p. 196.

no intention that I should have any power or freedom; they have freely copied or adapted all my plans and inventions all over the place but never gave me the credit of them; all Indian correspondence which does not happen to come to me direct is concealed from me, and I have been given to understand that my principal duty is to be punctually at my post by 10 a.m. and never leave it till the closing down. . . . But the great mistake was made as regards India that while paid from Indian revenues I have no responsibility to India or the India Office and no power to correspond except on trifling details. I do frequently write (but privately) to Sir E. Buck and I know from him that the position I am in is not at all what he intended.[120]

The Government of India itself was fast losing its faith. In 1891 it had sanctioned an annual grant of £1000, much before the Institute started functioning;[121] yet in 1894 the latter demanded more supplementary grants. When the GOI protested, Sir Abel replied with a long memo describing how the Indian contributions had been spent and why the GOI should send more.[122] The account furnished by Sir Abel is quite revealing:

Amount received from India £114,528
Expenditure:

(1)	Total value of space exclusively occupied by India	£38,923
(2)	Proportion of value of space occupied in common with the UK and Colonies	£41,309
(3)	Amount of Indian donations invested in Endowment Fund	£34,296
	Total	£1,14,528

In early 1896 a distinct Scientific and Technical Department (STD) was opened at the Institute, and was put under the charge of a noted chemist, Wyndham Dunstan, who fancied to make it 'the Kew of Chemistry'. These changes, however, failed to satisfy Indian requirements. George Watt complained that most of the enquiries sent by him to the Institute remained unanswered. The services of the Imperial Institute were called upon in sixteen cases, of which only six were reported

---

[120] J.R. Royle to Thiselton-Dyer, 31 Dec. 1893, ibid., p. 162.

[121] *Parliamentary Debates on India*, 1893, IOR. V/3/1596, p. 517.

[122] Memorandum from the Imperial Institute, 28 March 1895, Despatches to India, 1895, IOR, V/6/333, pp. 215–25.

upon. Finally, only two cases led to practical business. Naturally, J.B. Fuller (Secretary, Department of Revenue and Agriculture) concluded that the Institute does not 'promise any practical benefit to the production or trade of the country'.[123] Moreover, within India, a few laboratories had been established. There were chemical examiners attached to provincial governments, and by 1896 the Calcutta University and the office of the Reporter on Economic Products got separate laboratories. The GSI already had a laboratory of its own. So there was hardly any need to look to London for all sundry enquiries. But the Government of India was making a heavy payment for facilities that were not needed. It was paying £1,525 annually for the upkeep of the Institute and only £200 of this amount went to the research wing.[124] By March 1906, this annual grant had accumulated to £24,077, of which only £1,450 had been spent on research into Indian products, the rest being sent for what the India Office described as 'the general purpose of the Institute', salaries, etc.[125] The desirability of this cornering of Indian revenue was questioned on the floor of the House of Commons. M Bhownaggree (a Member of Parliament) wanted the Institute to train a few Indian graduates in return for Indian contributions. The Institute could have compiled; in fact some of Dunstan's staff had earned D.Sc. degrees from the University of London for research done at the Institute. But the Secretary of State refused.[126] Indian money was welcome, but not Indian students or Indian claims. When the Imperial Institute lost its autonomy in 1900 and was placed under the control of the Board of Trade, the Indian Government was not consulted. The people who paid for the Institute were ignored when its character was being changed.[127] Later, addressing the BAAS at York in 1906, when Dunstan formulated 'Some Imperial Aspects of Applied Chemistry', he deliberately excluded the Dominions and India from his purview. In reality the Institute was more useful to the other Crown Colonies which had no facilities of their own.[128] The Institute was

[123] Revenue, Agriculture, Practical Arts and Museums, nos. 3–5, March 1902.
[124] *Parliamentary Debates on India,* 1904, IOR, V/3/1607, p. 239.
[125] Ibid., 1906, IOR, V/3/1609, p. 73.
[126] Ibid., 1904, pp 285–395.
[127] Ibid., 1902,IOR, V/3/1605, p. 281.
[128] Its utility to other colonies gradually diminished. In 1911 the Government of South Africa reduced its annual contribution from £800 to £300, and Dunstan remonstrated in vain. Viscount Gladstone Papers, BM, Add. Mss. 46073, ff. 200, 205.

never 'imperial' in the fullest sense of the term and this 'Kew of Chemistry' gradually became a 'blurred white elephant' of the empire.[129]

The Imperial Institute was not cut out to give scientific advice. The Astronomer Royal and the men at Kew looked after the astronomical and botanical interests in India, but there were several other areas which needed expert advice. Pondering over the questions of classification and systematization was not enough. The GOI looked for tangible results; it wanted research to 'increase the productive capacity of the soil', and to provide 'new employments for the rapidly increasing population'. Keeping this in mind, Elgin asked for 'the advice of leading men of science in England who would exercise a general control over our researches'.[130] The Royal Society agreed to oblige and constituted an Indian Advisory Committee (IAC). But the whole exercise remained only on paper. The GOI almost forgot its existence, and the IAC was not supposed to take any initiative on its own. Thiselton-Dyer, a member of the IAC, resented this inactivity and in early 1902 even threatened to resign.[131] This protest had the desired effect and the GOI had to apologise to the Royal Society and renew its pledge. The Society agreed to retain the Committee.[132]

Meanwhile a need was felt to have an advisory body within India itself which could serve as a bridge between different scientific departments, ensure a tighter control and avoid duplication. Curzon was not happy with the scientific works in India. He found them 'sporadic, chaotic and wholly lacking in co-ordination'.[133] In February 1902 J.B. Fuller (Revenue and Agriculture Secretary to the GOI) put forward the idea of a Board of Scientific Advice (BSA) to bring the heads of all the scientific departments together twice a year under his chairmanship. Curzon and his Council readily agreed. There were two reasons. As Denzil Ibbetson, a member of the Viceroy's Council, wrote:

I think the Board an excellent idea. We are in need of all the help we can get in controlling and co-ordinating our scientific work in India. The Board will greatly strengthen the hands of the Secretary (J.B. Fuller) in coercing the Geological and

[129] M. Worboys, *Science and British Colonial Imperialism (1895–1940,* Ph.D. thesis, University of Sussex, 1979, pp. 153–7.

[130] Elgin to Hamilton, 22 Dec. 1898, Kew Papers, Misc. Reports, India Advisory Committee, f.2.

[131] Thiselton-Dyer to the Secretary, Royal Society, 18 Jan. 1902, ibid., ff. 15–16.

[132] Ibid., f. 20; Proc. of the IAC, Royal Society, 25 Nov. 1903.

[133] Curzon's note of 12 May 1902, Revenue, Agriculture, General, nos. 4–11, Jan. 1903, file no. 164.

Botanical Departments, who are the recalcitrant members, to give us some practical work, in addition to pure science; for with scientific assessors at the table, he will feel far less diffidence in pressing his views on points on which he has no technical knowledge.[134]

The latter consideration was perhaps more formidable. Curzon's government regretted that 'undue prominence had been given in the past to pure science, to the neglect of its economic application'.[135] The new Board was supposed to give the desired tilt a bit more firmly. It had roped in the heads of the meteorological, geological, botanical, forest, survey, agricultural and veterinary departments. Zoology was left out and Major A.W. Alcock, a noted zoologist, protested at this exclusion. The Surveyor-General was reluctant to join the Board and did not want his work to be reviewed by it on the ground that his department had no direct economic interests.[136] But the GOI insisted on placing at least the scientific portions of the survey and meteorological works (e.g. the magnetic and climatological studies) under the Board's jurisdiction. The issue was reopened in mid-1904, only to be dropped again.[137] So though the Board had academic interests, in practice, practical considerations were to prevail. The procedure was simple. Programmes and reports were prepared by the different departments and sent directly to the Board for its opinion. But the Board had no authority to issue orders for the execution of the programmes.[138] The GOI was the final authority to accept or shelve recommendations, or make alterations. Yet the Board's advice carried a lot of weight, as its members were top government officials. For a small expenditure of less than Rs 2500 a year on its secretariat, it was worth seeking advice.[139] The agenda of the Board reflected diverse interests and pulls. In its early meetings very few items were of a scientific nature; industrial or trade items dominated discussions. For example, in the very first meeting of the Board on 6 February 1903, the agenda listed a problem on how to regulate the flavours of *ghee* (a kind of butter-oil). This was not something worth the consideration of the central board.[140] In the second meeting of the Board on 23 March 1905, four sub-committees were

[134] Note by Denzil Ibbetson, 9.4. 1902, ibid.
[135] Resolution of the GOI, no. 22, 28 Aug. 1902, ibid.
[136] Revenue, Agriculture, Land Surveys, nos. 22–3, Oct. 1902, pt. B, file no. 130.
[137] Revenue, Agriculture, General, nos. 28–35, July 1904, file no. 180.
[138] Revenue, Agriculture, General, nos. 4–9, June 1905, file no. 130.
[139] Revenue, Agriculture, General, nos. 41–2, June 1905, pt. B, file no. 213.
[140] Revenue, Agriculture, General, nos. 14–22, April 1903, file no. 69.

appointed to look into specific areas of investigation.[141] Sub-Committee 'A', with the Surveyor-General as chairman and the Meteorological Reporter and the Director of GSI as members, looked into questions related to meteorology, terrestrial magnetism and cognate subjects. Sub-Committee 'B' had the Director of BSI as its chairman and the Superintendent of the Industrial Section of the Indian Museum and the IG of Agriculture as members, and it examined the agricultural products. Sub-Committee 'C' was chaired by the IG of Agriculture, and the Director of GSI and the IG of Forests were its members. It dealt with questions relating to soils and manures. Sub-Committee 'D' was meant for forest products. The IG of Forests was its chairman, and the Superintendent of the Industrial Section of Indian Museum and the Director of BSI were its members. Groups 'A' and 'C' had something to do with problems of a scientific nature, while the other two were product-oriented. The application of science to economic and agricultural development remained a major concern; the BSA was indeed meant only for that. And it did make a sound start, cutting across departmental boundaries. It was also unique in the sense that no other colony had a parallel to it, not even the metropolis.[142]

The establishment of the colonial BSA did not, however, render the metropolitan IAC redundant. The former had to send its reports, etc. to the India Office which in turn sent them for criticism and approval to the latter. The IAC had finally got a job, and it took it pretty seriously. Two problems surfaced almost immediately. The very nature and objectives of these two bodies were different. The Royal Society was a voluntary academic association par excellence, while the BSA was a mere government body. One looked for new knowledge, the other preferred its application. But, interestingly enough, the IAC was not enthusiastic about scientists in India revelling in basic research. The Fellows wanted pure science to be left to Britain alone, and the Government of India fully shared this 'colonial' view. This was resented by some of the scientists in India who were also members of the BSA and for whom original and

[141] Revenue, Agriculture, General, nos. 4–7, Sept. 1903, file no. 69.

[142] In England demands for a scientific advisory body (official and centralized, unlike the Royal Society) were occasionally raised throughout the last century. The Devonshire Commission had also asked for it in 1875. The establishment of a similar body in India raised hopes that the home government would also follow suit. The metropolis was now getting inspiration from the periphery. Yet the former had to wait for another twelve years, until 1915, when an Advisory Council on Scientific and Industrial Research was created for them.

clinical studies were vital. A conflict was inevitable.[143] Another problem was that of distance. There was no direct dialogue. In between the two came the GOI and the India Office, and this meant considerable delays in communication. Due to official involvement, the reports were written in a style too dry and formal to involve a fair mutual understanding.[144] Personal rapport (which had existed since the days of Banks) was the first casualty of the new system.

Serious differences arose over bacteriological, geological and agricultural works. The Indian Government had concentrated bacteriological research at one place under the charge of Dr Lingard (the Imperial Bacteriologist). His experiments with rinderpest sera were severely criticized by the IAC and the subsequent failure of his serum added grist to the mill. The IAC asked for empirical investigations in the field, in more than one centre, with the help of 'more skilful pathologists and bacteriologists'.[145] The BSA found these suggestions impractical. The IAC remonstrated and called the veterinary reports 'unscientific, ambiguous' and full of 'half-statements and generalities'.[146] As for geological investigations, the Royal Society regretted that the BSA had lain emphasis only on economic geology.[147] It wanted the GSI not to neglect its prime job of making a general geological survey and map of the whole country. T.H. Holland (Director, GSI) agreed with the IAC.[148] But the Board and the Government of India were united in giving primacy to economic considerations. Holland's survey was, in turn, described as 'sporadic and fitful'. These bodies crossed swords on several occasions, and this sapped the vitality of both.[149] One was abolished in 1910; the other

[143] A very comprehensive account of the uneasy relationship between the two is given in R.M. MacLeod, 'Scientific Advice for British India: Imperial Perceptions and Administrative Goals, 1898–1923', *Modern Asian Studies*, 9, 3, 1975, pp. 343–84.

[144] To quote MacLeod: 'Many Fellows of the Royal Society also saw themselves in the role of *arbiter elegantiarum* and used language more at home in an Oxbridge tutorial than in an exchange of views *inter pares*,' ibid., p. 359.

[145] Revenue, Agriculture, General, nos. 4–8, Aug. 1904, file no. 183.

[146] Revenue, Agriculture, General, no. 15, April 1907, file no. 80.

[147] Revenue, Agriculture, General, nos. 7–10, May 1905, file no. 127 (emphasis as in original).

[148] Revenue, Agriculture, General, no. 3, Aug. 1905.

[149] In May 1909, T.H. Holland launched a vigorous attack on the IAC. He claimed that the BSA itself was not an advisory body but only an annual conference of the heads of scientific departments, and as such was not obliged to look to the Royal Society all the time. The IAC members also felt sick of controversies. One of them,

limped into 1923, and there was no remorse felt when both came to an end.[150]

Dichotomies

Science administration in Victorian India mirrored colonial administrative policies in general; a top-heavy structure, inner contradictions (for example, imperial vs provincial claims) and professional jealousies were reflected everywhere. At the top was, of course, the Secretary of State for India, who enjoyed sweeping supervisory powers and total financial control. Every appointment and expenditure had to be examined and approved of by him, and he would not shed a fraction of his powers even in cases where his sanction was a mere formality. An interesting case arose in early 1880 when the Education Department in Bombay made some savings and out of that money arranged for a Professorship in General Biology. The Government of India happily approved of it (as it entailed no extra expenditure) and, while forwarding it to the India Office for confirmation, asked that the final power of sanctioning any similar expenditure (compensated by some saving) in future be delegated to it. The Secretary of State promptly refused on the ground that 'every local government and administration, whenever a reduction of expenditure becomes feasible, would feel itself at liberty to bring forward some scheme for disposing of the sum saved'.[151] The Secretary of State exercised his

---

A.R. Shipley (of Christ's College, Cambridge) wrote:

It is obvious the India Office and the Indian authorities are pretty mad with the IAC. I do not at all wonder at it. My view is that we should be thoroughly reorganized, that we should have a small committee, consisting of people who really know . . . for instance, I am not of much use on such a Body, and there are several who practically never attend. I am sure, at present our efforts are 'too sporadic and fitful'.

A.R. Shipley to D. Prain (Director at Kew), 23 May 1910, Kew Papers, Misc. Reports, IAC, f. 298.

[150] In July 1923, the President of FRI (DehraDun) refused to write a separate report for the BSA. In December all other scientific heads called the Board's report 'useless' which 'might, without loss, be discontinued'. In Jan. next, B.S.D. Butler (President of the BSA) noted, 'no one has a good word for the Board', and the curtain fell on it.

Education, Health and Lands Dept., Agriculture Branch, no. 43, Feb. 1924, pt. B.

[151] Finance, Accounts, nos. 489–91, May 1880.

powers rigorously, yet he was not always impervious to pressures. Pressure from 'home' worked wonders. The important scientific workers in India often used this pull through their mentors at Kew or the Royal Society, and the Government of India had to comply.

The Government of India had its own limitations. Financial considerations as well as the demands of administrative expediency hindered the growth of a well-knit and integrated scientific department. Except for a few branches which were of military and economic significance (e.g. the Survey of India and the GSI) none others were able to develop. On the whole, efforts remained ad-hoc, sporadic and provincialized. Buck's attempts to create an Imperial Scientific Department failed.[152] Later, in early 1904, the Secretary of State wanted scientists to be represented (particularly medical men, perhaps under pressure from the IMS lobby) on the Governor-General's Council in the form of a Sanitary Member, but Curzon refused. Curzon refuted the idea that scientific advice received insufficient regard from the GOI. 'If anything', he argued:

the reverse is the case. In our overwhelming work and admitted lack of technical knowledge, we rather bow down to the man of science, whether he is a meteorologist, or surveyor, or botanist, or an—ologist of any other description; with the result that scientific people like nothing better than serving under the Government of India, where they are treated with so much deference. When we do not take their advice, it is only because administrative considerations of vastly greater importance, as in the case of plague, have a superior claim. Council is an administrative body. The place of the expert is to advice, not to administer, and his place is accordingly outside it, not on it.[153]

The local civilian administrators wanted the scientific staff to be provincial rather than imperial; they wanted practical results rather than research papers. This was true even for those whose designation carried the prefix 'Imperial'. Lingard was called the Imperial Bacteriologist; his work was certainly imperial in scope but he was put under provincial control. The provincial governments would jealously guard their interests. When the Indian Museum was proposed for Calcutta, the Bengal Government insisted on total control.[154]

[152] *Selection from Despatches to India,* IOR V/6/323, pp. 261–2.
[153] Curzon to J Broadrick (the Secretary of State), 28 Jan. 1904, Curzon Papers, IOL, Mss. Eur. F.111/163.
[154] A contemporary scientist complained: 'The Bengal Government imagines all the museums at South Kensington to be under one management, and what is good for South Kensington must be good for Bengal.' C.B. Clarke (Supdt. Calcutta Botanic Garden, 1869–71) to Thiselton-Dyer, 16 May 1884, Kew Papers, India Letters (Bengal), 1863–1900, vol. 153, f. 180.

Excessive administrative control exercised at different levels (and that too by more than one centre of power) ensured that colonial scientists would always dance to the official tune. However, this bred a sense of dissatisfaction and demoralization among the scientists. Some of them resented the administrative responsibilities which hindered research, while others resented their total dependence on the bureaucracy for every minor favour. They had little faith in civil administrators and often felt insecure.[155] The administrators, on their part, seldom understood the complexities of scientific investigations, and, in their zeal for immediate tangible returns, bungled with the opportunities and funds they had at their disposal. Dyer found the reason for failure in the fact

that Indian officials, whose training is almost entirely literary, pay scientific men the compliment of thinking that they are competent to attack any problem. They forget how highly technical and specialized science is now becoming. That they are disappointed with the results is therefore not surprising. Thus the investigation of the poppy disease was entrusted to a gardener, that of the sugar-cane in Southern India to a zoologist, and that of tea to the reporter on economic products. No one of these men in his own line was second-rate, quite the contrary, but they were unfamiliar with the nature of the tasks set to them. The result was that in each case the subjects had to be referred to Kew for final discussion.[156]

The metropolitan scientists had an important role to play—counselling, advising, often assuaging the feelings of one party and sometimes mediating between the two. But this had to be more in a private capacity than in an official one, with the result that in many cases it lacked in strength and effectiveness. In its correspondence, the Government of India would give the advice due respect, but in practice would follow only what its officials considered 'rewarding' and 'remunerative'. In October 1903 the Government of India asked Dyer to supervize the abridgement of Watt's Dictionary. Later the Royal Society demanded its revision by Watt himself. The GOI was not prepared to allow a revision (because of the expense involved) but at the same time it could not blatantly

---

[155] In a private letter to Dyer (3 Nov. 1884), George Watt pointed out: 'Indian matters are liable to very considerable fluctuations, and paltry financial considerations are allowed often to upset the most valuable works and arrangements. . . . In India we have learned to put no faith in the promises of a Secretary to Government of India since one Secretary may abolish any part of his predecessor's scheme by saying that the public service demands that it should be done.'

Ibid., ff. 464–6.

[156] Thiselton-Dyer to the Under-Secretary of State, 6 Feb. 1899, Kew Papers, Misc. Reports, Indian Agriculture, 1869–1921, ff. 104–6.

disregard scientific opinion. The resultant confusion led Dyer to comment:

I am much disappointed. I find myself baffled by obstacles which however plausible seem to me to be animated by an oriental subtlety on the part of the Government of India which scarcely veils the desire to make the execution of the enterprise impossible while still professing a desire to see it accomplished.[157]

The Great Exhibitions and the Imperial Institute catered mostly to commercial needs and to British industrialists. The Royal Society made a few valiant attempts which ended in acrimony and failure.[158] This confusion in science administration was amply reflected in matters of science education and research to which we now turn.

[157] Thiselton-Dyer to India Office, 17 March 1905, Kew Papers, Misc. Reports, Watt, Commercial Products, ff. 114–17.

[158] MacLeod finds the history of these bodies (the IAC and the BSA), at least in part, 'a study in failure—failure in vision, in organization and in objectives'. R.M. MacLeod, p. 345, note 143.

CHAPTER 4

Science in Education

India is in an exceptional position, and just as the Revival of Letters preceded the development of modern science, so education in India must not be sacrificed by a premature attempt to foster science and original research before education itself is general and set on a firm basis.

DPI Report, Bombay, 1868–69, p. 227

For the development of her natural resources, India's most crying need is not higher technical education, but private enterprise and private capital.

Calcutta Review, 104, 1897, p. 238

The Raj was founded on 'the right of conquest', but its continuance depended more upon how much acquiscence and consent it was able to elicit from its subjects. Naked aggression alone was 'medieval' and sometimes repulsive to 'modern' susceptibilities 'at home'. It had to be masked. The Raj was more than just the army or the bureaucracy; it represented a culture and a way of life substantially different from the way of life of those it had subordinated. How was this difference perceived at both ends?[1] The colonial imperative was to make this difference more visible— more transparent—in order to show the 'utility' of one and the 'absurdity' of the other. Education was an important tool in this project. It masked the intentions and controlled the results.[2] It established Western hege-

---

[1] On this complex relationship Edward Said's *Orientalism* (New York, 1978) remains a masterpiece. He builds upon Gramsci's *direzione* and Foucault's discourse. The arguments were later sharpened and, to some extent, reconstituted by Homi K. Bhabha, 'Difference, Discrimination and the Discourse of Colonialism', in F. Barker, et al. (eds), *The Politics of Theory*, Colchester, 1983, pp. 194–211; S.P. Mohanty, 'Us and Them: On the Philosophical Bases of Political Criticism', *Yale Journal of Criticism* 2(2), 1989 and Edward Said, *Culture and Imperialism*, London, 1993.

[2] Gauri Viswanathan, *Masks of Conquest: Literary Study and British Rule in India*, New York, 1989.

money in a way no army could have done. Macaulay influenced the destiny of the empire more than any Governor-General (he can still be 'felt' in the South Asian corridors of power).

A recent work argues that the controversies around 'what to teach' and 'how to teach' were resolved by 1836 but 'whom to teach' and 'why teach' remained unresolved.[3] In my opinion, what was resolved was 'why teach' and 'how to teach'. The *raison d'etre* was of course the desire to consolidate power and control the people. This was to be done through the medium of English which was more than a language. The 'how' included secularization of instruction which was then not in vogue in the metropolis itself. Notwithstanding the war-cries of the evangelicals, the government opted for a policy of non-interference in religions matters. It meant 'a dramatic disavowal of English literature's association with Christianity' and a shift from religious to intellectual control.[4] But debates continued on 'what to teach' and 'whom to teach'. As seen in Chapter II, the Court of Directors would not give clear-cut instructions on the content of education. It had to be evolved at the local level. Yet the consensus was on the diffusion of European knowledge and values. Since every European traveller and observer harped on the deficiencies of the native character, 'character formation' was projected as the main purpose of education and the content of education was expected to revolve round it. The details of the curriculum were worked out by each presidency according to its own exigencies and opinions. Perhaps for the first time in Indian history, the state emerged as the producer of knowledge; the sole arbiter of what was to be delivered and to whom.

Western medicine, for example, was projected as a humanitarian rationale for British rule and, at the same time, was systematically used for subordinating a rival or 'other' system of medicine. While doing so care was taken not to gravely offend social sensitivities and people of higher castes were encouraged to join the new medical and engineering colleges. They themselves opted to enrol once they found that it brought them better jobs and more money. When the Grant Medical College was opened in Bombay (1845), initially the Brahmins did not come forward from fear of 'polluting' dissections, etc. Its Principal asked the Director of Public Instruction (DPI) for permission to open the College to lower castes. The latter replied in panic that 'we have come here to rule and not to cause social upheavals'.[5] The policy was English education for the elite

[3] Anil Kumar, *Development of Medical Science in India 1835-1911*, Ph.D Thesis, University of Delhi, 1991, p. 180.
[4] Gauri Viswanathan, pp. 94-117, note 2.
[5] Report of the Board of Education, Bombay, 1850-51, pp. 10-15.

and vernacular education for the lower classes. Thus continued the time-tested strategies of hierarchization and marginalization, which accentuated linguistic and social stratification to a hitherto unknown level. This, coupled with what Fanon calls 'cultural mummification', became the hallmark of colonial education.[6]

This is not to suggest that there was not enough space or opportunity to test new ideas. Secularization of education was one such experiment. The establishment of professional colleges and later universities, the formulation of the curriculum, etc. provided an opportunity to try what had not been done in any other colony or even in the metropolis. The development of colonial education and the debates over it do not suggest a uniform pattern or linearity. Even within official circles there were sharp cleavages, of course not on the broad objectives, but on the different modes and options. In 1845, when J.F. Mouat (DPI, Bengal) floated the idea of establishing a university, he was supported by the local government.[7] Cecil Beadon, then Secretary to the Government of Bengal, wrote to the GOI:

The establishment of an university will doubtless be a great boon. It would tend to call into existence a vast amount of talent which now lies dormant from the absence of such a stimulus, and to produce an efficient indigenous class both of scientific and learned men and of public servants.[8]

But this was not viewed favourably in London. Later the success of Indian medicos and 'the requirements of an increasing European and Anglo-Indian population' made the Court accept the proposal in the famous Wood's Despatch, but not without the emphatic declaration that 'the education which we desire to see extended in India is that which has for its object the diffusion of the European knowledge, having practical objects'.[9] It was not clear as to what these practical objects were. Contemporary opinion did look for a 'sound and fertilizing instruction in Natural Science', but future events were to belie such hopes.[10]

The University of Calcutta was patterned on the London model. It was to be merely an examining body, and so could not give a boost to science education as such. In its curriculum the Oxbridge tradition was apparent in the exclusion of science. Most of the professors were Oxford or Cambridge graduates, who sought to impart to Indians the type of

---

[6] Quoted in H.K. Bhabha, p. 205, note 1.

[7] F.J. Mouat, *Proposed Plan of the University of Calcutta*, Calcutta, 1845, pp. 57–63.

[8] Home, Public, nos. 2–3, 7 March 1846.

[9] IOR, E/4/826, Bengal Despatches, LXXXVII, 19 July 1854.

[10] *Calcutta Review*, 26 (51), March 1856, p. 221.

education they themselves had received.[11] The system led to the acquisition of literary rather than scientific knowledge—'tastes which were best satisfied by the profession of the lawyer teacher or the government official'.[12] Instead of giving incentives, the Syndicate of Calcutta University resolved in 1858 to oppose the introduction of a subject like geology into the academic curricula.[13] In July 1859, Oldham (Supdt. GSI) submitted a memorandum 'on the most effective and at the same time most economical means of teaching geology and its collateral sciences in Calcutta'.[14] This was of no avail. At the Presidency College the government could make some provisions for introducing courses in natural philosophy and geology, but non-governmental colleges, where the majority of students received instruction, had no means of appointing qualified science teachers and establishing laboratories, and, therefore, were unable to offer science courses.

The three presidency universities deliberated upon and developed the curriculum in their own way. In 1855 Elphinstone asked the government 'not to be guided by Home precedents', and to use the experience of England 'rather as a warning than as a guide'.[15] Hence the curriculum of London was not adopted. For the B.A. examination at Calcutta, the compulsory subjects were classical languages, history, mathematics, moral science and physical science. In 1863 physical science was removed from this list of compulsory subjects and students were asked to opt for one of the following courses: (1) geometry and optics, (2) elements of inorganic chemistry and electricity, (3) elements of zoology and comparative physiology and (4) geology and pysical geography.[16] This was perhaps in keeping with the growth and break-up of science. The Madras University offered only one paper in physical sciences and that too an optional one; its area was so extensive as to include physics, inorganic chemistry, botany, geology and zoology. Naturally no one ever passed in this branch.[17]

[11] *Nature*, v, 25 April 1872, p. 510; J. Murdoch, *Educational Reform*, Madras, 1893, p. 2.

[12] Note by E.C. Buck, 10 Jan. 1886, Home, Education, nos. 14–88, Oct. 1887, pt. B.

[13] Home, Educational, no. 4, 27 Aug. 1858.

[14] Home, Education, no. 1, 20 May 1859.

[15] *Papers on the Establishment of Universities in India*, Calcutta, 1856, p. LXXIX.

[16] *Selections from the Records of GOI*, LIV, Calcutta, 1867, pp. 11–12.

[17] S. Sathianadhan, *History of Education in Madras Presidency*, Madras, 1894, pp. 80–104.

The authorities in India looked to western models but they seldom incorporated their good points, and in the name of adapting models to local conditions often made a mess of them. Accepting the London model, the universities were only meant for examinations, but the fact that London University granted science degrees also was conveniently ignored. In Britain itself science education had attracted a good deal of attention during 1830–50. This was in consequence of the 'decline of science' agitation led by Babbage, Brewster, Herschel and others who raised a bogey that the Germans and French had stolen a march over British knowledge of science and that the supremacy of the nation itself would be in peril if the links between science and industry were not properly understood and strengthened.[18] Even Oxford offered Honours Schools in mathematics and natural sciences, and Cambridge a Tripos in natural sciences. But Calcutta, Bombay and Madras had to remain content with only one optional paper in science. The time, it seemed, was ripe for learning the languages of Rome and Athens, but not their science.

In 1868 the Asiatic Society of Bengal tried to do what the BAAS had been doing in England for so long. It petitioned the Viceroy on 'the absence of the means of teaching natural and physical sciences in Bengal'. It quoted extensively from the resolutions of the British Association and concluded that 'if this be the case in England, how much more needful is it to carry out these recommendations in this country'! The Society urged that 'no one should be allowed to pass the Matriculation examination in the university unless he has proved himself to have such general acquaintance with the elementary and fundamental truths of natural history and physics as will enable him to take up the further prosecution of these subjects in his after course not as an entire novelty or as a new language'.[19] The government took shelter behind their favourite excuse of circumstances not being favourable and opined that the students must pay more attention to learning English than to anything else.

Some of the officials also were fairly critical of government policy and did not conceal their feelings even in official communications. In 1872, for example, H Woodrow, the DPI of Bengal, wrote:

The Calcutta University by insisting on Sanskrit or a classical language as an obligatory subject at the First Arts examination, practically causes the postponment

---

[18] D.S.L. Cardwell, *The Organization of Science in England,* London, 1980, pp. 59–62.
[19] Home, Education, no. 7, 27 Feb. 1869.

of scientific instruction and shuts out the great majority of students from all knowledge of the natural and physical sciences—a result unfortunate in itself and antagonistic to the views announced in convocations by several vice-chancellors. I think that in the long run science, if well taught, and if encouraged at first by rewards, will have a better chance of success than Sanskrit. The real patriots like Ram Mohun Roy and M.L. Sarkar hold similar views.[20]

Woodrow wanted physical science to be taught even at the school level and asked for rewards and financial incentives to induce school teachers to first study and then to teach science. Woodrow then discussed at length the importance of drawing in the telegraph course, in medical studies, in geography, etc. and urged the government to encourage drawing instruction in schools. This gave Campbell, the then Lt. Governor, a lever and he dwelt upon only drawing in his reply, very cleverly ignoring the other points raised by this well-intentioned DPI.

The insistance on science education, however, did not go in vain. In the early 1870s there was uneasiness with the existing curricula and a growing realization of the importance of science education. In 1875 Gray called for a chair of science in Bombay.[21] The same year Madras University decided. to examine its matriculation candidates in geography and elementary physics in place of British history. Next year it remodelled the M.A. syllabus and the science course was divided into two, one comprising physics, chemistry and geology, and the other, botany, zoology and physiology.[22] Mathematics of course remained a separate paper. Calcutta divided its B.A. course into two branches—course A (i.e. literary) and course B (i.e. science). Course B had English, mixed mathematics, inorganic chemistry and geography as compulsory papers and one optional paper from physics, zoology, botany and geology.[23] This course B was the maximum the authorities were then willing to concede. E.C. Bayley, the same man who in 1870 had called for 'the erection of national schools of science', became cautious in 1874 and said:

we have now gone as far as is safe and justifiable in this direction, even in the interest of the physical sciences themselves. I feel assured that any greater and especially any earlier encouragement of such studies can only be given at the expense of the general training and discipline of the mental faculties, which it had been, and I believe always should be, the essential policy of our examinations to secure.[24]

[20] WBSA, General, Education, 37–9, Sept. 1872.
[21] *DPI Report*, Bombay, 1874–75, App. A, pp. 143–4.
[22] S. Sathianadhan, pp. 103–4, note 17.
[23] A. Croft, *Review of Education in India in 1886*, Calcutta, 1886, p. 28.
[24] *University of Calcutta Convocation Addresses*, I, Calcutta, 1914, pp. 231, 305–6.

Science courses received particular support in Bengal. In 1879, out of 373 candidates, 191 took up course B and the pass rate was also better than in course A.[25] But Bombay was the first, thanks to Richard Temple, to grant science degrees in 1880.[26] Two examinations were held for the B.Sc. degree. The first examination had mathematics and natural philosophy, inorganic chemistry, experimental physics, and general biology. The second examination asked for three out of the following nine papers: pure mathematics, applied mathematics, logic and psychology, experimental physics, chemistry, botany, zoology, animal physiology and physical geography and geology. This was quite an exhaustive course and the existing staff could not have done justice to it, even though a new professorship for biology was sanctioned.[27] In the absence of laboratory facilities, this degree was to remain both theoretical and unpopular. By 1895 only forty-three persons had taken this degree.[28]

The Bombay example was not emulated by Calcutta and Madras, but it did lead them to at least revise their syllabus. In 1882 Calcutta University simplified its course B and in place of five papers set earlier, now only three were asked. They were English, mathematics, and one out of physics and the elements of chemistry, chemistry with the elements of physics, physiology with either botany or zoology, or geology with either mineralogy or physical geography.[29] The B.A. course at Madras was a bit archaic. It had three papers, English, a classical or vernacular language, and one out of pure mathematics and natural philosophy, physical and inorganic chemistry, general biology with either botany, or zoology or geology, mental and moral science, and history and comparative philology. At the M.A. level, the revision in 1882 separated physics and chemistry, but zoology, geology and geography were lumped together and offered as an alternative to botany and physiology.[30] Punjab University, established in 1882, created a separate faculty for science, and had mathematics and physics-chemistry as compulsory subjects and one optional subject out of physiology, zoology, botany and geology.

The science course in Bombay, Lahore and Calcutta was similar, except

[25] *Calcutta Review*, 69(137), 1879, p. 65.

[26] R. Temple, *India in 1880*, London, 1881, p. 152.

[27] The GOI asked the Government of Bombay to first prune its expenses and from the savings this professorship was established. Finance, Accounts, 484–92, May 1880.

[28] E.M. Gumperz, *English Education and Social Change in late 19th Century Bombay, 1858–1898*, Ph.D. thesis, University of California, 1965, p. 163.

[29] A. Croft, p. 146, note 23.

[30] S. Sathianadhan, p. 196, note 17.

that English, which formed a compulsory subject in Calcutta and Madras, was altogether excluded from the two former universities and mathematics, optional in Bombay and Madras, was a compulsory subject in Calcutta and Lahore.[31]

The changes in the syllabus seem to have disturbed the students. In 1887 A.W. Croft, then DPI of Bengal, lamented that 'a few years ago there were more students of the science course than of the literature course, but in B.A. exam. of 1885 there were 308A candidates and only 120B. In that of 1886 the disproportion was even greater, nearly 675A candidates to 212B, and the percentage of success was 60 in A against 37 in B'.[32] But the response gradually improved. In 1892 the DPI reported that the percentage of successful candidates was 20 in the literature course and 46 in the science course.[33] During 1892–96 the percentage of those taking the science course rose from 16 to 35.4, and during these years 329 candidates passed their M.A. examination from Bengal, out of which 146 had English, 42 had philosophy, 33 had mathematics and 81 science.[34] A Science Degree Committee was formed in 1898 and the following year Calcutta University decided to institute the degrees of B.Sc. and D.Sc.[35] The science course continued to grow in popularity, even though this was not liked by many of the people. To quote a contemporary insinuation:

The scientific experts have been lowering the standard of science examinations, apparently to heighten their attraction, and they have been reducing the literacy science. There seem to be no literacy experts of sufficient weight who are able to counteract the mischief.[36]

Since science education involves a variety of subjects with varying degrees of practical utility (such as geology, botany, medicine, engineering, agriculture, etc.), they will be discussed separately.

Education in Geology and Mining

The Court was able to have the Professorship for Geology revived in mid-1858, but even so the post remained vacant because no suitable person

[31] A. Croft, p. 147, note 23.
[32] WBSA, General, Education, nos. 15–16, June 1887, pt. B.
[33] *DPI Report, Bengal*, 1891–92, p. 5.
[34] *First Quinquennial Review of Education in Bengal,* 1882–93 to 1896–97, pp. 34–5.
[35] Members of the Committee were J.C. Bose, E. Lafont, M.L. Sarkar, A. Pedler and P.C. Ray, *Minute of the Calcutta University,* 1898–99, para 331.
[36] N.N. Ghose, *Higher Education in Bengal,* Calcutta, 1901, p. 23.

could be found.[37] Moved by the destitution of the Presidency College in all matters of natural science, Oldham, in 1862, suggested a substitute for the vacant chair by establishing in GSI itself a centre or nucleus of a school of applied science.[38] This was not acceptable to the government. It was feared that a substantial teaching load would make geologists deviate from the original objectives of geological operations. In 1869 Oldham renewed his offer, and wanted at least two students to be trained every year on state support.[39] This was turned down by the Finance Department which, in its wisdom, made the rather innocuous suggestion that instead of fellowships, the universities should allow extra marks in the subject by way of inducement.[40] To this, the university rightly replied that 'such a mode of holding out inducement to the study of geology would be destruction of the scheme upon which the university examinations are conducted'.[41] And the matter rested there.

In September 1880 the government, however, resolved 'to encourage amongst the better educated classes of the native community a scientific habit of mind', and called upon the governments of Bengal, Madras and Punjab to offer suggestions on the best means of providing for geological training.[42] In his reply, Bengal's DPI preferred to see the students instructed in physics or in botany in which the means of practical illustration were ready at hand rather than in geology in which these means were necessarily wanting.[43] Moreover, it was found imprudent to incur the expense of procuring from England a special professor to teach geology, unless there existed some probability of employment being given to the natives who chose geology as their subject of study.[44] The bogey of the Bengali character and physique was also raised,[45] and it was held unlikely that the Bengalis would take up an abstruse subject like geology for mere love of the science.[46,47] The DPI of Madras thought the same way

[37] IOR, E/4/851, Bengal Despatches, CXII, pp. 601–8.
[38] Home, Education, nos. 2–4, 11 March 1863; *Calcutta Review*, 39(78), 1864, p. 430.
[39] Home, Public, Surveys, no. 65, 10 July 1869.
[40] Home, Public, nos. 63–4, 28 Aug. 1869.
[41] Home, Public, no. 148, 12 March 1870.
[42] Revenue, Agriculture, Surveys, no. 33, Sept. 1880.
[43] Revenue, Agriculture, Surveys, no. 8, Dec. 1880.
[44] Revenue, Agriculture, Surveys, no. 4, Sept. 1882.
[45] Revenue, Agriculture, Surveys, no. 8, Dec. 1880.
[46] Revenue, Agriculture, Surveys, no. 4, Sept. 1882.
[47] Among the Bengalis who made significant contributions to the study of Indian geology during 1880–1913, mention may be made of P.N. Bose, P.N. Datta, Sarat Chandra Das, Hemchandra Dasgupta and A.M. Sen.

as his counterpart in Bengal, albeit for different reasons. They were that firstly, an applied science like geology demanded considerable antecedent knowledge of the sciences of chemistry, physics and biology which could not be arranged for, and secondly, the chief educational centres in that presidency were situated in localities where the geological structure was incapable of affording much scope for field demonstration and instruction.[48] The Government of Punjab also expressed its unwillingness to establish lectures in geology until the standard of knowledge in physical science was improved.[49]

In sharp contrast to these pessimistic reports was the memorandum submitted in December 1880 by A.S. Harrison, Principal, Muir Central College, Allahabad. He wanted a distinct professorship for the geological classes, and what is more, pleaded for ten or twelve stipends of Rs 40 or Rs 50 a month to be given during the two years' course of instruction if talented students were to be attracted.[50]

H.B. Medlicott, successor to Oldham, was strongly against the introduction of special courses in geology and its branches like palaeontology, minerology and petrology. The pretext was the same—intellectual bankruptcy of the Indians.[51] Moreover, he pointed out that so far as the GSI was concerned, special provision for geological training in Indian colleges was not required, because the Department mustered only fifteen graded officers, 'a number too small to offer any inducement to the opening of special classes'. The Governor-General agreed with these views.[52] Geological education thus remained in the doldrums and it was not until 1893 that a Professorship of Geology was finally established at the Presidency College, Calcutta, and some students began to appear at the B.A. examination with Honours in Geology.[53]

In England a school of Mining and Geological Instruction had been established in 1851, and yet in 1857 the Mining Association of Great Britain called for a central mining college and mining schools in every mining district.[54] In India no provision whatsoever was made prior to 1904, even though striking developments had taken place in the working of the mineral resources of the country. The Mines Act of 1901 had made

[48] Revenue, Agriculture, Surveys, no. 2, Sept. 1882.
[49] Revenue, Agriculture, Surveys, no. 30, Sept. 1880.
[50] Revenue, Agriculture, Surveys, no. 2, Sept. 1882.
[51] Ibid., no. 3.
[52] Ibid., no. 5.
[53] *Calcutta Review*, 103 (205), 1896, p. 336.
[54] Institute of Geological Sciences, London, Mss. GSM. 2/12.

some sort of mining education obligatory for mine-managers and supervisors, but from where were they to receive such training? The Indian National Congress asked the government to establish a distinct school of mines.[55] A committee was appointed by the Bengal Government to discuss the proposal.[56] The Director of GSI, T H Holland, and the two inspectors of mines, Stonier and Grundy, favoured a special school but the more influential civilian officials rejected it on the ground that 'no real demand' existed for it.[57] As a consolation a mining class was opened at the Shibpur Engineering College.[58]

Education in Botany and Forestry

Due attention was paid to botanical investigations right from the early years of the Company in India, but the educational aspects were ignored. In Calcutta botanical education figured only in the medical curriculum. In 1863 when the physical sciences were made optional at the B.A. examination by Calcutta University, botany did not figure at all. Later, in its course B, botany was made an optional paper. The universities of Madras and Bombay also gave the subject some recognition. But the Madras medical course did not have even a single lecture on botany.[59]

In 1873 George Watt was sent to Calcutta to take up a professorship of botany. On arrival he found that the local government had abandoned the idea of such a professorship and he was asked to teach chemistry in place of botany, which he did for eight years. Even after botany had been made an optional paper in course B, the colleges did not encourage it, and they found it easier to provide for teaching physics and mathematics than botany or geology which involved working in gardens, going on tours, etc. In 1875 the principal of Dacca College pleaded for the removal of the subject of botany, but Richard Temple, then Lieutenant Governor of Bengal, did not agree.[60] He, however, reduced the number of botanical lectures delivered at Calcutta Medical College from forty to twenty.[61] He

[55] A.M. Zaidi (ed.), *The Encyclopaedia of the Indian National Congress*, IV, New Delhi, 1978, pp. 158, 263.

[56] *Record of the Progress of Mining Instruction in Bengal*, 1905–9, p. 2.

[57] Commerce and Industry, Geology and Minerals, nos. 7–9, March 1905.

[58] Home, Education, no. 64, June 1905, pt. B.

[59] K.R. Kirtikar, 'The Study of Natural Science in the Indian University', *The Modern Review*, I, 1907, p. 477.

[60] WBSA, General, Education, nos. 1–3, Sept. 1875.

[61] WBSA, General, Education, no. 1, Nov. 1876.

did this to oblige George King, the Superintendent of Calcutta Botanic Garden, who used to lecture there but did not want to devote much time to it.[62] The DPI and the Council of Medical College protested, without success. At other colleges, such as in Patna, Rajshahi, Hughli and Cuttack, botany was quite popular. But in 1882 the addition of physiology made the subject more difficult and its popularity waned. At Patna and Rajshahi, botany classes were stopped. At Hughli college only five students opted for botany in 1887 and the principal asked for its abolition. The DPI, Allen Croft, took a larger view and wanted to retain the subject on the ground that the government ought to offer a variety of subjects.[63] But his arguments fell on deaf ears, and the GOI abolished these classes again on the usual pretext of 'no real demand'.[64] The result was that except in Presidency College, botany never received a fair trial in any college in Bengal.

Botany was of great importance to forestry. In fact it formed the intellectual basis of both forestry as well as agricultural expriments. Richard Temple realized this and wanted foresters to be trained in botany.[65] In 1878 D Brandis proposed to establish a Central Forest School at Dehra Dun and the GOI accepted the proposal.[66] The school was an instant success and theoretical instruction was started in 1881. The papers taught were sylviculture, mathematics, surveying, botany, and elements of chemistry, physics and law. This course was open only to the natives working for the Forest Ranger's certificate. The superior officials were all Europeans who were trained at the French Ecole Forestiere at Nancy.[67] Theoretical instruction could be arranged at the Cooper's Hill College in England but for practical training they had to go to Nancy. Even on a limited scale, the Dehra school proved to be a good experiment and induced foresters to make observations and experiments in a more systematic manner.[68] But the professional botanists wanted a still better treatment of their subject. J. Wilson, the Revenue Secretary, conceded: 'It is a reproach to the Government of India that with its splendid and

[62] This was acknowledged by G. King himself in a personal letter to J.D. Hooker, 8 Dec. 1876. Kew Papers, Indian Letters, 1860–1900, vol. 155, ff. 380–3.
[63] WBSA, General, Education, nos. 15–16, June 1887, pt. B.
[64] WBSA, General, Education, nos. 1–5, Dec. 1888.
[65] *Bengal Administration Report*, 1875–76, p. 30.
[66] *The Indian Forester*, XI (12), Dec. 1885, p. 557.
[67] WBSA, Revenue, Forest, no. 152, Oct. 1885, pt. B.
[68] *The Indian Forester*, VII (4), April 1882, p. 397.

valuable forests so little has been done in this country towards the development of Forest Science as distinguished from Forest Practice, and it is high time that this reproach were removed.'[69] In early 1906 the status of the school was raised to that of a college, and what is more, named the Imperial Forest Research Institute, signalling a shift to research.

Agricultural and Veterinary Education

The British Government in India was perhaps the largest estate holder in the world and its sole beneficiary. Even so it ignored any scientific instruction in agriculture for a very long time. The Company did practically nothing in this regard and when the Crown took over a few hopes were raised. In the 1860s the Landholders' Association and the British Indian Association repeatedly asked for such education.[70] But the officials paid only lip service and were profuse in their regrets. A breakthrough was made by the Government of Madras which, in 1865, established an experimental farm at Saidapet 'to afford facilities for testing the merits of little known machines, implements, manures, crops, systems of culture, livestock, etc., believed to be suited more or less to the requirements of Southern India'.[71] When its Superintendent, Robertson, pointed out how literary education had robbed agriculture of her best men and called for the establishment of an agricultural college, the Madras Revenue Board would not agree on the ground that even the so-called 'degraded' agriculture of India fed 30 millions people, yielded a considerable surplus of rice, cotton, sugar, oilseeds and indigo for export, and paid 4 1/2 million sterling of land revenue.[72] But Robertson persisted and an agricultural college was opened at Saidapet in 1876.

That very year another agricultural college was proposed as an adjunct to the Patna College in Bihar. The then Bengal DPI admitted that as he had no precedents to guide him, he had to emulate European and American models. Cirencester was thought to be too theoretical and Massachusetts heavily experimental. So he preferred a Belgian example which could be partially imitated—partially because he had no hope 'of seeing any sufficient imitation of the beautiful laboratory, of the spacious

[69] Revenue, Agriculture, Forests, nos. 18—20, Jan. 1906.
[70] J. Long, *Introduction to Adam's Report*, Calcutta, 1868, p. 36.
[71] Revenue, Agriculture, nos. 13-14, Jan. 1884, pt. B.
[72] Madras Revenue Board Proc. 2410, 25 Nov. 1873, quoted in R. Ratnam, *Agricultural Development in Madras State prior to 1900*, Madras, 1966, pp. 390-2.

economic museum, and of the scientific equipment which the little kingdom of Belgium thinks it right, regardless of cost, to give to its colleges'.[73] But not even partial imitation took place; instead an industrial school was opened at Patna and the scene shifted to Pusa where the government was experimenting with tobacco culture.[74] The Lieutenant-Governor, R. Temple, wanted that farm to be kept permanently and made use of for scientific experiments connected with physiological botany and agricultural chemistry. But the failure of tobacco expriments there made the officials despondent, and so on the convenient pretext of 'the inutility of spending money in attempts to teach the Indian peasants the agricultural sciences of Europe', the Pusa estate was leased to Ms. Begg. of Dunlop Co. for tobacco cultivation.[75]

Richard Temple carried his enthusiasm for agricultural science to the Bombay Presidency. In 1879 an agricultural class was opened at the Poona Civil Engineering College, and such classes were also opened at six of the principal zilla schools.[76] Temple thought of native revenue officers, like Mamlatdars and Karkuns, as the most effective agents for the diffusion of agricultural knowledge and wanted the Bombay University to conduct a degree course in agriculture.[77] But the university refused to concede more than a diploma. Later Voelcker found the diploma course 'a sort of half-way house better than nothing, but not the equal of a degree'.[78] All these were meant for the natives to be placed in lower grades. There was need to secure in the higher grades a proportion of officers with a scientific and practical training in agriculture. The Famine Commission of 1881 proposed first a theoretical training of ICS probationers, and then the practical training of a few of these at an agriculture college.[79] The Government of India rejected the proposal on grounds of 'impractibility' and the Secretary of State claimed that 'a superficial theoretical training would not only be useless but would be injurious to a civil servant, for it would divert his time and attention from the study of those subjects which are necessary to his efficiency as a judicial or administrative officer in India'.[80]

Meanwhile, the Government of Bengal had created two special schol-

[73] BSAP, Revenue, Agriculture, nos. 17–19, April 1876.
[74] Bengal Administration Report, 1875–76, p. 165.
[75] Ibid., 1876–77, pp. 138–9.
[76] Home, Education, nos. 27–30, March 1879.
[77] Ibid.
[78] J.A. Voelcker, Report on the Improvement of Indian Agriculture (2nd cdn), Calcutta, 1897, p. 384.
[79] Home, Public, no. 89, July 1881.
[80] Home, Public, no. 182, June 1884.

arships of £200 a year to be held for two and a half years by science graduates of the Calcutta University at the Royal Agricultural College, Cirencester. The aim was to have a team of experts as a prelude to the establishment of colleges and schools 'for grafting on eastern practice, as far as may be found possible or desirable, the ascertained results of western research'.[81] It was a significant step and that too taken by a local government and not by the Government of India. When the Agriculture College of Salisbury requested the Viceroy to establish scholarships there for his Indian subjects, he simply declined.[82] Obviously, the Bengal example had failed to inspire the Imperial Government. But even this experiment did not prove a happy one. A few Cirencester scholars diverted their attention to legal and other studies. The Bengal Government issued warnings.[83] The scholars themselves felt unhappy when on coming home they found they could get nothing more than deputy collectorships.[84] So, in 1887, the government decided to discontinue the scholarship.[85] This was an extreme step, akin to chopping the head off as a remedy for headache.

The enthusiasm of the mid-1870s was thus on the wane by the mid-80s. Except for the Poona unit which conducted important experiments in cotton and plough-technology, all other centres were collapsing.[86] Saidapet, for example, notwithstanding the value of the theoretical and scientific instruction which it gave, could not turn out 'practical' farmers, and was therefore pronounced a failure by W. Wilson, Director of Agriculture, Madras. Wilson distinguished between the art and science of agriculture and wanted the former to be given precedence which Saidapet was not doing. Naturally he felt indignant and condemned the Saidapet system as 'a process very similar to that of attempting to acquire a language from its grammar instead of learning the grammar from the language'.[87] This, however, does not mean that agricultural education and experiments had come to a dead end. Thanks to E.C. Buck's insistence the issue was very much alive.[88] In June 1888 the Government of India resolved to encourage the introduction of studies 'inclining to the application of

[81] *Bengal Administration Report*, 1879–80, p. 495. The catch, however, lay in the paranthesis which made the whole thing dependent upon 'possibility' or 'desirability'.
[82] Revenue, Agriculture and Horticulture, nos. 34–5, Jan. 1881, pt. B.
[83] WBSA, General, Education, nos. 3–4, May 1886, pt. B.
[84] BSAP, Revenue, Agriculture, nos. 76–80, March 1887.
[85] Ibid., nos. 111–13, April 1887.
[86] Revenue, Agriculture, no. 20, March 1883.
[87] Revenue, Agriculture, nos. 13–14, Jan. 1884, pt. B.
[88] E.C. Buck's note, 10 Jan. 1886; Home, Education, nos. 14–88, Oct. 1897, pt. B.

natural science'. Madras Government promptly took up the clue and in December 1888 appointed a committee to report on the working of Saidapet Agricultural College. The Committee declared Saidapet to be absolutely essential and recommended provisions for instruction in veterinary and forestry. But the Madras Government did not agree and instead sanctioned five agriculture schools.[89]

Sagging spirits, however, got a boost from Dr Voelcker. His criticisms brought the whole issue into sharp focus. He found that no encouragement was being given to the pursuit of scientific investigation in India and that men who might have been original workers in science had to abandon it for the duties of school inspectors.[90] For the 'unsettled state' of Saidpet College, he held 'the constant change of policy pursued by Madras Government' responsible. He found the syllabus to be modelled on English and not on Indian experiences. Thus, practices such as 'paring and burning' and 'warping of land' were mentioned; manures such as sulphate of ammonia, dried blood-soot and artificial manures, none of which had any place in Indian agriculture, were introduced; the requirements for 'fattening animals' were supposed to be learnt and this in a country where no fattening of animals whatever was carried on. On the other hand, many subjects of special interest to Indian agriculture were omitted, such as canal and well irrigation, oilcake refuse, ghee, etc.[91]

In order to take advantage of Dr Voelcker's presence and advice, in October 1890, the Government of India called for a Conference of Directors of Agriculture. The conference discussed *interalia* whether special teaching in agriculture was desirable, and whether it was to proceed from above downwards or from below upwards.[92] Buck favoured the latter, i.e. not to insist on degrees from agricultural colleges and schools but only to teach in an elementary fashion as to how a crop grows. This was what he called 'the first process in the cultivation of the agricultural intellect, the first manuring of the mental field in which agricultural progress is to be developed'. It seems he was driven to this conclusion by financial contraints. Otherwise he very much desired the early establishment of high class agriculture colleges and the education of natives at Saidpet and Dehra instead of at Cirencester.[93] Another view was put forward by the Agriculture Commissioner of Madras, H.F.

[89] Home, Education, no. 6, Oct. 1895.
[90] J.A. Voelcker, p. 331, note 78.
[91] Ibid., p. 39.
[92] Revenue, Agriculture, nos. 7–11, May 1891, file 26.
[93] Ibid., Presidential Address, 6 Oct. 1890.

Clogstoun, who opposed Buck and advocated percolation downwards. He said: 'The source from which the teaching power springs must be put on a higher level and we can never look to the stream to flow healthily throughout the country unless the fountain head be pure.'[94] The conference finally recommended that higher agricultural education should not be provided by special institutions but should be grafted to the existing ones, and that the claims of men with training in scientific agriculture should be freely recognized in the revenue and congnate departments.

These proposals were considered in another conference in 1893, where again Buck's views prevailed and the very next year the Government of India resolved that 'the question is one which cannot be forced, but should be dealt with gradually, and that greater success is to be expected from making instruction in the rudiments of agriculture part and parcel of the primary system of instruction than from teaching it as a subject apart from the general educational programme'.[95] It was a carefully chosen dilatory move on the part of the government. Instead of being a specialized area of study where enquiry and experimentation were followed by application, agriculture resulted in becoming a part of the general school curriculum and that too as an optional subject.[96]

Until 1897, the Government of India held the view that high class educational institutions professing to teach agriculture should be national rather than provincial, and that one or two national colleges would meet the needs of the whole of India. But the provincial governments were clamouring for college level teaching in their respective areas. They argued that education in an agricultural college provided as good training for state officials in the land revenue and cognate services as did training in an arts college. The Imperial Government was attracted to this logic and in 1897 proposed the establishment of four high class colleges at Madras, Bombay, Calcutta and some other city in northern India.

One of the results of this resolution was the opening in 1897 of agriculture classes at the Shibpur Engineering College.[97] Agriculture classes were begun in some zilla schools also but they soon collapsed as agriculture was not included in the syllabus for the university entrance examination. The new regulations of Calcutta University did not provide for degrees in agriculture. It was suspected that the affiliation of agricul-

[94] Ibid., Proc. of Agricultural Conference, 10 Oct. 1890.
[95] Educational Resolution, 7 Sept. 1894; Home, Education, no. 6, Oct. 1895.
[96] J. Mangamma, *Technical, Industrial and Agricultural Education in Madras Presidency, 1854–1921*, Ph.D. thesis, University of Delhi, 1971, p. 80.
[97] Revenue, Agriculture, nos. 21–6, June 1897, file 48.

tural colleges to universities would accentuate the tendency to regard these institutions as avenues leading to government service and to place them still further outside the practical scope of persons who would turn the training received to profit in private farming or estate management.[98] This was the feeling of the Simla Conference on Agricultural Education held in 1901. It even recommended vernacular to be substituted for English as the medium of instruction and thereby the Government hoped to 'popularize the study of agriculture amongst those who live by its pursuit and are not debarred from it by their ignorance of English'.[99]

In Kanpur there grew an agriculture school out of a training school for Qanungos.[100] Taking advantage of the resolution of 1897, in 1901 the NWP Government asked the Allahabad University to raise the status of this school to that of a college and the University Senate readily agreed.[101] The basic objects were threefold: to train teachers, to give instruction to landowners and to maintain a supply of revenue officials. Research work was to be encouraged and the award of post-graduate studentships, even doctoral degrees, was contemplated.[102] Earlier in 1897 the Government of India had acknowledged the desirability of an agricultural college in NWP, but when the proposal came officially, it developed cold feet. J.B. Fuller, the Agriculture Secretary to GOI, agreed that such a college would have a good chance of attracting a fair number of students, but at the same time condemned it as being 'premature'.[103] He held that its course of study would be of no value in training men to manage landed estates, and that the revenue officers did not need acquintance with science so much as with practical agriculture. So in place of the college, Fuller advocated, and the Viceroy concurred to, the establishment of what he called a 'Zamindari School'. Similarly, in 1903, when the Madras Government asked for Rs 11,700 per annum to create rural teaching posts for the students of Saidapet as an incentive for agricultural education, the Government of India refused.[104]

Though the Resolution of 1897 talked of four high class colleges, the Government of India subsequently narrowed its choice to only one

---

[98] *Third Quinquennial Review of Education in Bengal,* Calcutta, 1907, p. 92.
[99] Home, Education, nos. 47, Nov. 1901.
[100] Revenue, Agriculture, no. 8, Sept. 1902, pt. B.
[101] Home, Education, nos. 57–9, July 1903.
[102] Revenue, Agriculture, no. 3, Dec. 1901, file 89.
[103] Ibid.
[104] Home, Education, no. 116, July 1903.

institute which would serve as the nucleus for agricultural education and research. But who was to foot the bill? The Imperial Government was reluctant while the provincial governments spent on agriculture only in a piecemeal fashion. The windfall came in the shape of Phipps' donation of a hundred thousand dollars. This money was utilized to create an agricultural college cum research institute at Pusa which would furnish a complete course of agricultural instruction from the elementary to an advanced stage.[105] At about the same time the establishment of a similar institution for forestry was also being considered. Research had begun to claim some attention.

Veterinary education was important for two reasons: it was of vital importance in an agro-based economy and the army also required a continuous supply of veterinary surgeons and assistants for its horses. In the mid-1880s veterinary classes were conducted at the Poona college of Science for purposes of the army.[106] Lahore had a veterinary school where the medium of instruction was the local language.[107] It was fairly popular and successful. But more important was the opening of a veterinary college in Bombay in 1886. It was the first to provide veterinary instruction in English by entirely European methods.[108] The proposal had originated from the Society for the Prevention of Cruelty to Animals. D.M. Petit donated the land. The Government of India was approached for some assistance but this was refused.[109] Later, the Bombay Government and the Bombay Municipality agreed to bear the burden.[110] Initially the College attracted a good number of students. The incentives were free professional education and fair prospects of employment. In 1887 the number of students was 80, but in 1891 it slumped to 42. The principal asked the government to guarantee a few appointments every year to the students of his college.[111] Though this was not possible, the college managed to survive. In 1890 it even acquired a patho-bacteriological laboratory. Its first principal, J.H. Steel, started a quarterly journal of veterinary science in India. On the whole, this college shaped well, and later earned the

---

[105] Home, Education, nos. 36–7, Dec. 1903, pt. B.
[106] Revenue, Agriculture, CB&CD, nos. 1–2, April 1885.
[107] This school prepared several books in Urdu. Revenue, Agriculture, CB&CD, nos. 35–6, May 1886, pt. B.
[108] Annual Report of the Bombay Veterinary College, 1886, p. 8.
[109] Revenue, Agriculture, CB&CD, nos. 2–3, Dec. 1883.
[110] Revenue, Agriculture, CB&CD, nos. 9–11, Sept. 1884.
[111] Revenue, Agriculture, HB&AS, nos. 17–18, Dec. 1891, pt. B.

patronage of the Government of India. Elgin, for example, on hearing that the Bombay Veterinary College was short of staff, suspended the CP veterinary department and shifted the surgeon there to Bombay.[112]

It is evident that agriculture began to claim a good deal of attention in the wake of Voelcker's visit. A bacteriological laboratory was launched at the Poona College for Science in 1890. The original idea was to make it a laboratory for the distribution of Pasteur's anthrax vaccine. The funds set for veterinary instruction in this college were diverted to the new laboratory, and it could now even think of original research.[113] It was named the Imperial Bacteriological Laboratory and was entrusted with an imperial survey of veterinary diseases throughout India. The original intention of veterinary instruction at Poona thus receded into the background.

Bengal lagged behind Bombay and Lahore. In 1883 a veterinary college in Calcutta was proposed but the Bengal Government refused to finance it. It was reconsidered in 1886, only to be dropped again. In 1890 the mercantile community of Calcutta established a *pinjarapole* (an asylum for animals) and this prompted a fresh discussion on the necessity of veterinary instruction.[114] Meanwhile, because Poona and Bombay had shown the lead, in 1893 Calcutta also got a veterinary school. Three years later its status was raised to that of a college.[115] Though the number of students was large, the pass rate was considerably lower than at Lahore and Bombay. Unlike the Calcutta Medical College, the veterinary college was not very successful. In Lahore just the reverse was the case. As Captain Trydell (Superintendent, Civil Veterinary Department, Bengal) explained:

The students at Lahore belong to the zamindar class, three-fourths of them are Muhamaddans, thoroughly accustomed to handle animals from an early age and a fair percentage of them are sowars from Native Cavalry Regiments. Conditions are very different in Bengal, where scarely 1 per cent of the candidates who gain admission can ride or have ever handled an animal, and many are at first too timid to approach a horse.[116]

Madras had an almost similar problem. Though the proposal had been around since 1893, a veterinary college was established only in 1901, and that too on Curzon's personal intervention. In a letter to the Governor of Madras, Curzon wrote:

[112] Elgin Papers, IOL, Eur. Mss. F84/64.
[113] Revenue, Agriculture, HB&AS, nos. 27–34, Aug. 1892.
[114] Revenue, Agriculture, HB&AS, nos. 16–72, Nov. 1891.
[115] WBSA, Revenue, Agriculture, nos. 6–19, June 1897.
[116] Revenue, Agriculture, nos. 1–8, Aug. 1906.

I will give you an illustration of the sort of things that makes our Secretariat howl with despair. They say that the Madras Government will do nothing to promote agricultural knowledge or improvement, and that any suggestion from us is met with active, or passive, resistance. . . . I remember the name of one Govindas, who offered you Rs 10,000 for veterinary schools, and was refused.[117]

The veterinary schools at Lahore and Ajmer served their purpose better. Their medium of instruction was Urdu and this made them more acceptable to the people. But later it was found more expedient to strengthen the Lahore school by making it a college and diverting the funds from Ajmer to Lahore. Hence, in 1905, the Ajmer school was abolished.[118] These veterinary schools and colleges were not geared for service on a large scale. They did prove useful to the civil veterinary departments, but the army benefited the most.

Medical Education

Medical colleges were fairly well established by the time the new universities assumed control over them. In 1856 the Calcutta Medical College had ten chairs in anatomy, physiology, zoology, chemistry, botany, materia-medica, medical jurisprudence, midwifery, surgery, medicine and opthalmic surgery. Around 900 bodies were utilized annually for purposes of study.[119] In 1860 the students were divided into four classes: the Primary class, the Apprentice class, the Hindustani class and the Bengali class. The Primary class students had the full course of five years in English and were eligible to sit for the Licence in Medicine and Surgery (LMS), the Bachelor of Medicine (MB), and the Doctorate of Medicine (MD) examinations of Calcutta University. The Apprentice class was for the Eurasians and was like the other two classes of three years' duration. In 1864 the Bengali class was subdivided into two: the Native Apothecary class which trained students for hospital assistantship, and the Vernacular Licentiate class which gave a more extended clinical training in order to fit the students for independent practice among the poor people.[120] This sort of divided system of education effectively met the most pressing local needs, particularly that of the army.

The medical colleges at Bombay and Madras were also doing well. In 1856 the course of instruction given at Madras was recognized by the

[117] Curzon to Ampthill, 21 Feb. 1901, Curzon Papers, IOL, Eur. F 111/203.
[118] Revenue, Agriculture, CVA, no. 51, Feb. 1905.
[119] *Centenary Volume of the Calcutta Medical College*, Calcutta, 1935, p. 39.
[120] *DPI Report*, Bengal, 1864–65, p. 19.

Royal College of Surgeons, London, and within a decade it had eight professors and five assistant professors.[121] But Madras had no provincial schools to impart medical education in the vernacular while Bombay did have them in Poona, Ahmedabad, Hyderabad (Sind), and NWP at Agra and Lahore. A medical school was opened in Agra in 1855. This was meant to relieve the pressure on the Hindustani class at Calcutta Medical College. But the very next year the proposal to establish a similar school at Benaras was turned down by the Court.[122] Its chairman, Charles Wood, preferred the elevation of the Agra school to the college level. The Lieutenant Governor of NWP jumped at the idea and formally asked for it in early 1862. The original function of these medical schools was to educate the natives as native doctors *only*, and with this in mind, the IG of the Medical Department refused any upgrading. Only the number of students, their scholarships and the salaries of the professors were increased.[123] The consolidation of British rule over Punjab and the needs of the army there called for the establishment of a medical school at Lahore also. This was accomplished in 1858. The pattern was the same—one English class in which Europeans, Eurasians and natives were trained as sub-assistant surgeons in five years, or as apothecaries in three years; and another, a Hindustani class which produced native doctors in three years.[124]

In the mid-seventies medical schools were established at Dacca, Patna and Cuttack to cater to local needs in the vernacular. Some of the officials were quite sceptical about these schools. Dr Wise (Supdt., Mitford Hospital, Dacca), for example, regarded the whole scheme of establishing vernacular medical schools as

a most ill-judged and retrograde one. . . . In a small provincial city, where everything that occurs is talked of and often misrepresented, the introduction of a study which is repugnant to the feeling of all classes of natives must afford for many years to come a fruitful subject for exaggeration.[125]

But there was certainly no dearth of students. At Patna, for example, about 80 boys applied for admission, out of which 31 were taken. The real problem was that of finance and incentives. Many of the students were too

[121] D.G. Crawford, *A History of the IMS*, II, London, 1914, p. 448.
[122] IOR, E/4/837, Bengal Despatches, XCVIII, 1856, pp. 37–8.
[123] IOR, V/6/291, Despatches to India, 1863, pp. 227–32; Home, Education, no. 21, 28 Feb. 1863.
[124] Home, Education, nos. 9–13, 7 April 1863.
[125] WBSA, General Education, nos. 113–15, June 1874.

poor to pay fees and many of them would leave before the completion of their studies. The principal of Patna Medical School asked the Bengal Government to double the stipendiary grants, and was supported in this by the Deputy Surgeon-General of Danapur circle. But the Surgeon-General was not in favour of such incentives and the plea fell on deaf ears. As a result the number of students gradually declined; in 1885 it fell from 151 to 92.[126]

The products of these schools did labour under certain disadvantages. They were almost entirely debarred from improving their professional knowledge, being unable to consult English works, while those in the vernacular were few and elementary. They had to spend one year at a military hospital or a civil dispensary before taking the professional LMS course. There was no uniformity in the curriculum. At Sealdah, Patna and Cuttack medical jurisprudence was taught both in the second and third year while it found no place at the Agra school which preferred arithmetic for the first year students. The Agra syllabus was more practical oriented. Every session of the three years' course had lectures on what was then called practical anatomy, practical pharmacy and practice of medicine. For final year students it prescribed even clinical surgery and clinical medicine. This was not done at the schools in Bengal. The schools at Sealdah and Cuttack introduced separate courses on anatomy and physiology, but at Patna. the course were combined. Again, midwifery was made optional at Patna and was not taught at Agra at all, while at Sealdah, Dacca and Cuttack it formed part of the second and third year curriculum. Barring these differences, the common subjects were anatomy, chemistry, materia-medica, medicine and surgery.[127]

Botany and other natural sciences were not taught in these schools but the medical colleges did pay some attention to them. Before 1880 the difference between the MB and LMS examinations in Calcutta consisted only in the absence of zoology in the course for the latter. This was often resented. Some thought that the medical colleges were wasting time on the natural sciences and should concentrate only on the 'professional skill, tact and practical knowledge' of the 'art' of medicine. The Sanitary Commissioner for Madras, Major Cornish, was opposed to what he called 'a meagre smattering of a few scientific subjects'.[128] But there were others who wanted them to be taught even at the LMS level. In a petition

---

[126] *Bengal Administration Report*, 1885–86, p. 317.
[127] These comparisons have been deduced from Home, Medical, no. 23–53, July 1880.
[128] Home, Medical, no. 42, Aug. 1877, p. 101.

(20 September 1886) to Lord Reay, Governor of Bombay, an assistant surgeon of Poona, V.R. Ghollay, complained that at the Poona Medical School botany was not taught; pathology was taught in connection with medicine and surgery and not separately; and that diseases of the eye were taught cursorily with surgery and not separately. Hence he called for the establishment of lectureships in these subjects.[129] This was never done. Rather, in 1888, the LMS examination of Calcutta University was made somewhat easier, though the course of study for both MB and LMS was allowed to remain the same. The result of this lowering of standards was that the annual percentage of candidates appearing at the LMS examination during 1890–1900 shot up to 27.8 from 9.4 of the previous decade, while the average percentage of candidates for the MB examination went down from 9.8 to 3.5. Later, during 1900–10, the percentage for the LMS examination went up to 58.4 but that for the MB examination remained as low as a mere 4 per cent.[130]

Medical education laboured under severe constraints. Candidates thronged at medical institutions, but there were not enough teachers to meet the demand, at least in the *mofussil* schools. At the medical schools in Ahmedabad and Hyderabad (Sind) the materia-medica classes were conducted by hospital assistants while at the Bombay Grant Medical College a professor had to take a class of more than 100 students and was hardly able to moniter their progress. Lack of laboratory facilities and scholarships, and problems regarding the medium of instruction compounded the difficulties. And to crown it all, the universities had fixed a very long period of study and a fairly high literary and professional standard for its MB and MD degrees. Until 1890 the University of London required only four years of study after matriculation for the MB degree but the Indian universities which had to deal with poor students and a population unable to remunerate medical men at a high rate, wanted its students to put in five years after an FA for an MB degree. Naturally most of the students sought only the LMS degree.

Engineering and Technical Education

Consolidation of the Empire brought in its wake certain building activities which in turn required a continuous supply of a less expensive but skilled and trained class of overseers, assistant engineers, mechanics,

---

[129] Reay Papers, Mss. 254560(3), SOAS, London.
[130] These figures were stated by Dr Nilratan Sarkar, Minutes of the Calcutta University, pt. III, 1913, p. 974.

surveyors, etc. Naturally, the government paid more attention to engineering than perhaps to any other branch of science. Students also flocked to this branch of study in view of the immediate and guaranteed employment it afforded.

The college at Roorkee was an instant success. But from the official viewpoint, the college at Calcutta could not stabilize. It had a staff of only three persons, and its students were found good in the book work of Euclid but not in geometry, etc. Roorkee was favoured and considered superior. Meanwhile, Calcutta University had set a very high standard in its examinations—in its Master of Civil Engineering (MCE) and its Licentiate in Civil Engineering (LCE). In 1862 it made an FA the minimum qualification for appearing for the LCE examination. The result was that no candidate could be found eligible for the LCE.[131] So in November 1864 this college was abolished and its classes were transferred to the Presidency College. This proved to be a mistake. In a petition to the government, one of its ex-students, B.N. Das, refuted the official charge of local apathy. He argued that the system of guaranteed appointments had such a salutary effect that 'many students had their scholarships transferred from the colleges of general education, and the very best student of the Presidency College did not hesitate to throw away his metaphysics and law for the exact sciences'. This was not liked by the heads of the general colleges, and hence the amalgamation of the Engineering College with the Presidency College, or rather 'its reduction from a position of rivalry to one of subordination'.[132] The system of guaranteed employments was subsequently withdrawn and engineering education in Bengal began to show signs of decline.

In England three years were found sufficient to train an engineer; at Roorkee even two years sufficed; whereas in Calcutta the students were asked to undergo two years practical training after completing the three years' theoretical course. This had a detrimental effect. As Das wondered: 'It cannot be that the Bengalis are so slow in acquiring a scientific education that they would take five years while the alumni of the sister college in NWP would take only two.'[133] In early 1878 a committee was appointed by the Government of Bengal to look into the shortcomings of engineering education. This committee recommended the removal of

[131] *DPI Report, Bengal,* 1862–63, p. 10.

[132] Memorandum by Bhola Nath Das on the causes of the decline of the Calcutta Engineering College, 5 Feb. 1876, in F.J.E. Spring, *Technical Education in Bengal,* Calcutta, 1886, pp. 27–9.

[133] Ibid.

engineering classes from the Presidency College. A separate engineering college was revived again, this time with a workshop and more facilities. Its educational aspects were to be looked after by the Educational Department while practical training was placed under the PWD. Four classes were opened, those for civil engineers, mechanical engineers, civil overseers and mechanical overseers. The courses were revamped in 1882. An entrance examination had to be cleared for getting admissions. After a two and a half year course the students appeared in what was called the first Examination in Engineering. The papers were on mathematics, natural science, engineering construction, geodesy and drawing. One more academic year after this examination made them eligible for the LCE examination. An FA graduate, after passing the first engineering examination, could sit for the BCE examination. The syllabi for both the LCE and BCE examinations were the same, and an LCE degree-holder, once he passed the FA examination also, could be admitted to the degree of BCE without further examination. The civil engineering branch had papers in mathematics, natural science, engineering construction and drawing. Mathematics had differential calculus, integral calculus and hydrostatics, while the paper on natural science concentrated only on geology, mineralogy and metallurgy. Engineering construction called for a knowledge of the construction of buildings, bridges, roads, canals and machines like turbines, steam engines, etc. The mechanical branch had a paper on machinery in place of the natural sciences and it dealt with different types of machines and workshop appliances.[134] The whole course stretched over four years and this was followed by one year of practical training. An apprentice department was also opened to train foremen, overseers, etc. A class on photography was added in 1892, a mining laboratory was sanctioned in 1894, and the next year a full practical course in electrical engineering was introduced.[135] With these openings the Shibpur college picked up rapidly. Of the twenty-eight candidates selected for admission in 1891, two had already passed the BA examination, twenty-two the FA examination and only four were students who had passed the entrance test.[136] Later the demand for admissions grew so great that the minimum qualification had to be raised from passing the entrance examination to being an FA degree holder.

[134] Home, Education, nos. 3.–5, June 1882.

[135] First Quinquennial Review of Education in Bengal, p. 106, note 34. To meet th expenses of electric installation students were levied one rupee each for ten months of the year.

[136] *DPI Report*, Bengal, 1891–92, p. 81.

In its enthusiasm for revenue and cadastral operations, the Bengal Government wanted its executive wing to learn at least the rudiments of surveying and engineering. The *mofussil* colleges at Hughli, Berhampur, Patna and Dacca were asked to arrange for such a course. The principals of these colleges expressed their inability and the DPI himself pointed out that their object was to provide for 'a liberal general education, and not for the requirements of any special occupation or profession'.[137] So three survey schools were opened at Dacca, Patna and Cuttack. Simultaneously several industrial schools had also sprung up all over the province. The principal of Shibpur college was authorized to visit and supervize these schools. His college now functioned as a central technical institution to which these *moffusil* schools sent their best boys for final training and from which they received their supply of teachers. The survey and industrial schools in Patna were amalgamated in 1896 to form a new engineering school, aimed at filling in the vacancies in the sub-overseer grade. The curriculum included a study of elementary mathematics, training in survey work, drawing, elements of road-making and constructing simple buildings, and workshop practice with hand-tools.[138]

In Bombay Presidency the most important was the Poona College of Science which had arisen out of a school established in 1854 for the purpose of educating subordinates for the PWD. This college was not an exclusive engineering institution; it held classes in agriculture and forestry also. The result was a hotch-potch of various types of instruction, and that too, without adequate staff. In a memorandum to the Governor of Bombay, a teacher of the college complained that

most of the professors sent out from England when here have no other object in view than of teaching what they have learnt in their days but never or rarely of indulging in the luxury of keeping pace with the advancing science or engaging themselves in scientific research, with the result that their teaching becomes deplorably old.[139]

Only the inferior class of matriculates joined the engineering or agricultural classes, looking for guaranteed appointments in government departments, while the arts colleges attracted more and better students. The workshop of the college did not get any financial aid; rather, it earned money for the government by executing different types of works assigned by the PWD and private firms. So its original function of instructing

---

[137] WBSA, General, Education, no. 53, June 1872.
[138] Indian Engineering, XXI, 5 June 1897, p. 424, and XXVI, 1 July 1899, p. 8.
[139] Reay Papers, note 129.

students was lost. Later a journal wrote: 'At the Poona Civil Engineering College, judging from results, the popular system of education or the teaching of smatterings seems to have reached a high stage of development. But not one single LCE, we venture to say, is fit to practice in the profession in which he possesses a licence.'[140]

The curriculum, the instruments and the very organization of these colleges were geared to meet the requirements of only subordinate grades. Seldom did private firms of repute touch them. And for the recruitment of superior grades in government departments, there was an apex college at Cooper's Hill in England. This college was established in 1869–70 by the Secretary of State without consulting the Government of India; rather it was contrary to its wishes. But the whole expense had to be borne by India without the advantage of any Indian benefiting from this education. Many of the officials did not like this superimposition of a 'super' class of engineers. The Lieutenant-Governor of NWP viewed Cooper's Hill as detrimental to the healthy growth of the Roorkee College. He asked for its abolition, and in its place preferred only a limited import of European engineers as and when the situation demanded.[141] The practical aspect of training at Cooper's Hill was found ineffectual in Indian conditions and its syllabi too non-professional and academic. This college turned out foresters also and the IG of Forests, who absorbed them, held the training there to be 'inferior to the best continental education'.[142] But the Home Government would not budge and the Government of India had to acquiesce. This college was abolished in 1903 after much hue and cry.[143] Its existence for more than thirty years nevertheless symbolized the supremacy of metropolitan institutions over colonial ones like those at Roorkee and Dehra.

The logic of the metropolis–colony relationship did not favour the latter getting a higher form of scientific or technical education. What the colony got was a sort of hybrid education emerging out of a careless fusion between industrial and technical education. It meant different things to different people, and even in the official hierarchy, its connotations differed. E. Buck (Secretary, Revenue and Agriculture), for example, treated it as the equivalent of practical training.[144] G. Watt (Reporter on

---

[140] Indian Engineering, XIII, 25 March 1893, p. 221.

[141] Home, Education, nos. 12–15, Dec. 1877.

[142] Finance and Commerce, Salaries and Establishment, nos. 149–55, April 1890, K.W.

[143] *Royal Indian Engineering College Committee Report*, Simla, 1903, p. 13.

[144] Curzon tended to agree with Buck. E.C. Buck, *Report on Practical and Technical Education*, Calcutta, 1901, p. 28.

Economic Products) discussed it as if it meant the general development of economic products, combined with research and practical training in particular industries, while Chatterton (Principal, Madras Art College) regarded it as a machinery consisting chiefly of workshops and a system of sales for developing manual industries.[145,146] For Campbell, the teaching of drawing and surveying appeared most important. 'The first of the technical sciences to be taught in schools', he urged, 'should be a good handwriting. In former days Bengalis were celebrated for their English handwriting'.[147] However odd this observation may appear now, he and Richard Temple had done a great deal to focus attention on scientific and technical education during the early 1870s.

But it was the Madras Government which thought of technical education in a big way. In 1885 it offered grants-in-aid for the establishment of applied science classes, laboratories, demonstration farms and workshops in connection with existing recognized colleges and high schools.[148] It even instituted what it called a higher examination in Science, Art and Industries. This test was meant for those seeking employment as science or technical teachers, engineers, foremen, foresters, veterinarians, or in different posts in the revenue, public works and other departments. This was rather a scheme of technical examination, not of technical instruction, and it was not so much concerned with teaching as with testing the results of teaching. This examination differed from the university ones in the sense that it contemplated 'not so much abstract science or science studied merely for the extension of knowledge and enlargement of the mind, but sience with a view to its application to various manufacturers and other industries'.[149] The choice was clear.

During the last two decades of the nineteenth century technical education overrode almost every other educational issue. Technical education was hailed as the panacea for all ills and every fund raising opportunity was utilized for establishing technical or industrial schools. The 1887 Jubilee celebration of Queen Victoria's reign was one such occasion, and Rs 80,000 was raised in Bombay alone. This, along with Rs 150,000 of the Ripon Memorial Fund, went into the establishment, in 1888, of the Victoria Jubilee Technical Institute (VJTI).[150] The Insti-

---

[145] G. Watt, A note on Technical Education in India, Home, Education, nos. 47–61, Nov. 1901.

[146] Ibid.

[147] WBSA, General, Education, no. 6, July 1872.

[148] Home, Education, nos. 30–2, July 1886, pt. B.

[149] Home, Education, no. 58, Sept. 1885.

[150] *VJTI Platinum Jubilee Volume*, 1962, p. 27.

tute was to supply textile technologists and mechanical engineers to the numerous cotton and other mills of Bombay. Expenditure on it remained static at around 50,000 rupees a year for almost a decade. In the initial years the attendance of students was good—about 350 in 1888–89—but it fell to 150 during 1895–96. In 1896 there was a sort of stock-taking and a thorough revision of the courses in order to make it 'more practical and more acceptable to the employers of skilled labour'. A series of popular lectures (imitating the Royal Institution perhaps) was started and several additions like metal-working, enamelling and electrical engineering were subsequently made. But all this did not increase the percentage of students who passed. In 1891, when the first three years' course in cotton manufacture was completed, eighteen students got the full Technological Certificates. But there appeared a distinct fall-off the very next year which turned out only four successful students. This trend continued till 1903—a period of thirteen years during which the annual average of successful students did not exceed seven. Interestingly enough, during the same period, the cotton mills alone required the services of over 5000 engineers, and mechanics.[151] Popular lectures had to be discontinued, the sheet-metal stamping plant (which cost Rs 45,000) broke down, while the enamelling department proved a failure from its commencement. The good students were drawn towards literary education, and VJTI had to content itself with matriculates who had not passed the examination. To teach them both the theoretical and practical sides of a profession was no easy task for the Institute.

Such a state of affairs did not dampen enthusiasm for technical education. People wanted technical colleges and institutes to be erected as memorials to the Queen. Even though money poured in through public donations, Curzon was not in favour of setting up such colleges.[152] He insisted upon a 'personal or structural' memorial. Arguments that such colleges would help vitalize Indian economy did not impress him. He thought of India's economic problems as being eternal, 'not to be solved by a batch of industries or a cluster of polytechnics'. 'They will scarcely produce a ripple in the great ocean of social and industrial forces', he added.[153] Engineering and technical education thus had to remain

[151] *The Indian Textile Journal*, XIV (165), June 1904, pp. 270–1.

[152] 'I regard this as using the Queen's name for an altogether improper and illegitimate purpose . . . and I predict with the utmost confidence, that the majority of these institutes, for which we are not ready in this country, will be still-born.' Curzon to Hamilton, 28 Feb. 1901, Curzon Papers, IOL, Mss. Eur. F 111/160.

[153] Curzon's speech at the Asiatic Society, Calcutta, 26 Feb. 1901; Home, Education, nos. 11–12, May 1901, pt. B.

confined to lower forms of instruction geared only to produce overseers, surveyors and mechanics of various hues, just as literary education produced clerks and pleaders.

The Medium of Instruction

The medium of instruction was a problem which perhaps was subjected to the most animated discussion, and which has still not lost its relevance. Although Macaulay's verdict was final, subdued voices were often heard in favour of either translation or instruction in local languages. European terminology not only limited a local student's perception of a particular scientific phenomenon, but also impelled him to first learn it by rote. The question of understanding—the most important thing in science education—came next and in many instances did not figure at all. In some cases at least local terms would have conveyed the meaning better. Ballatyne, for example, argued in favour of 'properly constructed Sanskrit terms'. For basalt he used the term *krishna-prastara* (i.e. black rock) and for iodide of potassium he used *laghutam ka arunaj* (potassium, being the lightest metal, is *laghutama*, and iodine is *aruna*, named after the colour of its vapour). This sort of adaptation, he believed, would have led to the percolation of science education (thereby scientific temper) down to the masses. He further explained: 'In a small provincial school we need not undertake to teach Berzelius' System of Chemistry, in its five thick octavo volumes, but why should the name of chemistry or science scare us from attempting to teach such matter of the chemistry of daily life as the fact that air is necessary to the burning of a lamp, the theory of rusting and why rust is prevented by oiling, the process of solution, filtration, evaporation and crystallization, etc.'[154]

For the purpose of general education the Court accepted the role of the local language, but 'for the conveyance of a high order of education in the science and literature of Europe', the Despatch of 1854 held it 'equally necessary that the English language should be the medium'.[155] English had certain inherent advantages, no doubt, but higher or professional education, whenever permitted in the vernacular, also proved successful. The British Indian Association of NWP cited the examples of vernacular classes at Agra Medical School and Roorkee Engineering College, and in 1867 submitted a memorandum pleading for science education

---

[154] J.R. Ballantyne, *A Discourse on Translation*, Mirzapore, 1855, p. 3.

[155] A.M. Monteath, note on State of Education in India, *Selections from the Records of GOI*, LIV, Calcutta, 1867, p. 113.

through the vernaculars. The memorialists gave very cogent arguments. They said:

The study of Sanskrit and Arabic is as valuable in India as that of Greek and Latin is in Europe. But we do not wish to see it overrated, as it was during the early part of this century. Neither do we wish to see the spirit of Anglicism riding so roughshod over it, as it has done since Bentinck's time. . . . English education may create a class of *Keranees* or Young Bengal but it can never by itself affect the life and growth of the nation. . . . We want education to be one that will not be foreign. The task of acquiring a foreign language like English is really a very difficult one, and consumes not only a great deal of otherwise precious time, but, even when accomplished is acknowledged by all English scholars to be but a *veneer*, and is furthermore injurious in its tendency of keeping knowledge in an *artificial* soil, and preventing its fruition.[156]

As expected, the memorandum met with a polite refusal.[157] Bereft of state support, the job of writing books in the vernacular or translating the standard English ones into local languages fell on individuals or certain societies. The Aligarh Scientific Society deserves special mention. Societies dealing with school books were doing a similar type of work at Agra and the presidency towns. But these were mostly esoteric attempts, confined only to primary or secondary levels. It is difficult to gauge whether they could make any major impact. There were no original or good works in the local languages while the English books were difficult to master; catechisms or help-books appeared in the market which were quite popular with the students. To quote a professor of City College, Calcutta:

The students are mostly young men and very imperfectly acquainted with the English language. They require help in the beginning and the best way of rendering help is to put before them a catechism of their text book, showing them the nature of questions that may be set in the examination, and the way in which they may be expected to answer. Having this view in mind I wish to prepare a catechism of your book.[158]

Those who took the trouble of writing text books in the vernacular

[156] The British Indian Association, NWP, publication no. 5–6, Aligarh, 1869, pp. 4–11 (emphasis added).

[157] Private Secretary to the GG dismissed it in a sentence: 'The question of the medium is quite subordinate to the task upon which Government is at present employed, that of educating youths up to a standard of fitness to receive higher education at all.' Home, Education, nos. 19–20, Sept. 1867.

[158] Rajendranath Chatterjee to T.H. Huxley, 22 May 1888, Huxley Papers, Mss. 12, ff. 174–5.

often found themselves in financial distress. George Watt, for example, while a lecturer at Hughli College, wrote a text book on botany and translated it into Bengali. This was done at the behest of the Bengal government. But by the time his book appeared (1876), a new Lieutenant Governor had taken over who abandoned the scheme of teaching botany in schools. Watt then had to sell 5000 copies of his labour as waste paper. Watt suffered again the very next year. The Calcutta Christian Tract and Book Society asked him to write a small text book on botany. This was to be translated into different Indian languages and used in missionary schools. But controversy arose soon after its publication. He had shown plants to be actually living beings and this was not liked by the Christian fundamentalists. So by the orders of the inquisition, the book was consigned to flames.[159]

At the official level there was hardly any vacillation. It was decidedly in favour of English. But in other forums and often at a personal level, officials did debate about whether something could be done to broadbase science education with the help of local languages and scripts. The Education Commission of 1882 pointed out how in Bihar the 'Sanitary Primer' printed in Kaithi (a local form of the Nagari script) was much more popular than that written in the 'High Hindi' of NWP.[160] Voelcker favoured teaching in the vernacular,[161] while Moreland rejected it as totally unsuited for scientific expressions. Moreland cited an example of such 'hybrid stuff' from a contemporary work on agricultural chemistry: '*Kul darkhtan nitrates, amonia or kuch amides kha sukh-te-hain, aur chunki amonia se nitrates juldi bante hain, galiban darkhtan upne kul nitrogen nitrates ke shakal main khate hain.*'[162] Final decisions invariably fell in favour of English.

Other Facilities

Language was not the only problem that dogged Indian students. Financial strain perhaps played more havoc. Most of them had to look for jobs long before completing their studies. Even while managing to continue in colleges, they often felt handicapped on account of high fees. While Oxford charged only £3.10 per annum, the Presidency College in

[159] The Indian Textile Journal, XIV (161), 15 Feb. 1904, p. 147.

[160] *Report of the Indian Education Commission of 1882*, p. 343.

[161] J.A. Voelcker, p. 388, note 78.

[162] Note on Agricultural Education by W.H. Moreland, Director, Agriculture, UP; Revenue, Agriculture, no. 8, Sept. 1902, pt. B, file 103.

Calcutta took £14.8 annually, and that was exclusive of fees for professional branches such as civil engineering.[163] Scholarships were few and future prospects so bleak as to leave scholars frustrated. It was only with the help of guaranteed appointments or scholarships that the medical or engineering classes could attract students. But these were professional courses, holding the prospects of some immediate return. Physical and natural science subjects held no such promise and did suffer for want of scholarship and laboratory facilities.

The task was not easy. Both official as well as private efforts were needed. The government always tried to shift the onus to private initiative. Right from the Woods Despatch, successive governments kept harping on the fact that scientific and technical education could progress only through private efforts. Very shrewdly the government limited its responsibility only up to the primary level. However, private initiative alone could not have succeeded in the given milieu and conditions. Such initiative was highly individualistic and often parochial. In 1887, for example, the Begum of Bhopal instituted a scholarship for Muslim students to study law or medicine in England.[164] The Bengal Government had instituted two agricultural scholarships at Cirencester. But when the demand was made for two identical scholarships for chemical or textile studies in England, the same government refused.[165] Even the agricultural scholarships were withdrawn in 1887 on the ground that the scholars felt unhappy when on returning to India they found they could get nothing more than deputy collectorships.[166] Later, in 1896, the prestigious Gilchrist scholarships were withdrawn for similar reasons.[167] It was only under mounting public pressure that the Simla Educational Conference of 1901 recommended ten technical scholarships.[168] The candidates were to be sent abroad 'to undertake definite courses of study in subjects connected with industrial science or research'.[169] In early 1904 an association was formed in Calcutta which, through donations and mass collections,

[163] Report of Proceeding of a Public Meeting held at Chapra (Bihar) in July 1870 to consider the Educational Policy of the GOI, *Tracts on Indian Education*, n.d. p. 14.

[164] WBSA, General, Education, nos. 6–11, April 1887.

[165] Home, Education, nos. 42–50, Feb. 1885.

[166] BSAP, Revenue, Agriculture, nos. 76–80, March 1887, and nos. 112–13, April 1887.

[167] Elgin Papers, IOL, Eur. Mss. F84/319.

[168] M. Bhownagaree raised the issue in the House of Commons on 15 July 1901; *Parliamentary Debates on India*, 1900–1, IOL, V/3/1604, p. 213.

[169] Home, Education, nos. 67–75, Oct. 1902; and nos. 14–18, Oct. 1903.

sent some students abroad for technical training.[170] The turn of the century thus saw a vast amount of attention focussed on scholarships and grants. The emphasis was obviously on sending students abroad. In the euphoria that ensued hardly anyone gave a thought to whether they would return better equipped than the other imported teachers; rather, as a contemporary critic pointed out, it only increased the mental distance between them and the local workmen whose training they were expected to influence.[171]

Other outlays such as that on laboratories were seldom considered till Campbell's regime in Bengal. Only the Calcutta Medical College could boast of a good laboratory, and that too perhaps because the office of the Chemical Examiner to the Bengal Government was associated with it. Other colleges had no specific laboratory arrangement. Sometimes a college would get azimuth and altitude instrument, or an achromatic telescope, or even a locomotive engine, but these were looked upon more as curios than instruments of instruction. From the 1870s things improved; more students were attracted to science subjects and a few colleges started getting around Rs 500 a year for apparatus and chemicals.[172] But the instruments had to come from England and indented through the Secretary of State. This led to inordinate delays, even of three to four years, and damages in transportation. Sometimes the DPI encouraged local manufacturers (like Murray and Co.) to supply the instruments.[173] But this was often condemned as 'nearly equivalent to giving away good money for bad instruments'.[174] It was J.C. Bose who imparted a certain amount of respectability to local talent by experimenting on self-made instruments. In 1896 he was given a special grant of Rs 1000 and subsequently Rs 500 a year for research purposes.[175] But the very next year when the Secretary of State recommended the establishment of a physical laboratory for 'advanced scientific training and research', the Government of India put the cost at around Rs 60,000 and backed away.[176] It was not prepared to establish even a research library for those who had received higher education in Europe and who wanted to continue work on their return to India. This demand was made by Ganesh Prasad, R.P. Raranjpye and forty-three other students then studying at Cambridge. They wanted

[170] Sumit Sarkar, *The Swadeshi Movement in Bengal*, New Delhi, 1977, p. 112.
[171] *The Indian Textile Journal*, XIV (160), 15 Jan. 1904, p. 103.
[172] WBSA, General, Education, nos. 1–3, Sept. 1875.
[173] WBSA, General, Education, no. 13, Aug. 1882, pt. B.
[174] K. Zachariah, *History of the Hooghly College*, Calcutta, 1936, p. 102.
[175] *DPI Report*, Bengal, 1895–96, p. 35.
[176] *Parliamentary Debates on India*, 1898, IOR, V/5/1601, p. 461.

this library out of the Queen Victoria Memorial Fund. But Curzon would not settle for anything other than personal memorials to the Queen—no institutes, no laboratories and no libraries.[177]

Major Problems

Five major problems plagued science education at the higher levels. The first was the very aim and character of the educational policy itself. Despite the increasing claims of science and materialistic philosophy, the education department clung to the notion that its goal was 'character formation'.[178] Moreover, universities were deliberately denied teaching jobs. K.M. Chatfield, the Principal of Elphinstone College, admitted that the institution of university professorships would indeed foster the development of knowledge through research but argued that not this, but the 'education of youth' was the aim of the system.[179] A recent critic noticed an interesting ambivalence on the part of the British. While on the one hand they complained that the Indians were too speculative, on the other they admitted that one virtue of having a contemplative, dreamy-eyed set of subjects was that it kept them from the pursuit of gain, allowing the British to step into that field.[180]

The second problem was the shortage of funds. For example, the total grant for the colleges of Bengal during 1871–72 was Rs 5,50,000, out of which only Rs 16,100 was given to the Presidency College for chemistry and physical science teaching, while Rs 29,500 and Rs 34,400 were spent on the engineering and law classses, respectively.[181] Since the state finances were gobbled up largely by the metropolitan college, the mufassil colleges could hardly afford to impart any science education worth the name. In 1906 Mr Jackson, Professor of Physics at Patna College, computed: 'In 1882–83 the four mufassil colleges of Patna, Cuttack, Hooghly and Krishnagar cost the Government Rs 1,05,142 and the Presidency College cost the Government Rs 78,850. In 1900 the share of the former declined to Rs 55,441 only, while the Presidency College soared to Rs 1,14,702.'[182] The Patna College, Temple Medical School and Bihar School of Engi-

[177] Home, Education, nos. 11–12, May 1901, pt. B, and nos. 17–18, July 1901, pt. B.

[178] E.M. Gumperz, p. 327, note 28.

[179] *DPI Report*, Bombay, 1868–69, p. 227.

[180] 'Baconian Philosophy Applicable to the Mental Regeneration of India', Calcutta Christian Observer, n.d, p. 125, quoted in G. Viswanathan, p. 161, note 2.

[181] WBSA, General, Education, no. 75, June 1872.

[182] Third Quinquennial Review of Education in Bengal, p. 19, note 98.

neering taught chemistry and physics in one common laboratory. Since the university had made science courses optional, several colleges discarded science classes altogether, and a few confined themselves to one or two science papers up to the intermediate standard. In 1897 the Government of India had acknowledged the desirability of an agricultural college in NWP, but when the proposal officially came, it developed cold feet. Behind this refusal perhaps lay the spectre of cost, which would have amounted to Rs 50,000 a year.[183]

Another allied problem was that so little laboratory work was prescribed that the course was almost worthless as a preparatory ground for entry into industry. Only a few of the big colleges could boast of even an improvised laboratory. The opening of a chemical laboratory in 1896 in Calcutta was sarcastically referred to by a contemporary as 'a Giant chemical laboratory for the use and abuse of its Giant classes, brimming full of sucking Newtons and budding Liebigs, who one day shall teach the West'![184]

The fourth problem was that the custodians of science education in India always looked to European models. C. Benson of the Saidapet Agricultural College, for example, justified the major emphasis on theoretical instruction there on the ground that Professor Jorgensen had done the same thing for the Royal Agriculture College at Copenhegen.[185] What was suitable in Denmark was thus held to be good for Madras also. In almost every field of activity, British institutions were looked upon as the *ideal* models. While proposing a scheme for the development of scientific and technical education in Madras in 1885, the DPI there first outlined developments in England.[186] Later, in 1894, E.C. Buck produced a 'Note on Technical Education in England',[187] before propounding his views on the desirability of such education in India. But while citing British examples, the publicists often tended to forget that one city alone—Manchester—used to spend £3,00,000 on one school, and this was about twice the annual provincial grant to education in the whole of Bombay presidency.[188] Revenues were not raised in far-off colonies for educational purposes because the colonizers were interested in filling their own coffers first. So, emulating western models seldom yielded construc-

---

[183] Revenue, Agriculture, no. 3, Dec. 1901.
[184] *Indian Engineering*, XIX, 11 Jan. 1896, p. 162.
[185] Revenue, Agriculture and Horticulture, no. 6, Oct. 1880, pt. B, file 45.
[186] Home, Education, no. 58, Sept. 1885.
[187] Revenue, Agriculture, nos. 11–13, Feb. 1894.
[188] A.W. Thomson, *Report on Technical Education in Primary and Secondary Schools*, Bombay, 1901, p. 2.

tive results. What is more, the adoption of English as the sole medium of instruction in science rather hampered its percolation down to the lower classes. In 1853 the Japanese language was not half as well advanced as Bengali or Hindi. It was argued that if useful scientific literature in Japanese could be cultivated, there was no reason why this could not have taken place in the Indian languages had they been properly encouraged.[189]

Finally, the most important problem was that of administration and management. This is evident from the treatment meted out to engineering and agricultural education. The Shibpur Engineering College, for example, was not under any institution. While its business was to train subordinates for the PWD, its teaching aspect was under the Education Department. The result was that as regards the theoretical aspect the students were under the Education Department, while as regards their manual work they were under the PWD.[190] The lack of an all-India policy led to certain management anomalies. The Saidapet Agriculture College and Poona College, for instance, were administered by the Education Department, while the Kanpur and Nagpur agricultural schools were under the Agriculture Department. But none of these provided a complete agricultural education.[191] The management of medical education was such that a professor of surgery often found himself transformed into a deputy surgeon-general and a professor of chemistry into a store-keeper for the government.

It may be argued that when Victorian England itself was lagging behind its continental competitors in science education and research, how could its government have thought of imparting higher scientific education to its dependencies? It is true that Germany and America provided greater state support to science education than did England. But the latter was certainly not oblivious to the need of providing science education. Britain saw an enormous growth in its number of scientific societies during 1770 to 1870. During the 1860s the government introduced vast changes in the university system through the royal commissions.[192] It marked a change from *polite* to *professional* education.[193] But in India this change did not occur.

[189] B.D. Basu, *Education in India under the East India Company*, Calcutta, n.d., p. 127.

[190] *Papers on Technical Education in India*, Calcutta, 1896, p. 45.

[191] Home, Education, no. 116, July 1903.

[192] G. Basalla, *Science and Government in England, 1800–1870*, Ph.D. thesis, Harvard University, 1963, p. 251.

[193] J.P.C. Roach, 'Victorian Universities and the National Intelligentsia', *Victorian Studies*, III, 1959, pp. 131–50.

CHAPTER 5

Research in Science

The Government of India is a *mule* as regards science. . . . It *won't* do anything unless driven.

Ronald Ross in 1898

India cannot, perhaps, afford to rank itself beside the more thoroughly developed European countries, where pure science is so richly endowed; and the practical difficulty here is to discover the profitable mean course in which scientific research, having a general bearing will, at the same time, solve the local problems of immediate economic value.

Thomas Holland in 1905

The question of scientific research in a colony brings to mind several reflections on its desirability, potentialities, ramifications (particularly economic), limitations, etc. Science in the metropolis had undergone a great deal of professionalization and specialization. But we have just seen that the Indian educational system was not geared to benefit from it. This fact put a severe constraint on the participation of Indians in research activities, which will be discussed in the next chapter. For a long time research was to remain an exclusive governmental exercise. And this largely determined the nature and scope of scientific research in India. A colonial scientist was supposed to not only discover new economic resources but also to advise on their wisest and most efficient use. No longer was he seen 'discovering' economic products and 'introducing' new crops but his role was to investigate, improve and safeguard agricultural production. The colonial scientist was no longer an 'explorer', he was now a 'settler'.[1] Colonial science primarily implied 'natural history' sciences and its main (if not sole) attraction was the exploitation of natural resources.[2] It was basically concerned with plantation research with

---

[1] M. Worboys, *Science and British Colonial Imperialism 1895–1940*, D.Phil. thesis, University of Sussex, 1980, pp. 68–9.

[2] Concern for industrial technology came much later, perhaps under the impact of the world wars.

emphasis on experimental farms, the introduction of new varieties, and the various problems of cash crops. The motive was economic, no doubt, but it had significant consequences for science itself. Next came survey operations in geology and meteorology. Another major area of concern was health. The survival of the army, the planters and other colonizers was at stake. The importance of medical research was always recognized but the degree of emphasis varied from time to time. In any case, pecuniary considerations warranted that research not be a leisurely activity. In 1875, the Curator of the Calcutta Botanic Garden, John Scott, was deputed to the poppy districts of Bihar and was asked to make two sets of reports— 'one to the Board of Revenue regarding practical (financial) points relating to the poppy and opium fields, the other to the Superintendent of the Botanic Garden, regarding the scientific points in physiological botany'.[3] This was the dual mandate.

Plantation Research

The provincial agricultural departments could seldom go beyond the collection of revenue data and famine-relief operations. No doubt, experimental farms had been established and in 1884 the Government of India enthusiastically resolved that 'in the management of the experimental farms, undue stress should not be laid on the financial results, no portion of the area being necessarily managed with the sole object of obtaining a net profit on the outlay'.[4] This generosity, however, was shortlived, for the very next year the same government declared that it 'cannot sympathize with any measures which involve the conduct of doubtful experiments entailing either lavish expenditure or a loose and vague record of results'.[5] For conducting experiments the government looked more to agricultural societies than to its own agencies.[6] The Society in Calcutta (AHSI) promptly agreed to conduct all experiments in economic products, and the Government of Bengal in return raised its 'grant from Rs 2400 to Rs 6000 per annum.[7] The AHSI accordingly experimented with plantain, muskmallow, rhea ochro, bowstring, hemp, etc.[8] The objects were:

[3] *Bengal Administration Report*, 1875–76, p. 173.
[4] Revenue, Agriculture, no. 56, April 1884, file 97.
[5] Revenue, Agriculture, nos. 12–14, Feb. 1886.
[6] In 1885 the Government of India emphasized 'the mutually supporting and harmonious relationship' between the agricultural departments and the agricultural societies. *Records of the AHSI*, 28 Jan. 1885, p. 157.
[7] Revenue, Agriculture, no. 12, March 1886, file 67.
[8] Revenue, Agriculture, Fibres and Silk, nos. 1–2, March 1886, file 10.

I. To obtain precise and trustworthy details as to the cost of cultivation and produce per acre of fibre-bearing plants of promising character, so that the Agriculture Department may be able to form a decisive conclusion as to the prospects of a profitable exploitation of the plants in question.

II. To secure a competitive trial of machines and processes for the extraction of the fibres, so as to raise a spirit of emulation which might, if not immediately, at all events at no distant date, induce competent and interested persons in either inventing or bringing out to this country a machine or process which would not only be efficient in its working, but cheap, portable and simple of construction.[9]

The official experimental farms were obsessed with cotton. Mounting pressure from British cotton tycoons had forced the Government of India to initiate a vigorous cotton improvement programme.[10] Cotton enthusiasts like Forbes, Ashburner and Rivett-Carnac were given charge of Dharwar, Khandesh and the Central Provinces, respectively. But, like the earlier projects of 1840s, the efforts of 1860s also failed mainly because of insufficient botanical knowledge or the necessary market research. Later, in 1890, Dr Voelcker specifically called for the association of an expert botanist in cotton experiments.[11] The cultivation and marketing of existing varieties produced a relatively stable and acceptable rate of return to ryots, money-lenders and dealers. New and untested varieties involved different methods of cultivation and greater labour input, without a higher level of output or profit, and with the risk of severe losses to each of these classes.[12]

The last quarter of the nineteenth century saw the closure of several experimental farms. But private farms patronized by cotton mills proved remunerative. For example, the Government's Mungeli farm at Bilaspur had to be abandoned,[13] whereas the nearby Kyragarh and Nandgaon cotton farms owned by Bengal-Nagpur Cotton Mills Company produced 6,00,000 lbs.[14] Why was there this difference? Perhaps, because the

[9] Revenue, Agriculture, Fibres and Silk, nos. 23–, May 1887, file 17.
[10] For details, see T.J.O'Keefe, *British Attitudes Towards India and the Dependent Empire, 1857–1874*, Ph.D. thesis, University of Notre Dame, 1968, pp. 134–6.
[11] Revenue, Agriculture, Fibres and Silk, nos. 11–17, Feb., pt. B.
[12] P. Harnetty, *Imperialism and Free trade*, Manchester, 1972, pp. 99–100.
[13] *Revenue Administration Report*, Central Provinces, 1903–4, para 6.
[14] F.G. Sly, *Report on the Development of Land Records and Agriculture*, 1900–1, para 61.

mills could procure cotton by advancing seeds to cultivators who employed traditional techniques, while the exercise in Mungeli failed because the government was enamoured of imported technology and ideas. The ever-growing Indian mills did not wish for a superior or long staple but only a pure one, and this indirectly discouraged improvement.[15]

Silk-manufacture was also quite remunerative and received some attention. Early research work in sericulture was conducted by W.M.H. Smith in 1814.[16] But the first clear statement of government policy came in June 1887, when it promised to aid sericulture 'in legitimate directions, such as instituting inquiries into diseases, difficulties, connected with the propagation of the mullberry and general administrative facts which are impending or may help to promote the silk industry of India'.[17] Next year, N.G. Mukherjee, a Cirencester scholar, was commissioned by the Government of Bengal to investigate silk-worm disease in Bengal.[18] That year he submitted a note on the decline of the silk trade in Bengal and pointed out that 'while European, Japanese or Chinese silk sells at 45 francs per kilo, Bengal silk sells at 32 francs. . . . If Pasteur's system is introduced in the country and healthy seed brought within reach of the peasantry, there will be no doubt an increase in the production of cocoons and silks. I believe the Government has been well advised in deciding to establish a sericultural laboratory'.[19]

But the Director of Agriculture of Bengal would not heed expert advice. He wanted Mukherjee to concentrate exclusively on the eradication or mitigation of the disease of pebrine. And only after that would he consider 'whether a laboratory for the investigation of other sericultural questions should be established'. Similarly, when in 1895 Mukherjee asked for the introduction of sericulture teaching in schools located in silk districts, the DPI of Bengal scorned him. A. Croft (DPI Bengal) wanted sanitation to be taught, not sericulture.[20] The Government of India obviously lacked a coherent policy and the will to act. Having permitted Mukherjee in 1888 to build a sericultural laboratory at Berhampur, the Imperial Government transferred its control to the Bengal Government which in turn handed it over to three silk tycoons. Mukherjee was made

[15] George Watt, *Memorandum on the Resources of British India*, Calcutta, 1894, p. 10.
[16] Home, Public, no. 33, 5 July 1814.
[17] WBSA, Revenue, Agriculture, nos. 11–16, May 1888.
[18] Ibid., nos. 48–51, May 1888, pt. B.
[19] BSAP, Revenue, Agriculture, nos 12–24, March 1890.
[20] WBSA, Revenue, Agriculture, nos. 7–9, March 1896.

accountable to a committee of merchants and not to the government. Such an abdication of responsibility by the government irked even the British Silk Association, whose president, T. Wardle, wrote: 'If the Government of France and Italy have for so many years seen the necessity of preserving their respective silk industries by State watchfulness and nurture, I feel certain that they are still more required in India. Were I in parliament, I would move for a Commission on this subject.'[21] But the Government of Bengal paid little heed to such remarks.

Henceforth sericulture was to remain a purely private concern. At the turn of the century Ms. Tata and Sons successfully started a silk farm at Bangalore for the introduction of Japanese methods of agriculture. After J.N. Tata's death in 1905, F.G. Sly, the Inspector-General of Agriculture, felt that the Bengalore farm would either be abolished or turned into a commercial undertaking and would not be available as an experimenting or training school.[22] Hence he sanctioned a similar farm with a Japanese expert at Pusa where an agricultural research institute had already come up.

Tea and indigo formed major items of export. In the 1870s, A.W. Blyth experimented with tea leaves and established a process by which it could easily be known whether the merest fragment of a plant belonged to the 'theine' class or not.[23] But nothing was done to control the scourge of blight. The AHSI made a slight attempt in this direction, but broke down for want of funds, being unable to procure from England, as they had desired, a skilled entomologist. By the late 1880s, a few progressive planters turned to science for qualitative improvement. They wanted to know the chemical basis of quality. A joint committee of the Indian Tea Association and the AHSI was formed and in 1891 M.K.K. Bamber was appointed its chemist. He gave advise on manures, drainage, insecticides, etc., performed chemical analyses of the soil and tea-leaves, and produced a book called *The Chemistry and Agriculture of Tea*. But these efforts lasted barely two years. Another attempt was made in early 1899 when the Indian Tea Association in London urged the Indian Tea Association in Calcutta to reconsider the possibility of establishing a scientific research organization.[24] The Governments of Assam and Bengal agreed to contribute Rs 5000 and Rs 2000 annually. Harold Mann arrived as the new expert. He aimed at studying (1) the relationship between the soil on the

[21] BSAP, Revenue, Agriculture, nos. 12–24, March 1890.
[22] Revenue, Agriculture, nos. 2–5, April 1905, file 38.
[23] *The Tea Cyclopaedia*, Calcutta, 1881, p. 34.
[24] P. Griffiths, *The History of Indian Tea Industry*, London, 1967, pp. 436–42.

one hand and the health and the yielding power of the bush, and the quality of tea obtained from it on the other; and (2) the nature of, and remedial measures to be adopted against, blister blight and thread blight. He was given every facility including the use of Watt's laboratory at the Indian Museum. His efforts met with success. In 1902 Mann asked for an organic chemist and an entomologist from England. What is more, he envisaged and succeeded in establishing a permanent tea research station at Toclai. All this was to push the cost of experiments to more than Rs 40,000 a year. The Government of India made an annual grant of Rs 15,000 for five years from 1 April 1905, and the quota from the Assam and Bengal Governments was raised to Rs 10,000 and Rs 3500, respectively. Thus the biggest chunk came from public revenue to help a private export industry.

Almost the same thing happened in the case of indigo. Indian indigo had remained unrivalled (and therefore, its scientific culture uncared for) till the end of the nineteenth century when Germany perfected its synthetic counterpart. This German feat sent shivers down the spines of British planters and brokers in India who hastily responded by forming an Indigo Improvement Syndicate. In early 1899, the Government of India deputed its agricultural chemist, J.W. Leather, to tour the indigo districts of north Bihar.[25] An indigo experimental farm was started at Dalsinghsarai in July the same year. The Syndicate frankly regretted that 'in more prosperous times no efforts were made in the direction of scientific improvements, and the urgency of prompt action, now that the synthetic indigo is making such rapid strides, is all the greater'.[26] George Watt, Economic Reporter to the GOI, called it 'a disgrace to the industry that so little should be known of the botany and agriculture of the plant upon which so much capital has been invested'.[27] The Syndicate annually spent Rs 28,000 on research and asked the government to contribute Rs 40,000 per annum for at least three years. The Lieutenant-Governor of Bengal agreed to give even more (Rs 50,000 a year) on the condition that the Syndicate would also contribute Rs 75,000 and the entire sum (Rs 1,25,000) would be devoted to further research. Bloaxam and Leake were sent to Dalsinghsarai in 1902, but they returned to England within two years as they found the Clothworkers' Research Laboratory of Leeds more appropriate for such research activities.[28] The interests of the indigo

[25] BSAP, Revenue, Agriculture, nos. 11–12, June 1899, pt. B.

[26] Ibid., nos. 1–8, May 1901.

[27] Ibid., K.W.

[28] *Report to the Government of India on Indigo Research Work Performed in the University of Leeds*, 1905–10.

lobby were also at work behind the creation of the Pusa Agricultural Research Institute. Pusa was selected as the site because of its proximity to the plantations of north Bihar. As the then IG of Agriculture noted: 'A research and experimental station in Bihar is likely to be more successful than in any other part of India, because useful results will be assimilated by the European farmers there and filter to the natives around them and then I hope to other parts of India.'[29] Alongwith this, B Coventery, who owned the Dalsinghsarai indigo experimental farm, was made the first director of the institute and also principal of the college at Pusa. Interestingly enough, he possessed no university degree in agriculture and his qualification was officially described as 'skilled practical acquaintance with agriculture, energy and administrative capacity, and sufficient general knowledge to be able to comprehend and utilize the results achieved by scientific experts'.[30] The government was quite aware of the possibility of being accused 'of doing a job for the sole benefit of European Industries (i.e. indigo)'.[31] Laying the foundation-stone of the agriculture college at Pusa on 1 April 1905, Curzon himself talked of the German synthetic indigo as a 'blue terror' and hoped that with the help of research the Bihar indigo would beat 'the finest product of Teutonic synthesis'; but he added, 'price, however, is the determining point. Science has got to help you to bring it down'.[32] Finance for such research posed no problem. Earlier the Bengal Government had granted Rs 50,000 a year to the Indigo Syndicate; another Rs 15,000 was saved by abolishing the agricultural unit at Shibpur; and now the entire money was pumped into Pusa.[33] Yet the 'Teutonic synthesis' was to win the indigo-war.

Botanical experiments were vital for plantation works. Though a centralized Botanical Survey of India came fairly late, a horde of both amateur and professional botanists had been active since the early years of the Raj. The works of Roxburgh, Wallich and Griffith have been referred to earlier. Later Jameson (1842–75) did good agricultural and technological work but practically no scientific surveys. Clarke (1869–71) did scientific surveys almost exclusively, while Lawson (1882–96) gave his attention almost entirely to practical work. Anderson (1859–70) and King (1871–98) rendered distinguished service to the scientific and economic aspects of botany. As late as in 1899 the Government of India complained that 'scientific research has hitherto been confined mainly to

[29] Home, Education, no. 116, July 1903.
[30] Revenue, Agriculture, nos. 14–18, Sept. 1905, file 193.
[31] Home, Education, no. 116, July 1903.
[32] Curzon, *Speeches*, vol. IV, Calcutta, 1906, pp. 116–17.
[33] Home, Education, nos. 36–7, Dec. 1903, pt. B.

the classificatory or systematic stage of science'.[34] It was looking for something more. The early history of botanical works in India shows that except for ferns, no special attempt was ever made to systematically survey the cryptogamic vegetation of the country, until Kurz (1864–78) joined the Indian botanical staff. For another two decades Drs Barclay and Cunningham of the IMS did this as their side work. And in 1900 when a cryptogamic botanist joined the BSI, he was immediately transferred to the agricultural service. So a systematic investigation of the cryptogamic flora of India continued to suffer.[35] The superintendents of the Royal Botanic Garden (Calcutta) had to handle a variety of jobs including the supervision of cinchona plantations and often the conservation of forests. Some, like the Saharanpur and Ganeshkhind garden superintendents were more interested in selling seeds than in scientific research.[36] The Calcutta garden lay emphasis on economic botany. Ceara rubber and the quick-growing guango trees were experimented with. Different types of grasses and bamboo-shoots were sent to Europe by Dr King for experiments on paper-making.[37] Ms. Routledge and Co. of London showed a keen interest in the raw materials needed for making paper.[38] In Bombay, Barber and Gammie were relieved of their purely scientific duties and allowed to devote themselves exclusively to economic ones. Of all the works the most beneficial was the introduction of cinchona. The Government of India was well advised, largely by C.R. Markham (who had brought cinchona from Peru), on the conditions required for its plantation and it flourished after some initial failures. The real problem was how to cheapen the cost of extracting alkaloids from the bark. The Superintendent of the Calcutta Botanic Garden, G. King, was made the quinologist. King chanced upon a process in Holland and shared it with the chemists, Wood and Gammie, who finally succeeded in perfecting the process. Cinchona experiments deeply affected the Calcutta garden. Teak and tea had broken into the superintendent's routine at different periods and for short times; but cinchona entered the routine and remained a long-time obsession.[39]

[34] Kew Papers, Miscellaneous Reports, India Agriculture, 1869–1921, f. 105.

[35] Kew Papers, Miscellaneous Reports, India Botanical Survey 1884–1920, ff. 123–4.

[36] *The Indian Forester,* XXIV, April 1898, p. 150; *Report on the Agricultural and Botanical Stations in the Bombay Presidency for 1905–6,* p. 115.

[37] *Report on the Administration of Bengal,* 1879–80, p. 28.

[38] *Journal of the Society of Arts,* XXV, 2 Nov. 1877, p. 1027.

[39] I.H. Burkill, *Chapters on the History of Botany in India,* Delhi, 1965, p. 136.

Geological Research

In spite of the fact that the works of the early geologists were intended to be primarily economic, by 1860 Oldham and his men had been able to lay the foundations of stratigraphical classification in Indian geology. Much was left to be desired, no doubt, yet many discoveries were made which influenced the course of geological science. For example, for the first time, there was world-wide recognition of the importance of deposits formed on land. It was proved that the Gangetic alluvium, formerly looked upon as a marine deposit, was a land deposit; that the stones and conglomerates of which the foothills of the Himalayas consist were formed, not in the sea, but on land, by rivers which were the historical equivalents of those now draining the Himalayas.[40] This was followed by the recognition of the Permian glacial epoch. H.F. Blanford was the first to establish that the Permian boulder-beds of India are relics of a bygone glacial epoch, and this opinion led to similar discoveries in Australia, Africa and South America. The other achievements were, of course, the preparation of a nearly perfect geological map of India[41] and a very significant study of the earthquakes in India.[42]

Even during military operations, the geologists were in the field. W.T. Blanford joined the military party to Abyssinia, collected almost 6000 specimens of geological importance, and expressed satisfaction that 'the campaign was not allowed, like so many others, to be entirely useless to science'.[43] Griesbach and W.T. Blanford took a keen interest in transfrontier explorations. Blanford regretted that Tibet still remained a forbidden ground and added that 'whatever may be the case in politics, it is certain that difficulties in science are not conquered by masterly inactivity'.[44] The virus of Tory jingoism had thus afflicted the geologists also. The Tibetan riddle had to await the arrival of Colonel Younghusband who later (on 13 February 1905) read a paper on 'the Geographical Results of the Tibet Mission' before the Royal Geographical Society.[45]

[40] *Nature*, LXII, May 31, 1900, p. 105.
[41] *Memoirs of the GSI*, vol. 19, pt. III, 1883, p. 163.
[42] After the great earthquake of 1897, the GSI promptly made deep scientific investigations and the findings were published in its *Memoirs*, vol. 27, pt. II, 1898, pp. 46–272.
[43] W.T. Blanford, *Observations on the Geology and Zoology of Abyssinia*, London, 1870, pp. 138–9.
[44] W.T. Blanford, Presidential Address, *Proceeding of the Asiatic Society of Bengal*, 1879, pp. 35–6.
[45] *The Geographical Journal*, XXV (5), London, 1905, pp. 481–98.

The geological works of several explorers would have been lost had someone not thought of preserving the collections, subjecting them to laboratory tests and making the results known through publications. The Asiatic Society of Bengal was the first to think of a museum as early as in 1796. In 1814, Dr Wallich offered specimens from his own rich collections in order to form the nucleus of a museum. The opening of the Raniganj coal-field and the reports of Dr Helfer and other scientific officers directed so much attention to mineral resources that it was resolved in 1856 to establish a Museum of Economic Geology.[46] Within two years, Dr Oldham and his men had been able to arrange and name altogether 6,800 specimens of fossils, 1550 simple minerals, 700 rocks and 1500 ore, making a total of about 11,000 specimens.[47]

Not only this, foreign collections were also looked for. In 1865 about 223 meteorite falls were bought from R.P. Greg of Manchester. In 1867, Oldham chanced to see at Giessen a very extensive and valuable collection of Prof. Klipstein who, to tide over his pecuniary problems, wanted to sell (for £3500) 1600 specimens of fossils, 9000 geological specimens, 12,000 geographical specimens plus a splendid head of dinotherium. Oldham promptly had the Secretary of State sanction the purchase of what he called 'a series so entirely unlikely to be met with again that it is extremely difficult to fix the true limits of its value and also admirably suited to the wants of India'.[48]

Geological works could not have been put on a scientific footing without a comparative study of Indian specimens with the standard collections preserved at Vienna, Turin, Geneva, Normandy, London, Munich, Moscow, etc. Oldham was so eager to make such a study that he and Stoliczka (the palaeontologist) were willing to proceed to Europe at their own expense, provided they be considered 'on duty'. As he put it: 'Our object is to facilitate the progress of the Survey, and we are willing to do this at considerable cost to ourselves, seeing that the desired end cannot be attained, excepting by such personal examinations.'[49] This is quite illustrative of their zeal.

Closely connected with the works of the museum were that of the laboratory and the determination of minerals, rocks and fossils sent by

[46] *Transactions of the Mining and Geological Institute of India*, vol. V, pt. 1, 1910, pp. 31–2.
[47] *Annual Report of the Superintendent, GSI*, 1858–59, p. 5.
[48] *Selection from the Records of the GOI*, LXIV, 1868, pp. 175–7.
[49] Ibid.

Robert Kyd
(1746–1793).

William Roxburgh
(1751–1815).

Nathaniel Wallich
(1786–1857).

Sir George Watt
(1851–1930).

W.B.O' Shaughnessy (1809–1889).

Standing: F. Stoliczka, R.B. Foote, W. Theobald, F.R. Mallet, V. Ball, W. Waagen, W.L. Willson. *Sitting*: A. Tween, W. King, T. Oldham, H.B. Medlicott, C.A. Hacket 1870.

Madhusudan Gupta (1800?–1856).

The first four Indian medical students (B.N. Bose, G.C. Seal, D.N. Bose and S.C. Chackburty) who went to England for higher studies.

Radhanath Sikdar,
1813–1870.

Master Ramachandra,
1821–1880.

M.L. Sircar, 1833–1904.   P.C. Ray in 1896.

J.C. Bose photographed in 1897 with his short electric wave apparatus.

Ronald Ross with Md Bux at his Hyderabad Laboratory

A group photograph taken at Mukteswar on the occasion of the visit of Professor Koch in 1897

P.C. Roy seated at the centre and his favourite students
M.N. Saha and J.N. Ghose, 1916

Dr S.S. Bhatnagar, Dr H.J. Bhabha, and Pandit Jawaharlal Nehru
with Sanjay and Rajiv Gandhi, New Delhi, 1952

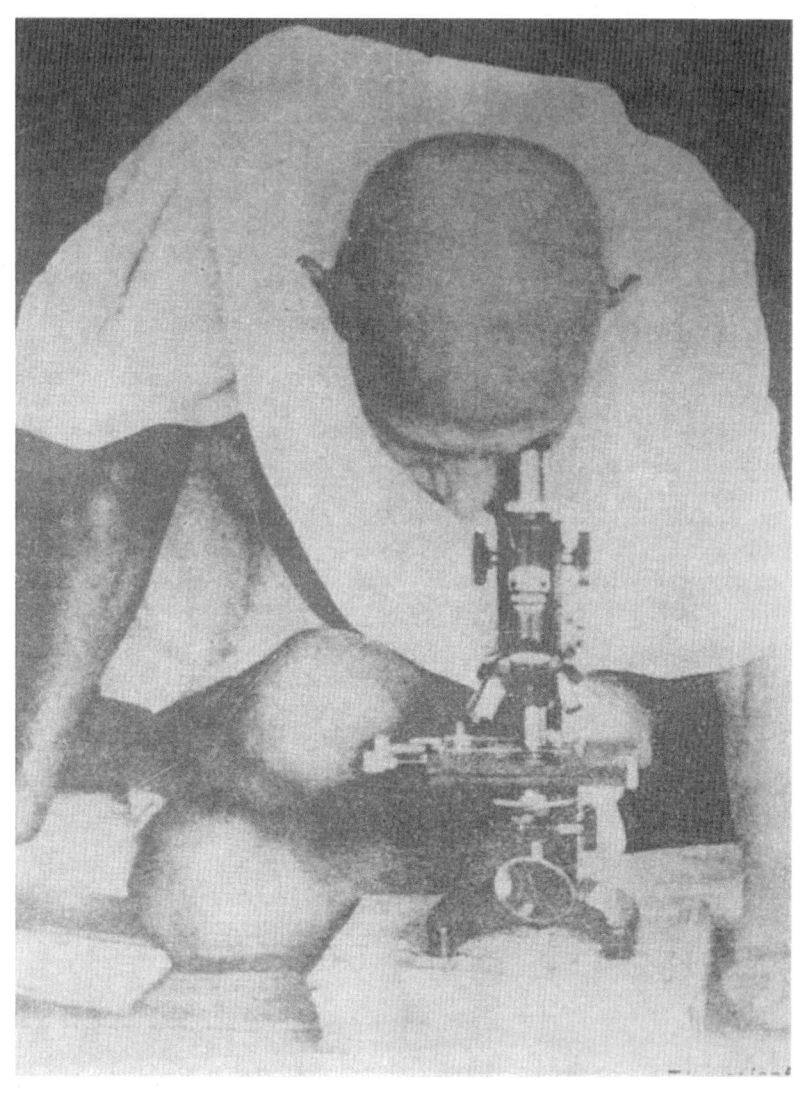

Mahatma and the Machine
(Empire under scrutiny!)

private individuals as well as those collected by the GSI staff. A tea-specu-lator would ask about tea-soils while the PWD would enquire about iron sand stones from Rangoon. Commercial interests thus provided the motive force. However, under T.H. Holland's curatorship, laboratory work touched new heights; while during 1882–90 the average was about 50 tests a year, during 1903–8 it was 790 tests.[50]

All these explorations and research activities could have reached discerning quarters only through publication. When the staff of the GSI was increased in 1856 and its labours systematized, it was felt that the reports on different districts, examined geologically by its officers, should be published in one continuous and uniform series, not, as previously, in various journals and in different forms.[51] The result was the *Memoirs of the GSI* catering to the need for detailed and comprehensive accounts of specific economic minerals of a region. In 1862 was launched the *Palaeontologia Indica* with the prime objective of publishing the result of palaeontological research carried out by the GSI on fossil collections. The region-wise volumes gave place, in 1899, to the New Series permitting diversity in the region as well as in the subject treated within the same volume. The third important publication was the *Records of the GSI* started in 1868 to publish short and brief scientific papers prepared by the officers of the Survey.[52] This annual publication usually formed a four part volume, part 1, being mostly the Annual General Report of the Department, part 2, devoted to mineral production satistics, and parts 3 and 4 to scientific papers.[53] The scientific spirit of the GSI was also reflected in its attempts to forge links with numerous scientific (geological in particular) societies and organizations throughout the globe, and the publications of the GSI were put on the circulation lists of almost all the leading libraries of the world.[54]

Besides, there was prepared a 'Manual of the Geology of India' in two volumes by H.S. Medlicott and W.T. Blandford, published in 1879, to which was subsequently (in 1881) added a volume on 'Economic Geo-logy' by Valentine Ball and one on Minerology by P.R. Mallet. These volumes not only contained much information collected by the survey,

[50] Transactions, p. 37, note 46.
[51] *Memoirs of the GSI*, vol. I, Calcutta, 1856, p. v.
[52] Home, Public, nos 26–8, June 1868.
[53] *Records of the GSI*, vol. I, Calcutta, 1868.
[54] See *Index to Proc. Department of Commerce and Industry, Geology and Minerals Branch, 1858–1905*, Calcutta, 1921, pp. 3–10, 22–3, 61–2.

which it had not been possible to publish previously, but for the first time, by collecting scattered information into one general review, they made the geology of India generally accessible and intelligible. It was Ball's work which stole the limelight, obviously because of its economic value.[55]

Oldham talked of government pressure and the GSI had to contend with what he called 'the force of circumstances to take up in detail the examination of special districts'.[56] As the GSI was a government organization, it had to work mostly along those lines which brought economic benefits to the Raj. Oldham himself tried to rationalize this:

In a new country, the mineral wealth of which is only now becoming known, the natural and inevitable tendency of such geological enquiries is more practical and economical than they would be where the energies and skill of many, interested in such pursuits, had been for generations brought to bear upon them.[57]

The geologists, as seen earlier, were expected to perform a dual role—as explorers (thereby as agents of imperialism), and as scientists busy in advancing the frontiers of geological knowledge. Oldham, no doubt, kept a very close watch on all geological happenings in every part of India, and made a practice of personally examining new discoveries and advances whenever possible.[58] Yet there was no clear-cut distinction as to who should work on pure research and who should be engaged in survey work. In fact both were considered inseparable, but emphasis was always laid on the economic aspect. This will be clear from the following chart submitted by Dr W. King, Director of the GSI, to the government outlining the programme of work of the geologists during 1893–94.[59]

Name	Area of work	Nature of work
C.L. Greisbach	Superintendent Baluchistan,Punjab	Scientific, Economic to be done by his subordinates Mr Smith and Edwards
T.D. LaTouche	Offg.Superintendent Sukkur	Economic and Oil
P.N. Bose	Dy. Superintendent Rewa	Scientific, Economic as found
F. Noetling	Palaeontologist Burma	Scientific and Economic

[55] *Calcutta Review*, 75(149), 1882, p. XIII.
[56] Revenue, Agriculture, Surveys, no. 1, Jan. 1872.
[57] *Memoirs of the GSI*, vol. I, 1856, p. VI.
[58] L.L. Fermor, *First Twenty Five Years of the GSI*, New Delhi, 1976, p. 67.
[59] Revenue, Agriculture, Surveys, no. 1, Oct. 1893, file 174.

C.S. Middlemiss	Dy. Superintendent Madras	Scientific and Economic
P.N. Dutta	Asstt. Superinten- dent,Headquarters	Scientific, Economic as found
T.H. Holland	Asstt. Superinten- dent, Headquarters	Scientific

Thus, only Holland had been assigned purely scientific work. By the 1890s the government had begun to show some genuine interest in scientific research and therefore wanted to divide the GSI into scientific and economic sections with two-thirds of the staff given to the scientific section and one-third to the latter section. The Secretary to the Government of India, E.C. Buck, accordingly asked the GSI to henceforth adopt the following division:[60]

Scientific section	Economic section
1. C.L. Griesbach	1. T.D. LaTouche
2. P.N. Bose	2. F. Noetling
3. P.N. Dutta	3. C.S. Middlemiss
4. F.H. Smith	
5. W.B.D. Edwards	
6. T.H. Holland	

The idea was thus the decentralization of that section of the GSI which dealt with practical aspects as divorced from purely scientific enquiry.[61] But the issue remained far from resolved. For example, the Annual Report of the BSA for 1902–3 put emphasis only on Economic Geology and absolutely ignored the scientific side.[62] Blanford criticized this at a meeting of the Advisory Committee of the Royal Society (on 12 January 1905) and it was resolved that 'the Committee, while sensible of the necessity of giving attention to the geological survey of areas of economic interest, desires to urge the importance of completing as quickly as may be practicable a general geological survey of India and Burma'.[63] In reply, T.H. Holland (Director, GSI) prepared a memo which deserves to be quoted in full:

In the preparation of our annual programme, we are not wholly free to follow the line which we think is best for the country in the long run; questions of economic

[60] Ibid., no. 2.
[61] Revenue, Agriculture, Surveys, nos. 11–13, July 1894, file 128.
[62] Revenue, Agriculture, General, nos. 4–7, Sept. 1903, file 69.
[63] Revenue, Agriculture, General, nos. 7–10, May 1905, file 127.

importance sometimes cannot be postponed without immediate loss and inconvenience, whilst an unsurveyed area, about which nothing certainly is known, is neglected from one field season to another . . . the study of pure palaeontology, with no end immediately in view but the forms and zoological relationship of fossils, has established the data by which isolated exposures of strata have been correlated in a stratigraphical scale, which, in the case of coal, for instance, forms the first guide to prospecting operations; the study of microscopic petrology conducted by a research worker, interested only in the physics and chemisty of crystals, has resulted in establishing the laws of the mode of occurrence and habits of mineral lodes. It is for reasons of this kind, not always expressed in words, that all enlightened countries now maintain institutions of pure science, and facilities for scientific research which may, or may not, be of benefit to the particular country in which the work is done. India cannot, perhaps, afford to rank itself beside the more thoroughly developed European countries, where pure science is so richly endowed; and the practical difficulty here is to discover *the profitable mean course* in which scientific research, having a general bearing will, at the same time, solve the local problems of immediate economic value.[64]

The pure light of science had nevertheless been slightly dimmed by the smoke of commercialism. For instance, the *Memoirs* were started in 1856, and up to 1901, 34 volumes had been published, containing 82 parts, of which about 30 parts were devoted primarily to the study of various coalfields in the country, apart from the numerous references to it scattered in almost all the volumes. By contrast, gold received full attention only in three parts. Obviously the geologists were more concerned with black gold. It is, however, significant that out of these 82 parts not less than 40 were devoted to the study of geological structures and physical features of several areas encompassing the length and breadth of the country. The rest dealt with isolated topics like the occurrence of laterite, lignite and ammonites, the geology of several rock systems, mineral statistics, earthquakes, thermal springs, etc.[65]

So great was the economic value of geological works that when Valentine Ball wrote part III of the *Manual of Geology of India* (London, 1881) dealing with economic geology, it was hailed as sufficient proof to 'justify the existence of the Survey were any justification needed', and the government granted an honorarium of Rs 1000 in recognition of the

[64] Revenue, Agriculture, General, no. 3, Aug. 1905, file 127 (emphasis added).

[65] In sharp contrast to this Victorian phase, during 1902–47 one finds only 7 articles on coal, 8 on earthquakes and seismic problems, about 12 on various minerals and ores, and about 7 on petrology—all this out of 43 volumes containing about 75 parts. The rest deal with structural and physical studies. This information has been culled from the *List of GSI Publications*, Calcutta, 1975, pp. 1–16.

expeditious and satisfactory manner in which he brought about the book.[66]

A more concrete example can be found in the exploitation of manganese ores. By 1865 the manufacture of ferro-manganese had started in England which led to a tremendous increase in the demand for its ores. Supplies from continental Europe fell far short of the demand. Indian mineral exploitation then became commercially viable. At Vizagapatanam, Mr H.G. Turner, previously the Collector of this district, formed a syndicate in 1891 to work the deposits of managanese in Kadur mines. In 1895 the syndicate was converted into the Vizanagaram Mining Company Limited, which, in search of new 'pastures', started prospecting works in Chikvadi and Nagpur districts following up the references given in Ball's *Economic Geology*.[67] The result of this was that managanese production in India rose from a meagre 685 metric tons in 1892 to 14,498 in 1900; and the export of manganese from India rose from 1000 tons in 1893 to 95,225 tons in 1900. On top of this, these exports were made in raw conditions resulting in heavy revenue losses, but the loss to India meant a corresponding gain to Britain and her manufacturers. The total value of manganese production during 1892–1906 was about £2,437,461. Had this been converted into ferro-manganese in India itself, the minimum value to India would have been £11,448,582. India thus suffered a loss of £9,011,119 as a price for not manufacturing ferro-manganese in the country.[68] The thread which bound together the discoverer and the manufacturer (i.e. science and industry) in Victorian England, was sadly missing in India.

Medical Research

India was the largest natural disease repository in the British empire, yet its medical services (particularly the IMS) were not geared to meet this task. Its primary responsibility was to look after the troops and the European civil population. Some of its members also shouldered the burden of imparting medical education. No wonder co-ordinated research was not found feasible. There were no doubt scientifically trained persons and most of the early investigations in natural history were made by them, but the Service, on the whole, had a fairly poor record of achievement in

[66] Revenue, Agriculture, Surveys, nos. 14–23, April 1882, file 46, pt. B.
[67] L.L. Fermor, 'History of the Indian Manganese Industry', *Memoirs of the GSI*, vol. 37, pt. III, Calcutta, 1909, pp. 420–3.
[68] Ibid., pp. 536–42.

its own field (i.e. medical research *per se*). The development of the public health system and of medical science in Britain had been an organic one. In India this was not to be so.

First of all, Indian medical traditions were completely ignored and the study of indigenous drugs found no place in the medical curriculum of Indian universities.[69] W.B. O'Shaughnessy had drawn the government's attention to this neglect, but in vain. Later Dr Waring worked on the same aspect.[70] In Bombay, Pandurang Gopal prepared a catalogue of indigenous drugs found there.[71] Kanny Lal Dey (Professor of Chemistry at Calcutta Medical College since 1862) was perhaps the most forceful crusader. In 1894 he had the honour of presiding over the section of pharmacology at the first Indian Medical Congress. He used this platform to focus greater attention on the study of indigenous drugs, their systematic cultivation and increased use in medical depots.[72] But the Surgeon-General of Bengal, J Cleghorn, did not think 'that the GOI would be justified in initiating experiments with regard to their efficacy; such investigations should be left to individuals.[73] Earlier the Principal of the Grant Medical College (Bombay) had vetoed the establishment of a chair of pharmacology on the ground that 'it would be much cheaper and much more satisfactory to pay skilled pharmacologists in Europe to investigate the Indian Drugs'.[74] The Government of India appointed a Committee to look into the points raised by Dey, which, in turn, appointed local committees.[75] Four years later it felt constrained to admit 'defects in the system of provincial committees', and recommended their replacement 'by one or more selected physicians in important towns who would be required to give the various inquiries entrusted to them their special and personal consideration'.[76] Finally nothing more than a fairly long-drawn debate emerged out of the proceedings of this committee.[77]

Bacteriological investigations were the next to attract attention. The cholera epidemic of 1861 gave a new dimension to the colonial health

[69] G. Mukhopadhyay, *History of Indian Medicine*, I, Calcutta, 1923, p. 94.

[70] D. Waring, *Pharmacopia of India*, London, 1868.

[71] P. Gopal, *A Catalogue of Drugs Indigenous to the Bombay Presidency*, Bombay, 1874.

[72] K.L. Dey, *Indian Pharmacology: A Review*, Calcutta, 1894, p. xxxi.

[73] Home, Medical, nos. 15–18, Dec. 1895.

[74] Ibid.

[75] Home, Medical, nos. 54–7, Aug. 1896.

[76] Home, Medical, nos. 48–62, March 1901.

[77] For details, see *Report of the Proceedings of the Central Indigenous Drugs Committee of India*, Calcutta, 1901.

policy in India.[78] If cholera was to be prevented, it had to be sought out, not waited for. Segregation on sanitary grounds was not enough. More knowledge about the disease itself was required. The Government of India promptly appointed a Cholera Commission—the first systematic enquiry into a major epidemic. In 1864 three presidency sanitary commissions were set up. In 1868, at the suggestion of the Army Medical School at Netley, two medical officers, T.R. Lewis and D.D. Cunningham of the AMD and IMS respectively, were sent to Germany for further training in research, and on their return to India were appointed special assistants to the sanitary commissions with the Indian government.[79] They failed to produce conclusive evidence on the cause of cholera, but their study was valuable in the sense that they stressed the importance of seeking other causes than the contagion theory. In 1883 Robert Koch discovered the cholera 'Comma' bacillus in Egypt and visited Calcutta the same year to confirm the discovery. This was an important contribution and helped to establish the germ theory of disease causation over the earlier miasmatic theories. This shift in focus in England and western Europe from sanitation to epidemiology and bacteriology had significant implications for India. Now inspection, isolation and attention to the water supply were added to the concern for sanitation.

Richard Strachey took a keen interest in bacteriological research as a member of the Secretary of State's Council. Strachey sought the advice of the noted biologist, T.H. Huxley, and informed him:

The proposal to send Klein (Dr E.E.) to India to investigate the supposed or actual cholera microbe will come on for discussion in the Indian Council. There is naturally a residuum which prefers doing nothing in such a matter and possibly the residuum in this case will be a majority of the Council and the proposal will be squashed. I personally have a belief that such an enquiry is very desirable . . . and it would strengthen me greatly if I could get an expression of your opinion. . . . The opposition in our Council talk of the expense. But certainly it would be worth many thousand pounds if we really knew how cholera was propagated.[80]

The Leprosy Commission of 1890–91 focussed attention on bacteriological problems.[81] Soon few small bacteriological laboratories were set up. Hankin (Chemical Examiner to the Government of NWP) estab-

---

[78] Radhika Ramasubban, *Public Health and Medical Research in India*, Stockholm, 1982, pp. 17–18.

[79] *Report of the Sanitary Commissioners to the GOI*, vol. I, Calcutta, 1868, p. 26.

[80] T.H. Huxley, Mss. 27, 111–12, Imperial College of Science and Technology, London.

[81] *Report of the Leprosy Commission in India*, 1890–91, Calcutta, 1893, p. x.

lished a laboratory at Agra.[82] About the same time Dr Lingard was looking into the bacteriological causes of the diseases of domesticated animals. But the most important step was the establishment of the Plague Research Laboratory in Bombay under Haffkine in 1896. It undertook several things—preparation of anti-plague serums and anti-cholera vaccinations, examination of pathological specimens, instruction in bacteriological work and research, etc.[83] Haffkine's anti-plague vaccination was most successfully employed on a large scale in India. It reduced the number of cases and considerably lessened the death rate in those inoculated. Of the 2197 persons inoculated out of a population of 6033 in an epidemic area, only 30 died of plague, while from the non-inoculated, 1482 died, a difference of 82 per cent in favour of inoculation.[84] One of his assistants, Dr Lamb (IMS 1893 and trained at Pasteur Institute, Paris, in 1897) discovered the existence of the Mediterranean or Malta fever in India, and was inclined to believe that many of the obscure cases of fever amongst native troops, hitherto indiscriminately ascribed to malaria, were really cases of this disease.[85] By the turn of the century two more important laboratories came up—one was the King Institute of Preventive Medicine in Madras and the other was the Pasteur Institute at Kasauli. The former supplied lanoline vaccine paste while the latter produced anti-rabic and anti-typhoid vaccines. Kasauli was expected to become the nucleus for investigations in tropical diseases and clinical bacteriological work.[86] But all these institutions worked under severe limitations. The research structure that eventually evolved was the result of a piece-meal and ad-hoc response to sudden epidemic emergencies, and there existed no enduring foundation for the growth of medical science in the country. Addressing the Indian Medical Congress in 1894, Ernest Hart (editor of the British Medical Journal) severely criticized the IMS for having neglected to lay the basis for the scientific exploitation of the largest disease laboratory in the British Empire.

A pestilence breaks out in an important English colony. The local physicians display—as I trust Englishmen ever will—unlimited courage and devotion to duty; but the scientific part of the work has to be done for us by outsiders. . . . Run over some of the names associated with most of the important recent discoveries, and judge. Cholera, Koch; leprosy, Hansen; malaria, Laveran; dysentery, Loesch;

[82] Home, Medical, no. 15, Dec. 1895.
[83] *Report of Bombay Plague Research Laboratory*, 1896–1902, p. 17.
[84] Patrick Hehir, *Hygiene and Diseases of India*, Madras, 1913, p. 843.
[85] Lamb Papers, IOL, Mss. Eur. D. 893.
[86] Home, Medical, nos. 103–6, March 1902.

liver abcess, Kartulis; beri-beri, Scheube and Balz; endemic haematuria, Bilharz; latest discovery of all, the bacillus of plague, Kitasato.[87]

There were, however, one or two cases of original discovery. The most significant was that of Ronald Ross (IMS 1881) who worked on the relationship between malaria and the mosquito. He received inspiration and initial guidance from Patrick Manson. Manson was a trend-setter in the sense that he was one of the first to speculate upon the possibilities of 'insect-born transmission of disease' while practising medicine at Amoy and Hong Kong during the 1880s. Later he shifted to London and won recognition for the clinical treatment of tropical diseases based upon parasitological studies. He lay emphasis on research and clinical control rather than preventive sanitary measures. This was a metropolitan concept of science in furtherance of which he sought the services of Ross. Here was a colonial scientist eager to carry out the necessary field-work with great diligence, and in the process brought Manson's work to its logical conclusion. Ross frankly admitted his intellectual debt to Manson.[88] But both the IMS administration and GOI were cold to Ross. In 1895 the Maharaja of Patiala offered to finance malaria investigations by Ross on a full-time basis. But the Lieutenant Governor of Punjab would not agree and even asked the Maharaja to withdraw his offer.[89] A few months later Ross was refused permission to work at Hankin's laboratory in Agra.[90] The reason was that Ross was on military duty and could not be easily spared. In early 1897 the GOI declined to put Ross on special duty for a few months for malaria research on grounds of financial difficulties.[91] Even a bacteriologist like Haffkine refused to accept Ross. He apprehended that if the mosquito-malaria theory failed, it would affect the reputation of his institute.[92] At home, Manson raised the question of national honour and lobbyed with Charles Crosthwaite (a member of the Secretary of State's Council) to put Ross on exclusive malaria duty. This seemed to work and soon the Director-General of IMS expressed his appreciation of Ross's work and promised to provide him 'with more favourable opportunities'.[93] But for another six months Ross continued to remain in the doldrums. In sheer disgust he wrote to Manson:

[87] E. Hart. *The Medical Profession in India,* Calcutta, 1894, pp. 73–4.
[88] R. Ross, *Ross-Manson Letters,* London, 1929, p. 4.
[89] Ross Papers, Mss. 01/072, London School of Hygiene and Tropical Medicine.
[90] E.H. Hankin to R. Ross, 23.4.1896, ibid., Mss. 49/134.
[91] Ibid., Mss. 01/133.
[92] Ross to Manson, 9.12.1897, ibid., Mss. 02/137.
[93] Ibid., Mss. 03/064.

You will exclaim 'the Government has promised to put you on malaria duty'. Yes, but *when?* I fear I know Government very well. After this frontier business, there will be the plague, after that the 'financial condition' will remain. Then one will pooh-pooh the whole theory. Then they will forget all about it; and lastly if I remind them of it, they will tell me mind my own business. . . . In asking to be put on malaria duty I ask in reality nothing at all. The investigation will not cost Government *one penny.* I am willing to undertake the research on reduced pay, namely on the bare salary of my rank, and to give up my regimental allowance to the man who does my work in the regiment. I am also willing to pay my travelling expenses myself. . . . You must remember that I cannot write all this frankly to my superiors as to you. Indeed I am scarcely allowed to write at all.[94]

In February 1898 Ross was put on special duty to investigate malaria for six months. But one more job was added. He was ordered to investigate kalazar in addition to completing his work on malaria. Ross asked to be relieved from the work on kalazar or to be given assistance but his request was ignored and an assistant was denied.[95] Yet he managed to complete his work on malaria. This led to some rivalry in the scientific circles in Europe. A few Italian scientists like Bignami and Grassi tried to seize the credit. In *The Lancet* (of 3 and 10 December 1898) Dr Bignami claimed to have discovered proof of the mosquito theory of malaria. *The Indian Medical Gazette* called it 'a new Fashoda incident'.[96] Ross was the first to point out (1) that the eggs of the Indian Anopheles are laid in a pattern on hard surfaces (e.g. stones) and not in water and (2) that the larvae of Indian Anopheles do not float head downward like those of culex but float on the surface like sticks. Ross gave credit for the second fact to his Indian assistant, Muhammad Bux.[97] Another interesting dimension was that Koch was also trying to seize the credit for malaria research for himself and tried to blacken Grassi and his work in the eyes of the Italian authorities

[94] Ross to Manson, 6 Dec. 1897, ibid., Mss. 02/136 (emphasis as in original).

Earlier in June 1897 Ross had seriously thought of leaving medical science for good and taking to literature and writing novels. Ross to Manson, 11 June 1897, ibid., Mss. 01/099.

[95] The Kalazar parasite was finally spotted by W. Leishman of the Army Medical School, Netley, and C. Donovan of Madras Medical College in 1903 with R. Ross acting as a liaison between them. E. Gibson, 'The Identification of Kalazar and the Discovery of Leishmania Donnovani', *Medical History*, vol. 27, April 1983, pp. 203–13.

[96] *The Indian Medical Gazette*, Jan. 1899, p. 15.

[97] Ross to G.H.F. Nuttal (editor of *Journal of Hygiene*, Cambridge), 28 April 1899, Ross Papers, Mss. 02/229.

and scientists.[98] Manson, however, did his very best to uphold Ross. This controversy gave a handle to Ross's detracters within the IMS. Surgeon-Colonel Lawrie of Hyderabad attacked him in *The Lancet* and also in the Indian papers under the name of Buggobutty Bose, MD. As Dr Bignami was a man of considerable eminence, the Indian authorities began to think that Manson and Ross had been quite mistaken. They remained by and large cold, and in early 1899 Ross left India for good. Back home he gained a professorship at Liverpool and world-wide recognition in the form of the Nobel Prize. But soon he drifted from Manson—his mentor. Manson treated tropical medicine as a metropolitan rather than a satellite (or colonial) activity. His programme was essentially reductionist and stressed on research rather than on active control.[99] In contrast, Ross developed interest in the practical implications of his scientific work, seeing malarial control as a matter of reducing the population of mosquitoes by destroying or treating their breeding areas.[100] This was a practical and preventive approach but it involved considerable expenditure and large-scale sanitary measures. No wonder the governments in India and at home preferred to ignore Ross. As Ross himself put it: 'Instead of using me for the large sanitary schemes as I desired, my countrymen offered me only three minor occupations—to teach students, to dissect parasites, or to prescribe pills. The British are a practical people; they seldom actually kill the goose that lays the golden eggs; they force her to lay goose's eggs!'[101]

N.C. Macnamara (Bengal Medical Service 1854) was devoted to the study of cholera. He found that the rice-water-like stools of cholera were infective when swallowed, and that a number of persons who by accident drank a portion of the watery cholera stool, developed the disease within a day or two. This convinced him that the disease was caused by an intestinal bacillus and led him to carry out animal experiments on the causation of cholera. In the course of these he contracted the disease and had to retire in 1876. Thereafter he studied bacteriology under Koch at Berlin at his own expense. In February 1883 he requested the India Office to depute him to Egypt, where cholera had broken out, to work on its

[98] G.H.F. Nuttal to Ross, 19.3.1899, ibid., Mss. 02/218.

[99] M. Worboys, 'The Emergence of Tropical Medicine: A Study in the Study of a Scientific Speciality', in G. Lemaine (ed.), *Perspective on the Emergence of New Disciplines*, Paris, 1976, p. 90.

[100] R. Ross, 'Malaria in India and the Colonies', *Journal of the Royal Colonial Institute*, vol. xxxv, Dec. 1903, pp. 14–25.

[101] R. Ross, *Memoirs*, London, 1923, p. 468.

bacteriology. His application was refused; yet eight months later Koch was given every facility for his investigations in Egypt which resulted in the discovery of the cholera vibrio. Later Macnamara wrote: 'The Secretary of State's refusal to accept my offer crushed any hopes I had of completing my life's work.'[102]

Haffkine was a little more fortunate. He was given every help to develop his prophylactic and his inoculation programme became popular. But this called for far more money, men and technical and administrative facilities than the GOI was willing to give. The AMD on its part saw Haffkine's authoritative handling of the epidemic problem and his apparent command over the new scientific knowledge as a threat to their professional monopoly in India. Misfortune struck him in 1902. In Punjab 13 persons inoculated from a single bottle contracted tetanus and died.[103] The AMD officials promptly asked for the suspension of his activities, and the government obliged. It was also resolved that henceforth only officials of the British AMD would be appointed to head research laboratories. Haffkine left India in disgrace. Later it was found that the tetanus germs had entered the bottle not in the laboratory but through an accident in transit in Punjab. But Haffkine was never officially exonerated by the British Government.

Meteorological Research

Climatology and meteorology attracted a good deal of attention. Ozone registration was started as early as in 1863. Dr Henry Cook, Medical Superintendent at Mahabaleshwar, noticed 'a decided connection between the absence or marked decrease of ozone and the presence of cholera'. Earlier, during the spread of cholera at Strasbourg in 1854–55, a German scientist, M. Bokel, found that ozone had disappeared at the commencement of the epidemic and reappeared after it was over.[104] Later H.F. Blanford carried out many investigations on weather problems and forcasting, and wrote a number of papers and memoirs. His successor, Eliot, wrote more than fifty original papers on various meteorological

[102] L. Rogers, 'A Tragedy: How Surgeon-Major N.C. Macnamara was Deprived of Priority in the Discovery of the Causative Organism of Cholera', *Transactions of the Royal Society of Tropical Medicine and Hygiene*, vol. 43, Jan. 1950, pp. 395–400.

[103] E. Chernin, 'Ross Defends Haffkine: The Aftermath of the Vaccine-Associated Mulkowal Disaster of 1902', *Journal of the History of Medicine and Allied Sciences*, 46(2), April 1991, pp. 201–18.

[104] Home, Public, nos. 5–7, 3 Aug. 1865. pt. B.

subjects, particularly on cyclones. His *Report on the Visakhapatnam and Backerganj Cyclones of October 1876* excited so much attention that requests were made in the British House of Commons for the report to be produced and laid on the table of the House.[105] His last publication was *The Climatological Atlas of India*. Walker, another noted meteorologist, discovered the southern hemispheric oscillation, and also worked on the relationship between equatorial pressure and monsoon rainfall, etc. Non-government agencies also contributed to the knowledge. The St. Xavier's College of Calcutta regularly took meteorological observations right from 1868 and sent them to the meteorological observatories in different parts of the world.[106] This college also conducted spectroscopic observations. Professor Tacchini, an Italian astronomer, visited India in December 1874, and was so impressed with the value of solar observations in India's cloudless sky that he persuaded Father Lafont to erect a spectroscope at the St. Xavier's.[107] Father Lafont remained the moving spirit behind it.

Solar research got a boost from Norman Lockyer's interest in meteorology and solar physics. He worked on the relationship between sun-spot activities and changes in atmospheric conditions. In a report to the Indian Famine Commission of 1880 he pointed out that the famines in Madras since 1801 could be correlated very closely with sun-spot minima, and called for the establishment of a Solar Physics Observatory in India.[108] This enquiry, incidentally, tried to give a scientific basis to the notion that famines were natural calamities and therefore, beyond the control of any human agency or government.[109] These works nevertheless succeeded in focusing attention on solar studies. The total solar eclipse in January 1898 generated a lot of interest. A party of Jesuit Fathers camped at Dumraon and photographed the spectrum and the corona.[110] The British Astronomical Association sent a party to Talni in the Central Provinces and the Astronomer Royal of Scotland, Dr Copeland, settled at Ghoglee near

[105] *Hundred Years of Weather Service* (1875–1975), New Delhi, 1976, p. 143.
[106] Rev. E. Francotte, *Meteorological Observations at St. Xavier's College*, 1868–1913, Calcutta, 1918.
[107] *The Xaverian*, vol. II, April 1912, p. 308.
[108] A.J. Meadows, *Science and Controversy: A Biography of Sir Norman Lockyer*, London, 1972, p. 127.
[109] *Indian Engineering*, vol. XXIX, 16 March 1901, p. 168.
[110] Rev. S.J. De Campigneulles, *Observations taken at Dumraon, Behar, during the Eclipse of 22 January 1898*, London, 1899.
It got a rave review in *The Athnaeum*, 20 May 1899, p. 629.

Nagpur. The party sent by the Royal Society and the Royal Astronomical Society, under Norman Lockyer, chose the old Viziadurg Fort. The Astronomer Royal, W.H.M. Christi, stationed himself at Sahdool in Rewa. Naegamvalla of Poona along with two Japanese professors, Taero and Hirmaya, made observations at Jeur. These investigations were addressed at the problem of what the corona is, what it is made of, its connection with sun-spots, whether it is an electric phenomenon or is due to the sun's magnetism, etc.[111] An interesting sidelight of this interest in the tropical sun was exploration into the possible uses of solar heat. As early as in 1878, the Deputy Registrar of Bombay High Court, William Adams, experimented with flat mirrors and concluded that even an ordinary vertical cylindrical boiler stationary steam engine, of any power, could be driven at an expense of £2.10 per horse-power in India from 8 a.m. to 4 or 5 p.m. He wanted the Bombay mills to use the steam generated by solar heat as an auxiliary to, and with, the steam generated by conventional fuels.[112]

Zoological Research

Zoology ranked quite low in the priority list of the colonial government, perhaps because of its limited economic potential. The chief progress in Indian zoology had been in the old-fashioned faunistic or 'natural history' direction. Hamilton-Buchanan evinced interest in ornithology and established a menagerie at Barrackpur.[113] Edward Blyth gave credence to these studies and published a number of papers in the *Journal of the Asiatic Society*. Amateurs, however, continued to dominate this field. During the 1860s and 70s, A.O. Hume of the Bengal Civil Service (and later one of the founders of the Indian National Congress) devoted his leisure hours to the study of ornithology. He fitted out expeditions with a staff of collectors and taxidermists to Sind, Coorg, Manpur, Malaya, Tennaserim, Andman, etc., and collected about 400 skins of mammals, 63,000 skins of birds, 300 nests and 18,500 eggs. He started an ornithological journal, *Stray Feathers* and published numerous papers. Later he turned to politics and in 1885 donated his entire collection to the British Museum.[114]

---

[111] S.G. Burrard, *Report on the Observations of Total Solar Eclipse at Sahdool in Central India*, Dehra Dun, 1898, p. 7.

[112] W. Adam, *Solar Heat: A Substitute for Fuel in Tropical Countries*, Bombay, 1878, pp. 92–3.

[113] Buchanan Papers, IOL, Mss. Eur. D. 541.

[114] A.O. Hume Papers, BM, Natural History, Mss. HUM, 1867–81; *Nature*, 24 Sept. 1885, p. 498.

During the 1870s Dr J. Anderson, Curator at the Indian Museum, and W.T. Blanford of the GSI, maintained a semblance of zoological research and sent some valuable specimens and reports to the Zoological Society of London.[115] When the HMS Challenger was sent from England for scientific explorations in distant seas and archipelagoes, Indian waters were excluded from its route. The Asiatic Society of Bengal resented this and in vain even tried to obtain a research vessel to carry out marine investigations in the Bay of Bengal.[116] On the whole, till the end of the century, Indian zoology had not graduated from being of passing interest to collectors and systematists to becoming an area of serious study and research which was the case in Europe.

One man who fought for this transformation and established Indian zoology on a sound scientific pedestal was Major A.W. Alcock. Alcock had a fascinating career. A school drop-out from England he came to Wynaad in 1876 to learn coffee-planting and later went to Purulia as an assistant to a cooly-recruiting agent. There he chanced to read Michael Foster's *Physiology Primer*, came into contact with the Deputy Sanitary Commissioner, resolved to become a doctor, and in 1881 returned to England to study medicine. In 1885 he got into the IMS. After a stint with the Punjab Infantry, he preferred to become the Surgeon-Naturalist to the Indian Marine Survey and worked at the survey ship 'Investigator' from 1888 to 1892. The next year he joined the Indian Museum as the Superintendent of the Natural History Section.[117] His works on deep sea fishes, the deep-sea Madreporaria and the deep-sea Brachyura brought him great fame and a Fellowship of the Royal Society in 1901. But he gradually became disillusioned. The Trustees of the Museum kept him in subordination 'through the medium of a clerk who was a mere Eurasian office-boy without a tincture of any academic culture'.[118] The last straw came in 1903 when Curzon asked him to dismantle the galleries of fishes and invertebrates and in their place house the exhibits illustrating the history of British India which had been collected as a tribute to the late Queen Victoria. Alcock later received the backing of the Royal Society, which reviewing the Report of the BSA for the year 1904–5, stressed upon the

[115] Home, Public, nos. 37–9, 20 Aug. 1870; Home, Education, nos. 106–8, May 1882.

[116] Presidential Address by T. Oldham, *Proceedings of the Asiatic Society*, 5 Feb. 1873, p. 62.

[117] Alcock Papers (1859–1933), London School of Hygiene and Tropical Medicine, Mss. FA: D1 (14) DON 12223.

[118] Ibid., f. 11.

usefuleness of zoological studies in relation to tropical diseases and strongly recommended 'the establishment of zoological laboratories and small teaching museums in all the large university colleges of Northern and Western India, with professors in charge, whose sole duty should be to teach and to conduct research'.[119] But this remained on paper and was of no avail. In May 1906 Alcock decided to resign. Most ruefully he wrote: 'We zoologists feel hurt at being treated like the musty old museum-mongers of a century ago, whose little lives were surrounded with stuffed skins and cabinets of butterflies and shells, and who in other affairs were as hapless as owls at mid-day!'[120]

Riding a Mule

Ronald Ross very aptly summed up the problems of a colonial scientist when he wrote: 'The Government of India is a *mule* as regards science . . . it *won't* do anything unless driven.'[121] The scientists at home (e.g. the IAC) and the colonial administrators consistently held that scientists in India should leave pure science to Britain and apply themselves to the applications of science. As Russel wrote in 1902:

General fundamental problems are best worked out here (England) or in Europe or America where the number of workers is greater and where it is easy to get into touch with those able to render useful assistance.[122]

Excessive government control of scientific undertakings often hampered the logical development of modern science in India. The government would always goad the various organizations to work only along economically beneficial lines. Most of them buckled under this pressure. Watt, for example, was asked in 1903 to prepare an abridged volume of his famous *Dictionary of Economic Products*. But he was not given a free hand in selecting the products. He was asked to include only those which were of commercial value. The result was that instead of a Dictionary of *Economic* Products, he produced a Manual of *Commercial* Products. Dyer (who was supposed to supervize this abridgement) was upset. He wrote: 'The omission of the unimportant products is most undesirable. These are precisely those one wants.' Watt replied:

[119] Revenue, Agriculture, no. 15, March 1907, pt. B.
[120] Commerce and Industry, Practical Arts and Museums, nos. 1–2, July 1906.
[121] Ross to Manson, 28 June 1898, Ross Papers, Mss. 02/159 (emphasis as in original).
[122] *Nature*, 24 Dec. 1908, p. 236.

I had never personally differed from your view. But neither your opinion nor mine had been accepted by the GOI. The Government of India gave me not only the list of subjects to be dealt with but the relative spaces to be devoted to each. . . . I certainly did set myself to write a manual of commercial products of India, since such may be accepted as the orders of the Government of India to me. If I had my way the book would have dealt mainly with the unimportant economic products.[123]

A few of the scientists, such as Ross and Alcock, preferred to resign.[124] There were some who wanted to make fundamental investigations in India itself, while the majority preferred to depend on Europe. Strahan, the Surveyor-General, for example, asked for an astronomical observatory to determine the slight movement of the polar axis of the earth. But Eliot, the Meteorological Reporter, opposed it on the ground that since this movement was common and affected the whole earth, work could be easily done within Europe. Eliot then added that 'any determination in India, if it agreed with their (European) results, would only be confirmatory, and if it disagreed would probably not be accepted'. To this Strahan reacted very sharply: 'I regret extremely that it should be in print . . . our observations of tides, pendulums, latitudes, longitudes, and azimuths have all been given the highest weight in Europe, and it is most invidious to suggest that any determination made in India would probably not be accepted.' Eliot and Strahan were temperamentally poles apart. Both were colonial scientists, but one was looking to the metropolis and the other was not willing to forego his own identity and the importance of his work.[125] Similarly, the agricultural experts looked to the European models despite Dyer's repeated warnings that 'all developments in India will have to be along *its own lines* and that it will only be attained by gradual experiment'.[126]

Colonial researchers often found themselves unable to distinguish between 'basic' research and 'applied' research. This was particularly true of the geologists and botanists. Holland's memo, referred to earlier, illustrates this.[127] The dilemma was fairly acute. On top of it, though the

[123] Dyer to Watt, 23.4.1908; and Watt to Dyer, 29.4.1908, Kew Papers, Miscellaneous Report, Commercial Products, ff. 364–6, 371–4.

[124] Revenue, Agriculture, nos. 22–3, Jan. 1907, pt. B.

[125] Revenue, Agriculture, Meteorology, nos. 1–12, March 1899.

[126] Thiselton-Dyer to Strachey, 22.10.1888. Kew Papers, Miscellaneous Reports, no. 5, India Agriculture, 1869–1921, ff. 16–17 (emphasis as in original).

[127] See reference no. 64. In later years Holland became more eloquent. In 1918, as President of the Indian Munitions Board, he claimed: 'Science is not the monopoly

colonial government would always recognize the importance of science, it would never approve of 'any large outlay upon them which must, however useful in its remote results, be immediately unremunerative'.[128] Some of the specialists (especially the botanists) felt slighted. A few received a great deal of attention while others none; for example, large sums were spent on geological explorations and nothing on the examination of agricultural soils.[129] George Watt thought it 'absurd to suppose that the Geology of India requires fourteen European experts, while the Agriculture and the Industries of India must be content with two or three expert investigators'.[130] Yet, despite difficulties and obstacles, their individual conduct was exemplary. Dr Lingard, the Imperial Bacteriologist, himself suggested to the government that he ought to be debarred from private practice.[131] Ross accepted less salary in order to continue with research. Rogers preferred to pursue kalazar investigations on the minimum pay due to a person of his rank.[132]

A significant feature of colonial research is the relative neglect of medical and zoological sciences. This is in sharp contrast to larger investments in botanical, geological and geographical surveys from which the British hoped to get direct and substantial economic and military advantages, while medical or zoological sciences did not hold such promise. Hence, even though a medical college had been established in 1835, the expected emphasis on research came quite late in the Plague Commission Report of 1904. Here for the first time was underlined the need for a Medical Research Department and the establishment of the Indian Research Fund Association for promoting research on medical problems.[133] The dynamics of research within medical science itself led to

---

of Europe, but we must do more than transplant the results, if it is to grow in India. We must undertake our own research work here.' *Proceedings of a Conference for the Organization of Chemical Research in India*, Simla, 1918, p. 2.

[128] Home, Education, nos. 14–88, Oct. 1897, pt. B.

[129] Revenue, Agriculture, nos. 1–2, March 1889.

[130] G. Watt to Thiselton-Dyer, 15 Jan. 1902 (marked private), Kew Papers, Indian Letters (Madras, Bombay, Bengal), 1901–14, vol. 160, ff. 366–8.

[131] Revenue, Agriculture, HB&AS, nos. 27–34, Aug. 1892, file 2 of 1891.

[132] L. Rogers, 'Note on the Difficulties Encountered by an IMS Officer in Obtaining Suitable Opportunities for Medical Research in India, 1893–1920', Rogers Papers, Mss., pp. 99–101, Wellcome Institute for the History of Medicine, London.

[133] The Bohr Committee Report on Medical Sciences, vol. I, Calcutta, 1946, pp. 23–4.

an important, perhaps unfortunate, consequence. The bacteriological advances of the late nineteenth century had put curative medicine on a scientific basis. Clinical treatment won ascendancy. This, in turn, encouraged the expansion of the private medical profession (both European and Indian), for a few medical colleges were a cheaper (but not necessarily an effective) alternative to expending government resources on sanitary reforms for the general population. Alcock worked for sometime as the Deputy Sanitary Commissioner of Bengal, and this experience convinced him that all talk about sanitary reforms was mere humbug.[134] The growth of preventive and social medicine was irremediably pre-empted and the rising medical profession made its spoils from the ever-expanding disease market.[135]

Another important feature is the almost total absence of pure or theoretical research. Research activities in sciences like physics and chemistry which had by then reached 'a professional stage' in Europe, were hardly noticeable in India. In the *Centenary Review of the Asiatic Society*, P.N. Bose apologetically wrote: 'Our chapter of chemistry at the Asiatic Society is near being as brief as the proverbial chapter on Snakes in Ireland.'[136] Till the advent of P.C. Ray, only one chemical paper had appeared—by A. Pedler on the volatality of some of the compounds of mercury.[137] There were chemical analyzers in every province but their job was confined only to medico-legal cases and the inspection of government stores. India was found suitable only for field research. She was in fact used as a 'vast storehouse' with exotic varieties of flora, fauna and minerals which were to flood the European laboratories for many years to come. The real research was thus to be done in the metropolis. India could get only ancillary units. And this happened in a century when England itself was undergoing a phase of transition wherein professional scientists, the government and industrialists who understood the full potentialities of science, were all attempting the very difficult task of integrating science into the English government, industry and education. In India the story was, however, different. Here scientific explorations brought the government, science and economic exploitation into a close relationship. But the Indians and India's interests were left largely in the cold.

[134] Alcock Papers, note 117.
[135] R. Ramasubban, p. 41, note 78.
[136] *Centenary Review of the Asiatic Society*, Calcutta, 1884, p. 101.
[137] *Annual Report of the Asiatic Society*, 1889, p. 96.

CHAPTER 6

Response and Resistance

A Hindu philosopher in Torricelli's place would have contented himself with simply announcing in an aphoristic *sutra* that the air had weight. No measure of the quantity of its pressure would have been given; no experiment would have been made with mercury; no Hindu Pascal would have ascended the Himalayas with a barometric column in hand.

Bankimchandra Chattopadhyaya on
'The Relation of Hindu Philosophy to True Science' in May 1873

The Mahomedans with their philosophy are exactly in the position of the schoolmen of Europe, that is they have travelled half way towards actual civilization; consequently when the modern reformed philosophy of Europe once gains an entrance to their minds, they will be able to make more rapid progress than their neighbour Hindoos. Among us a Newtonized Avicenna or a Copernicized Averroes may spring up who may be able to criticize even sons of Sina and Rushd.

Maulavi Ubaidullah, *Essay on the possible influence of
European learning on the Mahomedan mind in India*, Calcutta, 1877, p. 47

The psychology of colonization is superbly captured in Mannoni's interpretation of the three Shakespearean characters, Prospero, Ariel and Caliban. Ariel is a 'good' native, ever grateful to his master for small mercies (like the *bhadralok*?). Prospero has magical powers; the colonizers claimed a similar power in their rationality, heritage and tools. Prospero seeks to educate the 'bad', though hard-working, native, Caliban (the cannibal or subaltern?), but abandons him because he tries to rape the master's daughter. Caliban begins plotting against Prospero—'not to win his freedom, for he could not support freedom, but to have a new master whose foot-licker he can become. He is delighted at the prospect. It would be hard to find a better example of the dependence complex in its pure state'.[1]

The real colonial situation, however, is much more complicated than

[1] O. Mannoni, *Prospero and Caliban: The Psychology of Colonization*, New York, 1956, pp. 57–107. See also, L.D. Wurgaft, 'Another Look at Prospero and Caliban:

the Shakespearean drama. The colonial relationship is not always that of staunch adversaries, determined to fly at each other at the slightest provocation. Though the parties are often fierce adversaries, they negotiate their relationship, and tend to leave, consciously or unconsciously, some gap and some space for re-negotiation. These provide the basis for new strategies, re-alignments or experiments. As Volosinov said, the ideological sign is always multi-accentual and Janus-faced.[2] How did these signs appear in Victorian India?

Edward Said has drawn attention to the relationship between the different types of texts (which the colonizers and the colonized possess and profess), the relationship between the text and context, the representation of other cultures and societies and the role of the intellectual in shaping this relationship; and finally links them to the larger issues of power and knowledge relationships.[3] Can knowledge be non-dominative or non-coercive? This is where the role of context comes in, and the context of colonization precludes the possibility of knowledge being used in a non-dominative way. This can be seen in the paternalistic tones and invincibility (more apparent than real) of the colonizing power. It was a powerful ruse to inspire awe and elicit acquiscence. Under this relationship the colonizer would not only flaunt his civilizational 'superiority', citing the Greek and Renaissance achievements, but would also impose upon the natives (through force or pursuasion) his opinions, assessments or prejudices. It is a situation in which 'A knows B more than B either knows himself or A, and in which A tells B what he is or ought to be'.[4]

The colonial discourse thus presents distinct categories and powerful hierarchies. But in the very process of its enunciation, Homi Bhabha spots what he calls 'ambivalence', 'hybridization' and 'mimicry'.[5] Unlike Said, he believes that colonial power and discourse is *not* possessed entirely by the colonizer. He locates the kernel of colonial discourse in the conflict

---

Magic and Magical Thinking in British India', *The Psychohistory Review*, VI, (1), 1977, pp. 2–26.

[2] Quoted in Homi K. Bhabha (ed.), *Nation and Narration*, London, 1990, p. 3.

[3] Edward W. Said, *Orientalism*, New York, 1978; *The Word, The Text and the Critic*, Massachusetts, 1983; 'Orientalism Reconsidered', *Race and Class*, XXVII(2), 1985, pp. 1–15; *Culture and Imperialism*, London, 1993.

[4] T.O. Ranger, 'From Humanism to the Science of Man: Colonialism in Africa and the Understanding of Alien Societies', *Transactions of the Royal Historical Society*, XXVI, 1976, p. 126.

[5] Homi K. Bhabha, 'Signs Taken for Wonders: Questions of Ambivalence and Authority under a Tree Outside Delhi, May 1817', *Critical Inquiry*, 12(1), 1985, pp. 144–65.

of power/disempowerment, knowledge/disavowal, mastery/defence, absence/presence; the disavowal of difference, for example 'turns the colonial subject into a misfit—a grotesque mimicry or "doubling" that threatens to split the soul and whole'.[6] It is this ambivalence which allows the colonial power to produce the colonized as an unreadable 'other' and yet entirely knowable and visible at the same time. It is this ambivalence which explains the predicament of a native intellectual whether or not to own or disown a traditional notion or text, and how. Though produced as an *other* to provide a plausible pretext for the exercise of naked power, he is also a *producer*, provoking reaction in the master.[7] The binary rigidities are thus removed; history is relativized and seen more as a process, as a flux, than an assortment of categories.[8]

By arguing that the colonial power and discourse is shared by both the colonizer and the colonized, and by putting emphasis on 'ambivalence' and 'indeterminancy' in this relationship, is Bhabha trying to take the (political and economic) sting out of colonialism? How could the native, whose entire economy and culture was threatened, share or 'possess' colonial power?[9] One may not rule out the possibility of some 'genuine confusion' in perception of both the parties but one cannot present colonialism as 'a genuinely confused spectacle'. The history of material and discursive conflicts between conquerors and natives is too dense to permit that. This relationship is perhaps best captured in Fanon's description of it as a 'Manichean' struggle.[10] The story of this struggle is fascinating as well as instructive. But it cannot be studied in isolation. The imperium remains an important determinant. In fact the very question of response has within it a very close relationship with British policy. Both influenced each other.

Articulating Distance/Difference

Distance and difference are no new entrants in the imperial lexicon, though of late they have been discussed a great deal by cultural historians.

[6] Homi K. Bhabha, 'Difference, Discrimination and the Discourse of Colonialism', in F. Barker et al. (eds), *The Politics of Theory*, Colchester, 1983, pp. 194–211.

[7] Paul Brown, 'This thing of darkness I acknowledge mine: The Tempest and the Discourse of Colonialism', in S. Dollimore (ed.), *Political Shakespeare*, Manchester, 1985, p. 61.

[8] Gauri Viswanathan, *The Masks of Conquest*, New York, 1989, p. 97.

[9] A.R. Jan Mohamed, 'The Economy of Manichean Allegory: The Function of Racial Difference in Colonial Literature', *Critical Inquiry*, 12, Autumn 1985, pp. 59–87.

[10] Frantz Fanon, *The Wretched of the Earth*, New York, 1968, p. 41.

One can spot an articulation of distance and difference (psycho-social, epistemological, spatial) in the numerous travelogues and colonial writings of the eighteenth and nineteenth centuries. Guns and sails could win territories but the empire could be sustained only by emphasizing differences and operating from a higher pedestal. The Raj rested on prestige. A curtain had to be put across even where it did not exist and even when it was not required.[11] Racialism was an extremely important component of this project. Imperialism itself rested on ascendency or predominance, the rule of one race over another. A contemporary found imperialism consisting of two things—pride and greed.

With the capitalist class and mining magnates the latter spirit is more powerful as it probably was with Cecil Rhodes. With Mr Chamberlain, Lord Rosebery, and the average Englishman, the former spirit chiefly prevails. Both classes, however, are to some extent influenced by both sentiments, and it is merely a question of degree.[12]

Gradually racialism came to acquire a certain amount of respectability in academic circles. Darwin's theory of the survival of the fittest, for example, was warmly welcomed by the social scientists of the day, and they believed that mankind had experienced various levels of evolution culminating in the white man's civilization. In the second half of the nineteenth century it was accepted as a fact by a vast majority of western scientists.[13] Politically, many like Ellenborough, attributed English domination in the East to the right of conquest and to ascendancy of blood. As a corollary they violently attacked 'the insane principle of treating in India natives and Europeans on the same footing'.[14] Any sign of discontent or revolt on the part of natives irritated the rulers, leading to a wider chasm between the two. After the revolt of 1857, for example, Colonel Waugh, who had

---

[11] Raghavan Iyer (ed.), *The Glass Curtain between Asia and Europe*, Oxford, 1965.

[12] J.G. Godard, *Racial Supremacy Being Studies in Imperialism*, Edinburgh, 1905, p. 9.

[13] Some American medical researchers even argued that black males were primitive because the distance between their navel and penis remains small throughout their life while white children begin with a small separation but the span increases as they grow—the rising belly button was taken as a mark of progress. S.L. Gilman, 'Degeneracy and Race in the Nineteenth Century: The Impact of Clinical Medicine', *The Journal of Ethnic Studies*, 10, (4), 1983, pp. 27–50.

For several case studies of 'scientific' racialism, see H.L. Gates Jr. (ed.), *Race, Writing and Difference*, Chicago, 1986.

[14] *The Times*, 9 Dec. 1857, quoted in T.J. O'Keefe, *British Attitudes Towards India and the Dependent Empire, 1857–1874*, Ph.D. thesis, University of Notre Dame, Indiana, 1968, p. 48.

earlier tended to respect local talent like his predecessor, George Everest, felt that 'no Englishman can ever thoroughly know or understand a native'.[15]

Notwithstanding Waugh's frustration, it is true that the natives were very keenly observed by their rulers. Some would criticize them, some would discover a few positive points, while some would offer suggestions for their improvement. David Smith, an iron expert, appreciated the native steel-makers, but added: 'They must entirely change their habits and conditions; European dress must be adopted; they must eat more nourishing food, and give up their *Poojas* and holidays of long duration.'[16] As skilled workers they were found to be tolerably good. But where higher mental faculties were required, the British were sceptical. Professor A. de Morgan of Cambridge was impressed with Master Ramchandra's work on maxima and minima and wanted the education departments to foster original research among the natives. To this the DPI of Bengal replied: 'Indeed if I am asked what steps should be taken in furtherance of the Professor's views, my answer must be *none*.'[17] The new bhadralok had become so enamoured with the Oxbridge tradition of classical education that they themselves often derided the study of physical sciences.[18] This trend in fact continued for a long time. To quote a contemporary:

The establishment of science courses has not led to any scientific education properly so-called. The F.A. candidates hardly ever read the books prescribed, but depend upon notes published or dictated to them by their teachers. In several colleges the laboratory is most imperfect, in some possibly non-existent. For the B.A. examination also notes are very largely depended upon. And there are deeper reasons for the desultory study of science. The students look forward to no scientific career which may be a stimulus to their special studies. They aim at the same career as literary students, and if they select science as their optional group of subjects, it is because they expect to pass more easily in that group.[19]

Several instances were cited to show that the Indian students did not respond favourably to whatever rudimentary scientific course they were offered. In 1875, out of twenty-eight native students at the Lahore

---

[15] Col. Waugh was then Surveyor-General, and was so incensed with the revolt that he even prepared a long memo on it, describing the causes and suggesting measures. Waugh Papers, IOL, Mss. Eur. F. 181/6.

[16] *Selections from the Records of GOI*, LXIV, Calcutta, 1868, p. 83.

[17] WBSA, General, Education, no. 90, Aug. 1860 (emphasis as in original).

[18] H. Woodrow, DPI, Bengal to Secretary, Govt. of Bengal, 20 May 1872; WBSA, General, Education, nos. 37–9, Sept. 1872.

[19] N.N. Ghosh, *Higher Education in Bengal*, Calcutta 1901, p. 12.

Medical School, as many as twenty-one were dismissed for 'idleness and incompetency'.[20] Employment of a scientific nature was scarce, no doubt, but whenever opportunities arose, no suitable person could be found.[21]

It is not that the natives always met with reprobation. Saiyed Mohsin's theodolities were found accurate even after thirty years of heavy duty in the field;[22] Smyth and Thuillier frankly acknowledged that Parts III, IV and V of their *Manual of Surveying* were largely written by Radhanath Sikdar. But in the second edition of this manual, Sikdar's name was omitted; acknowledgement of valuable scientific work done by a native was probably considered inconsistent with British prestige.[23] Due to political reasons, British explorers could not enter Tibet and eastern Turkestan and so they sent a few natives, in the guise of *pundits* to Lhasa with mapping instructions.[24] One of them, Pt. Nain Singh mapped a distance of 1200 mils from Nepal to Lhasa through absolutely unknown areas, made 31 determinations of latitude with a sextant, and fixed the absolute altitude of 33 stations by means of a boiling-point thermo-meter.[25] The government rewarded him with the grant of a village,[26] and later the Geographical Society of Paris acknowledged his services with the award of a gold medal.[27] Its counterpart in London may have felt constrained to go to that extent for a native. In the field of medical and engineering education Indian performance on the whole was found impressive to the extent that some of the students of the relatively less important Nagpur Medical School were considered quite capable of

[20] Home, Medical nos. 96–101, Oct. 1877.

[21] In Dec. 1875, G. Watt and G. King, after interviewing five candidates for the post of Botanical Tutor at the Calcutta Medical College, reported: 'They indicated a very imperfect knowledge indeed of the theoretical part of botanical science, and a want of acquaintance with practical botany which impressed us both very painfully. We cannot recommend any of the candidates as fit for the post.' WBSA General, Agriculture, no. 6, Jan. 1876.

[22] *Account of the Operations of the GTSI*, II, Dehradun, 1879, App. I, p. 19.

[23] P.N. Bose, *A History of Hindu Civilization*, III, Calcutta, 1896, pp. 101–2.

[24] 'Wonderful men those *pundits!* The English have made extraordinarily clever use of their intelligence and fidelity, and in many cases the pupil has excelled the teacher. They work with the accuracy of self-registering instruments, and wherever their labours have admitted of control their trustworthiness has been fully demonstrated.' S. Hedin, *Scientific Results of a Journey in Central Asia* (1899–1902), IV, Stockholm, 1907, p. 531.

[25] *Jr. of the Royal Geographical Society*, XXXVIII, 1868, p. 129.

[26] J.T. Walker Papers (1876–81), Royal Geographical Society Archives, London.

[27] *General Report on the Operations of the Survey of India*, 1885–86, p. 9.

'passing many of the Examining Boards in the United Kingdom'.[28] In 1867, two students, after failing their LMS in Bombay, went to England and gained their diplomas in both London and Edinburgh.[29] The response was perhaps better in the field of engineering. Even the high-caste youths were not unwilling to study practical engineering and undertake mechanical work.[30] But the prospects were hardly lucrative. R.E. Fife, who had a long stint with the Royal Engineers in India, considered the PWD 'singularly inimical to the native civil engineer'. In 1879 he calculated that no native figured among the 18 chiefs and 54 superintending engineers; out of 516 executive engineers, only 9 were natives, out of 632 assistant engineers 79 were natives, out of 173 sub-engineers, 39 were natives; out of 341 supervisors, 103 were natives; and out of 768 overseers, 492 were natives.[31] Native subordinates, however intelligent they may be in engineering matters, had little chance of rising above the subordinate grades. As least there were a few officials who recognized this.

By the 1860s many changes had taken place. Universities had been established in the presidencies. The medical and engineering colleges were striking deeper roots. A good number of medical, physical and other societies had sprouted and some like the Students' Literary and Scientific Society in Bombay consisted exclusively of natives.[32] Some Indians who had opportunities for higher studies in England (e.g. S.G. Chakravorty, G.C. Seal, B.N. Bose) had distinguished themselves.[33] The years to come were to see more and more of what Richard Temple called intellectual restlessness and mental fermentation.[34] The government was only too acutely aware of this growing restlessness and sought to defuse it by clearing the lower decks for natives.

Moreover, the importation of skilled hands from Europe was becoming quite expensive. Even twenty years after the establishment of the GSI, its superintendent, Oldham, lamented the unavailability of good geologists. It took on an average nine and half months to procure a qualified

[28] *Report on Sanitary Measures*, XIII, 1879–80, p. 33.

[29] *Administration Report*, Bombay, 1867–68, p. 191.

[30] *Administration Report*, Bengal, 1879–80, p. 54.

[31] R.E. Fife, *The Civil Engineering Profession in India*, London, 1879, pp. 8–9; Fife Papers, IOL, Mss. Eur. D. 1015.

[32] *Jr. of the Royal Society of Arts*, XXIII, no. 1170, 23 April 1875, p. 517.

[33] Hodgson Pratt, *University Education in England for Natives of India*, London, 1860, p. 28.

[34] *Jr. of the Royal Society of Arts*, XXIV, no. 1250, 3 Nov. 1876, p. 3034.

geologist from England, and it became increasingly difficult to obtain a geologist because of the poor incentives offered by the Survey.[35] The medical service also faced similar problems. Out of the 74 students produced by the Calcutta Medical College during 1852–57, as many as 61 joined the government service, but only 43 continued in the service. Frosyth (Director-General, Medical Deptt.) reported that the natives 'who nearly invariably leave the college with as large an amount of knowledge as is to be met with in any medical graduate of any European school, in default of any great object or prize to aspire to, sink in a few years into the condition of tolerable routine practitioners, among whom not a single man of distinguished scientific attainment is to be discovered'. This happened because, as an official forwarding the Frosyth's Report, explained:

We expect too much for what we give. There is no high prize, no possibility of rising to any eminent position. There is nothing to attract the ambitious, no high reward for the industrious. We keep all the prizes in the hands of the Europeans and then expect the natives not to be less assiduous and enthusiastic than they would with a large field of usefulness and advancement before them.[36]

So far as the other scientific departments like the GSI, Survey of India and Meteorological Department were concerned, things were no better. At least in the Geological Survey there did exist a sharp cleavage between the high administrative officers of the government and the head of the GSI (Medlicott in particular) regarding the role to be assigned to the natives. A.O. Hume, Secretary to the Government of India, was determined to 'set on foot a plan for superceding, to a certain extent, European by native agency'.[37] He wanted to do this because it would have made the survey operations much more economical. He felt that some of the natives would be as useful as some of the Europeans then drawing Rs 1000 a month; and was so insistent that he even recommended the dismissal of the officers (Oldham and Medlicott in this case) who refused to implement his plan. The high-ups in the GSI, however, continued to fulminate over this policy decision. To Medlicott, the Indians appeared 'utterly incapable of any original work in natural science'. He wanted to wait until the 'scientific chord among the natives' was touched, and added, almost contemptuously, 'if indeed it exists as yet in this variety of the human race. So let us exercise a little discretion with our weaker brethren, and not expect them to run before they can walk'.[38] He was obviously oblivious

[35] Revenue, Agriculture, Surveys, no. 1, Jan. 1872.
[36] Home, Public, nos. 26–8, 7 Oct. 1859.
[37] Revenue, Agriculture, Surveys, nos. 15–19, Oct. 1872, K.W.
[38] Revenue, Agriculture, Surveys, no. 25, Sept. 1880.

of the fact that a few enlightened Calcuttans had already established an institution, the first of its kind in Asia, called The Indian Association for the Cultivation of Science, in 1876.

To such fiery outbursts, the governments's reaction was rather cool. The then Under-Secretary to the GOI, C.S. Bayley, wrote a brief note: 'Mr Medlicott writes, as usual, intemperately, hence no use in discussing with him a point which has been thoroughly thrashed out already'.[39] The motive was clear—why incur extra expenditure by importing qualified geologists from England. As an official had noted earlier: 'The natives are very much cheaper; if the work is to be really done thoroughly in future generations, it must be done by the natives of India.'[40] This was reaffirmed by Lansdowne in 1891 with the sanguine hope that an improved educational system would gradually enable the GSI to recruit better people.[41] But how could this hope have been realized, even partially, in the absence of a comprehensive system of science education, particularly in geology? This dichotomy was well brought out by a reviewer in 1864 when he wrote: 'The difficulties are those of our own making. The Egyptians wanted bricks made without straw, and became proverbial for their folly. The Indian Government goes a step further and wants its bricks made not only without straw but without brick-makers. It contemplates great things in a hazy way; and looks for great results.'[42] But such criticisms would not deter the men in power who, though inclined to urge the induction of natives into the geological staff, could not find 'sufficient reason to justify any special education for the purpose' nor found the 'native character likely, with any kind of training, to fulfil the requirements of the Departments'.[43]

Officials like Medlicott harped on the so-called intellectual bankruptcy of the Indians. The idea was to retain a European hold in areas which were vital to the colonial economy. Scientific knowledge was not for every one. Medlicott had one more convenient pretext. His department (GSI) was small with only fifteen graded officials, so why dilute it?[44] As late as the mid-1880s when a number of natives had already given a good account of their abilities in scientific pursuits, the government was chary of reducing the European element even in the junior branch of the Survey

[39] Revenue, Agriculture, Surveys, no. 3, Aug. 1886, K.W.
[40] Revenue, Agriculture, Surveys, nos. 25–33, Sept. 1880, K.W.
[41] Revenue, Agriculture, Surveys, no. 7, Sept. 1891, file no. 25.
[42] *Calcutta Review*, 39, no. 78, 1864, p. 438.
[43] Revenue, Agriculture, Surveys, nos 1–6, Sept. 1882, K.W.
[44] Ibid., no. 5.

of India. The maximum concession was one vacancy in every five or six to natives, just as an experiment, 'without any risk of endangering the efficiency of the Department'.[45]

Indians were thus treated only as minor partners, in case partnership was at all required. In topographical survey works particularly, and more especially in unhealthy portions of the country, where European life was at a greater risk, it was not possible to reckon on a high rate of progress without the employment of natives. J. Mulheran, who headed the Topographical Party of Hyderabad Survey, preferred natives as field-workers because 'they require less assistance from the local authorities, have greater influence with the people, are more rapidly replaced if they give dissatisfaction, and as a body are more easily managed and directed'.[46]

With the expansion of railways and irrigation works, there was an increasing need for the training of natives as mechanics, plate-layers, drivers, etc., of course, with a moderate amount of European supervision,[47] and so in 1875, Calcutta got its first School of Apprentices to feed the mint and other government establishments employing skilled mechanical labour.[48] The Committee constituted to report on Indian Surveys in 1904 recommended an increase in the proportion of natives in junior services to one-half, but for higher grades took recourse to the well-known cliche that the time was not yet ripe.[49] This was in the interest of the Raj.

A proper system of education in various branches of science may have tilted the balance in favour of the natives. This was not created except for a rather heavy emphasis on mathematics and mathematics alone. *Tatvabodhini Patrika* (No. 86, 1850) calculated that advanced school students had to spend fourteen hours per week on mathematics only, compared to four hours on history and five on literature.[50] This may have been done with a view to create a class of junior surveyors or calculators for assistance in the gigantic topographical and cadastral operations. Mathematics fitted well into the 'arithmetic of imperialism', but not the other exact or applied sciences. When in July 1883 the people of

[45] Home, Public, no. 224, Dec. 1884, pt. B.
[46] *General Report on the Topographical Survey of the Bengal Presidency*, 1860–62, p. 2–20.
[47] F.J.E. Spring, *Technical Education in Bengal*, Calcutta, 1885, pp. 85–6.
[48] *Report of the Committee for the Establishment of a School of Apprentices in Calcutta*, Calcutta 1875.
[49] *Report of the Indian Survey Committee*, 1904–5, p. 60.
[50] Benoy Ghosh, *Samayikpatre Banglar Samajchitra*, II, Calcutta, 1964, p. 423.

Backergunge requested the Lieutenant-Governor of Bengal to provide for two scholarships to enable Indian students to go to England to study chemical and textile manufacture, the reply was that the engineering colleges of Shibpur and Roorkee were more than adequate.[51] The memorialists then approached the Viceroy, Lord Ripon, seeking for an explanation of the fact that if scholarships could be provided to study agriculture at Cirencester, why could they not be provided for chemical or textile engineering, which were no less important. The DPI, Bengal, reacted very strongly by advising the agricultural scholarships to be stopped which had apparently whetted the natives' appetite.[52] Fortunately, the agricultural scholarships were saved but the proposal fell through.

Racialism was a tool, not an end in itself. The end was to legitimize a complete psychic subordination. This is most explicit in the use of gender-based metaphors such as feminine, rape, etc. This was done systematically and consistently for a very long period. In 1782 Robert Orme described Indians as 'the most effeminate inhabitants of the globe'.[53] In 1840–50, Dr James Esdaile performed mesmeric operations in Calcutta and attributed his 'mesmeric' success to the native physique and psyche which could be 'readily subdued'.[54] Thanks to the 'native psyche' and his own determination, he could project even mesmerism as a science, and his methods did attract people. This was not something unusual in a country where astrology enjoyed a high 'scientific' reputation, and where superstition and science perhaps mingled more freely than anywhere else.

Racialism had other, perhaps lighter, aspects. Even insane Europeans did not like to be attended to by natives. They would feel humiliated and degraded if coerced by native attendants. Hence the recommendation that in asylums 'restraint ought to be applied exclusively by European attendants'.[55] Racial segregation of European and Eurasian children was advocated by a Select Committee of the House of Commons in its Report on European Settlement and Colonization in India (9 August 1859). The Bishop of Calcutta immediately picked up the hint and demanded separate schools in the hill stations, away from the Heathen and Maho-

---

[51] Home, Education, nos. 44–5, Feb. 1885.

[52] Ibid., no. 49.

[53] Robert Orme, *Historical Fragments of the Mogol Empire*, London, 1782, p. 472.

[54] James Esdaile, *The Introduction of Mesmerism as an Anaesthetic and Curative Agent into the Hospitals of India*, Perth, 1852; also, Dr Elliotson, *Mesmerism in India*, London, 1850, p. 46.

[55] Report on the Asylum for European Insane Patients at Bhowanipore, 1856–57, *Selections from the Records of the Govt. of Bengal*, XXVIII, Calcutta, 1858, p. 14.

medan population, for European and Eurasian children who could be groomed as 'mechanics and practical agriculturists', then very much in demand by the British planters. Both the Governor-General and the Secretary of State approved of this idea.[56]

Here it is worth mentioning that the term, 'Natives of India', as defined in Statute 33, Vic. Ç-3, which enabled the Government of India to appoint such persons to posts reserved for the Covenanted Service, also included the Eurasians and Europeans who satisfied the conditions of being born and domiciled within 'the dominions of Her Majesty'.[57] Taking advantage of this, the Eurasians were getting into the medical and engineering services on a large scale. But their status was uncertain. Theoretically they would enjoy the same social status as the Britons recruited from Cooper's Hill College of Engineering but in practice would labour under disabilities in matters of leave, pension, etc. which put them on par with the natives. To create a better equation, successive Secretaries of State like Salisbury, Cranbrook and Hartington, urged the Government of India not merely to render the Roorkee College of Engineering available to the natives of India in the ordinary colloquial sense of the term, but to reserve the guaranteed appointments exclusively for such natives. The Government of India did not go that far, but (vide Resolutions of 11 November 1881 and 14 Feburary 1883) reserved a few appointments for natives of pure Asiatic descent. The Anglo-Indians felt affronted. This was the year of the famous Ilbert Bill. Some sort of a 'white mutiny' was in the offing.[58] They petitioned but the government refused to take it as a race or 'caste question'. It may be too much to read in this isolated instance an attempt to drive a wedge between the Indians and Anglo-Indians. Both were natives nevertheless, and hardly stood to gain much under the existing set-up.

The natives, on the whole, were not viewed favourably. Social Darwinism could not have got a finer expression than in what Lansdowne wrote to Cross: 'Original scientific research demands mental and physical qualifications which are not apparently found in races bred in tropical climate to the same extent that they exist in the more vigorous races of northern latitudes.'[59]

[56] *Selections from the Records of GOI*, LIV, Calcutta, 1867, pp. XLI, XLIV–XLIX.

[57] Home, Public, nos. 215–23, Aug. 1883, pt. B; Despatches to India, 1872, IOR v/6/300, p. 45.

[58] E. Hirschmann, *White Mutiny*, New Delhi, 1980.

[59] From Governor General-in-Council to the Secretary of State, 28 July 1891; Revenue, Agriculture, Surveys, no. 7, Sept. 1891, file no. 25.

Signs of Restlessness

A close contact with a vibrant, though alien, culture could not have failed to produce ripples in the minds of the local people. The native press reflected the stirrings, and a few individuals responded by publishing scientific books and journals, while some took to organizing scientific associations and institutions. Among the Bengali journals the important ones were *Vigyan Kaumudi* (1860), *Vigyan Rahasya* (1871), *Vigyan Vikas* (1873), *Vigyan Darpan* (1876), *Sachitra Vigyan Darpan* (1882), *Chikitsa Darshan* (1887), etc.[60] These ofcourse did not carry original papers but only translations, aimed at popularizing science. Even general periodicals like *Samvad Prabhakar,*[61] *Tatvabodhini Patrika,*[62] *Bengal Spectator,*[63] *Somprakash,*[64] etc., every now and then harped on the importance of science education and research. *Somprakash* (No. 19, 1868), for instance, observed that

in a country like France even at the primary level or at the very ordinary school, sufficient attention is given to science. In India the study of true science is negligible. It remains limited to the Roorkee or medical colleges. The Asiatic Society, reflecting upon so deplorable a condition of science, proposed that science should be studied properly at the university level right from the Entrance. But the Government refused saying that the time was not yet ripe. Is not the Education Department the cause of our scientific and technological backwardness?[65]

Apart from criticizing the government, *Somprakash* also chided the native medical students for not having made any new discoveries in medicine.

Their education is bookish. Somehow they get themselves passed in the examination and then concentrate mainly on their fees. . . . Some of the European doctors are eager to know about our books on medicine (*nidana*). But none of our medical students showed any such inquisitiveness. None of them, after knowing the system of therapy (*chikitsa*), have tried to examine chemically the medicinal matters (*dravya*) and make use of them.[66]

Such criticism served the dual purpose of imploring the government's

---

[60] B. Bhattacharya, *Banga Sahitye Vijnan,* Calcutta, 1960, pp. 138–44.
[61] Benoy Ghosh, I, Calcutta, 1962, pp. 94, 164–5, 306, 356–9, 386–8, note 50.
[62] Ibid., II, Calcutta 1964, pp. 420–9, 466–8.
[63] Ibid., III, Calcutta, 1964, pp. 179, 181–2, 184, 193.
[64] Ibid., IV, Calcutta, 1966, pp. 505–6, 530–2, 552–3, 566.
[65] Ibid., p. 520, also *The Athenaeum,* no. 2168, 15 May 1869, p. 672.
[66] Ibid., p. 505.

support on the one hand, and exploring new possibilities within the natives' structure on the other.

There was certainly no dearth of enterprising men who would translate articles of scientific interest and there was no dearth of journals or periodicals either to carry them. But most of them lacked consistency and had to be abandoned as financial disasters. Their subscription was limited to a very small section of the upper middle class. The government on its part would not lend a helping hand by subscribing to them for its schools and other institutions. In 1885, G.C. Bose, J.N. Dey and H. Patra brought out the Indian Agricultural Gazette.[67] It was an entirely Indian enterprise, but in the absence of any government support it was closed in 1889.[68] Likewise most of the scientific periodicals were short-lived. Their potentialities were appreciated but never properly exploited.

The number of publications was many. In 1857, 571670 copies of 322 books were issued in Bengali from 46 printing presses. But this included only 9 works on natural science, of which 12,250 copies were printed.[69] The ratio improved later as science education spread. In 1875 the Calcutta School Book Society compiled a catalogue comprising 1544 Bngali books which included 112 books on mathematical science, 61 on physical sciences and 99 on medicine. A further analysis suggests that arithmetic (51 books) and medicine (68 books) were the favoured subjects, while for natural history and surgery the figure was as low as a mere 2 and 3. Out of the 4122 books published in all the languages in the year 1874, 348 (i.e. 8.44 per cent) were in science. Only 219 were published in the provincial languages—Bengal had a share of 86, Punjab 49, the NWP 30, Bombay 18 and Madras only 14. The trend is clear. Science got much less than its due and technology even less. The only book in Bengali on engineering is Durga Charan Chakravorti's *Vishwakarma* written in 1886. Important Bengali writers like Aukhoy Dutt, Ramendra Trivedi and Bankim Chandra were fascinated more by natural sciences than by such prosaic subjects as engineering or mechanics.[70] In *Mayapuri*, Trivedi

[67] *The Indian Agricultural Gazette*, I, April 1885.

[68] Ibid., IV, March 1889. Even the influential editor of The Statesman, Robert Knight, who published also an agricultural journal (The Indian Agriculturist), complained of 'weak support of Government' and his request to the government for a raise in subscription met with refusal. Revenue, Agriculture, nos. 18–19, Sept. 1885, pt. B. file no. 124.

[69] D.P. Bhattacharya, 'Scientific Writings in some Major Indian Languages in the Nineteenth Century', paper presented at a Seminar on *Science and Empire*, NISTADS, New Delhi, 21–23 Jan. 1985.

[70] B. Bhattacharya, pp. 351, 388, note 60.

calls for 'science for the sake of knowledge and nothing else. Telegraph, telephone, dynamo, motor, electricity, steamships are very small, lowly and negligible in comparison with the sublime ecstasy which a truth seeker derives from pure science.'[71]

This, however, is not to minimize the importance of individual efforts in the dissemination of scientific learning. Master Ramchandra in Delhi and Bal Shastri Jambhekar in Bombay wrote hundreds of popular science articles. Aukhoy Dutt wrote on geography and chemistry in Bengali. Onkar Bhatt wrote on geography in Hindi. Geography and astronomy were made the first targets because the Pauranic myths were considered the strongest in these fields. Vyas, the author of *Srimad Bhagwat*, for example, talked about the ocean of milk and nectar. This is part of popular Hindu myth even now and this was attacked by persons like Onkar Bhatt, Bapu Deva Shastri and Aukhoy Dutt. Onkar Bhatt explained that Vyas was only a poet, not a scientist, and since his interest was merely to recount the glories of God, he wrote whatever fancied him.[72] Even laureats like Bankimchandra Chattopadhyaya wrote a series of articles on astronomy. Significantly enough, he does not shower blind praise on the Indian scientific tradition. He wrote:

A Hindu philosopher in Torricelli's place would have contented himself with simply announcing in an aphoristic *sutra*, that the air had weight. No measure of the quantity of its pressure would have been given; no experiment would have been made with mercury; no Hindu Pascal would have ascended the Himalayas with a Barometric column in hand. To take a parallel case, the diurnal rotation of the earth is shadowed forth in the *Aitereya Brahman*. Arya Bhatta distinctly affirms it. 'The starry firmament in fixed', says he, 'it is the earth which, continuously revolving, produces the rising and the setting of the constellations and the planets'. The only legitimate deduction from the combination of these three facts, viz. the diurnal rotation of the earth, the fixity of the heavenly bodies, and the apparent motion of the sun, was the heliocentric theory. But the heliocentric theory was never positively put forward—never sought to be proved. In modern Europe, the announcement of the Copernican theory rendered certain the future discovery of the laws of Kepler and the great laws of universal gravitation. In India Arya Bhatta's remarkable announcement rendered certain that nothing further would come of it.[73]

Similar sentiments can be noticed in what a noted Islamic scholar, Syed J. Afghani, said in a lecture in Calcutta in 1882:

[71] S.K. Chattopadhyay (ed.), *Ramendra Rachnavali Samgraha*, Calcutta, 1965, p. 87.
[72] Onkar Bhatta, *Bhugolsar*, Agra, 1841, p. 39.
[73] J.C. Bagal (ed.), *Bankim Rachnavali*, I, Calcutta, 1869, pp. 146–7.

There was, is, and will be no ruler in the world but science. . . . Our ulema at this time are like a very narrow wick on top of which is a very small flame that neither lights its surroundings nor gives light to others. . . . The strangest thing of all is that our ulema these days have divided science into two parts. One they call Muslim science, and one European science. Because of this they forbid others to teach some of the useful sciences.[74]

In later years Bankimchandra retreated into Hindu spirituality. He found Darwin's theory of natural selection closer to the Hindu concept of trinity. He also believed in the power of *mantras* and astrology. Positivists like Girish Chandra Ghosh, Guru Das Chatterjee and Dwarka Nath Mitter wanted to learn from western science and commerce but not at the cost of Hindu ethos and its social institutions. Jogendra Chandra Ghosh insisted that reform be carried out with due respect to traditional institutions and through traditional channels.[75] Aukhoy Kumar Dutt and Ramendra Sunder Trivedi believed in the scientific method and demonstrable proofs, but while propagating western sciences both betrayed streaks of revivalism and had a tendency to show that whatever was good in western science existed in ancient India also. For example, Trivedi's discussion on Darwin ends with the Gita and he concludes that ultimately this world is *Maya*—controlled by some supernatural force which is beyond the comprehension of science. Like Ram Mohun Roy, Trivedi starts with material perceptions but finally lands up in a metaphysical mirth. His first book *Prakriti* (nature) was published in 1896 and his last is *Jigyasa* (inquisitiveness or thirst for knowledge) in 1904. His *Jigyasa* could not be satisfied by science.[76] This dichotomy is a recurrent feature of the time. An average educated Indian found himself torn between two schools of thought. In 1882 Radha Govind Kar wrote a book, *Kobiraj-Doctor Samvad,* wherein he gave a graphic description of the tussle between the traditionalists and the modern medical men.[77] Contact between two different systems was bound to set off controversies—exploding old myths and often creating new ones.

Signs were, on the whole, propitious. Even the Urdu poets, devoted mainly to the romantic aspects of life, felt drawn towards western science and technology. Ghalib, for instance, talked about the achievements of western civilization based upon steam and coal power. Hali in his *Musaddas* described science as 'a precious jewel tested as yet by no one,

[74] Quoted in Pervez Hoodbhoy, *Islam and Science: Religious Orthodoxy and the Battle for Rationality,* London, 1991, p. 59.

[75] G.H. Forbes, *Positivism in Bengal,* Calcutta, 1975, pp. 63–72, 90–5.

[76] B. Bhattacharya, pp. 195–6, note 60.

[77] Ibid., p. 360.

with its great powers still hidden'.[78] The next logical step was to give some organizational shape to the growing yearning for modern science; the lead was given by Syed Ahmad, followed by more concrete steps taken by Mahendralal Sarkar a few years later.

In 1864 Syed Ahmad founded the Aligarh Scientific Society with the support of the leading *zamindars* and officials of the NWP. The Society called for the introduction of technology in industrial and agricultural production.[79] This was not possible unless the people were made aware of western scientific and technological advances in their own languages. So a massive translation programme was undertaken, a journal was started and often popular lectures were organized. The most striking thing about these efforts was its totally secular character. Religion was given no place at all. Sir Syed was not a blind follower of theological writs. He would not accept what appreared vague and mystic in Indo-Islamic traditions. Criticizing the establishment of an Oriental Faculty at Allahabad University, he wrote: 'No doubt, the Muhammadans cultivated other sciences also. But they cultivated the other science merely to make them serve theology, always considering them as subservient to divinity.'[80] Sir Syed felt that backwardness was a direct result of superstitious beliefs and the rejection of *maaqulat* (reason) in favour of blind obedience to *manqulat* (tradition).[81] The Aligarh Society preferred to lay maximum emphasis on mathematical studies. The Society was basically a landholders association and mathematics was important for revenue purposes. Seventeen books on arithmetic, algebra, trigonometry, geometry, etc. written by Tod Hunter, Bernard Smith, Galbraith and Haughton were translated. *Modern Farming* by R.S. Burns, *Electricity* by W.S. Harris and C. Tomlinson's work on mechanics were also translated and published. Apart from popular and demonstrative lectures (such as one on the sewing machine), even a model farm was laid. English and American pumps were fixed for raising water. Foreign varieties of wheat, cotton and vegetables were imported. Several government-aided horticultural societies were engaged in similar work in other areas. The interests of the landed aristocracy, which the Society represented, and those of the government coalesced in at least this field.

[78] A. Rahman, 'Science, Cultural Values and Historical Analysis', *New Orient*, I, Dec. 1960, pp. 19–22.

[79] A.L. Azmi, The Aligarh Scientific Society, *Proceedings of the Indian History Congress*, Varanasi, 1969, p. 418.

[80] Home, Education, no. 19, Nov. 1886.

[81] P. Hoodbhoy, pp. 55–6, note 74.

This Society did succeed in creating some awareness of the efficacy and benefits of science and reason. The orthodox *ulemas* and *moulavis* naturally felt unhappy. The Society members were called *Natury* (meaning those who study nature and believe in reason) and dubbed as heretics on this ground. The *mutawalli* (keeper) of the Holy Kaaba declared Sir Syed to be *wajib-i-qatl* (deserver of death). Sir Syed and his associates could not succeed against the traditionalists as they had neither a broad-based support nor was there any economic movement behind them as was the case in the West. Moreoever, the opponents of science were in a fortunate position. While the barrier of language prevented the protago-nists of science from taking their message to the people, the former could fully utilize the regional languages with all their emotive force.[82] But they did succeed in establishing off-shoots of the Society even in far-off towns like Muzaffarpur and Gaya.

Syed Imdad Ali, a sub-judge in Bihar, felt inspired and founded the Bihar Scientific Society at Muzaffarpur in 1868. This was not a branch of the Aligarh Society, but the two corresponded with each other and shared the same ideals. In 1867 the Aligarh Society submitted a memorial for the diffusion of European sciences through vernacular languages.[83] In 1868 the Bihar Society did the same thing.[84] Another memorial was sent to the government in July the same year. The result of all this was the GOI Resolution of 31 March 1870, accepting the demand. Meanwhile the Bihar Society engaged a translator on Rs 200 per month, but the emphasis here also remained on mathematical works. The same books, undertaken at Aligarh, were translated in Bihar again.[85] These efforts failed to invoke a positive response from the officials. In a representation to the Lieuten-ant-Governor, the secretary of the Society pointed out how in the selection of text books for schools, the publications of the society were unjustly rejected by the educational officers to make room for their own books.[86] The Society also founded schools at Muzaffarpur, Gaya, Sitamarhi

[82] A. Rahman, note 78.

[83] Home, Education, nos. 19–20, Sept. 1867.

[84] V.A. Narain, 'The Role of Bihar Scientific Association in the Spread of Western Education in Bihar', *Proc. of the Indian History Congress*, Varanasi, 1969, p. 421.

[85] V.A. Narain and B.K. Sinha in their articles say that works on astronomy, geography, natural philosophy, geology and botany were also translated, but they do not give the names of those books. Ibid., and B.K. Sinha, 'Syed Imdad Ali Khan— An Eminent Educationist of 19th Century Bihar', *Jr. of Historical Research*, XIII, no. 1, 15 Aug. 1970, p. 4.

[86] BSAP, Judicial Bundle no. 33, Letter no. 75, 14 June 1873, quoted in J.S. Jha,

and Chapra. It held periodical meetings in which lectures on scientific subjects were delivered and experiments in chemistry and physics were exhibited to the audience. The Society issued a bimonthly paper in Urdu called the *Akhbar-ul-Akhyar* in which articles on educational subjects were published. The trouble with these societies, however, was that they leaned heavily upon any one individual. At Aligarh the main figure was Sir Syed. In Bihar it was Imdad Ali whose retirement in 1875 greatly reduced the strength of the Society until it became defunct. They started with a bang only to end in a whimper. In 1870 the *Somprakash* compared them with the autumn clouds which thunder a lot but do not bring rain. (*Sharder Megh sadrasya arambhkale koto tarjan—garjan hoilo kintu bindupat matra hoilo na.*)[87]

But there was to be at least one exception. It was Mahendra Lal Sarkar's brainchild, the Indian Association for Cultivation of Science (IACS). Sarkar initiated the move in 1869. He criticized the Hindu ethos as 'a chaotic mass of crude and undigested and unfounded opinions on all subjects, enunciated and enforced in the most dogmatic way imaginable', and this, he thought, could be remedied only 'by the training which results from the investigation of natural phenomena'.[88] The universities were then not providing this type of training. He did not blame the students, but said: 'The true cause, why our educated youth have not hitherto turned to any substantial profit the knowledge they have acquired at school, lies in the fact of want of opportunity, want of means, want of encouragement, and not in a defective moral nature, nor in an easily spent precocity, nor in a badly developed physique, nor in inadequate food.'[89] Hence he wanted to establish an institution which would combine the character, scope and object of the Royal Institution in London and of the BAAS. Sarkar's scheme was fairly ambitious. It aimed at not only original investigations but at popularizing science as well.

We shall be able to institute two series of lectures on each subject, one general for the general public, and the other special for the instruction of few who would like to form themselves into a class to learn the subject. In this way we shall have in each section under the head worker, a few sub-workers as it were, who by virtue

---

'Origin and Development of Cultural Institutions in Bihar', *Jr. of Historical Research*, VIII, no. 1, 15 Aug. 1965, pp. 4–5.

[87] Benoy Ghosh, IV, Calcutta, 1966, p. 531, note 64.

[88] M.L. Sarkar, *On the Desirability of a National Institution for the Cultivation of the Sciences by the Natives of India*, Calcutta, 1869, pp. 3–6.

[89] M.L Sarkar, *The Indian Association for Cultivation of Science*, Calcutta, 1877, p. 29.

of the training they will receive, will soon become workers in science themselves, and will be of help to the institution as well as to the community in general. In this way a taste for science will soon be disseminated among the general community, and science will then count her votaries by thousands and hundreds of thousands. And then India of her own accord, unaided and unsolicited, will equip and send out scientific expeditions, as civilized governments under pressure are now doing. No part of the world requiring exploration will be without explorers from India. . . . Is this a dream? Yes, it is, but it is one of those dreams which can be willed into a reality. Give me money and I can show you that, though yet a dream, it can be made as much a reality as anything in nature.[90]

The press and the intellectuals in Calcutta welcomed the idea. But the real problem was that of funds; at least one lakh of rupees was needed and donations trickled in slowly. During the first six months of his efforts Sarkar was able to collect only Rs 13,500, and it took him six years to raise the required amount. There were other, perhaps more formidable, hurdles. The Lieutenant-Governor, Richard Temple, himself was inclined more towards technical education than original scientific research. This made the big zamindars reluctant to contribute to Sarkar's scheme liberally. Moreover, Sarkar had envisaged an autonomous native management. Though Temple was made the patron, no government aid was sought; and any aid, if acceptable, had to come without strings. This was unpalatable both to the high officials and their henchmen among the wealthy natives. A visit to Calcutta by the Prince of Wales in December 1875 provided the opportunity to float a few other schemes in the name of commemoration. The Lieutenant-Governor proposed a zoological garden and obviously had no dearth of funds. But technical education remained his major obsession, and sensing this, the zamindars of the India League hastened to call for the establishment of a big technical institute in Calcutta on the occasion of the royal visit. A subscription of one lakh and forty thousand rupees was raised within five days. This fund yielded an annual income of Rs 8000, and to this was added another Rs 8000 a year as government grant.[91] Temple had tilted the balance. For example, Laxmipat Singh (a zamindar) had donated Rs 40,000 for the restoration of BA classes at the Berhampur College. But on hearing that the Lieutenant-Governor intended to open a technical institute in Calcutta, he immediately shifted the money to the new fund.[92] Temple now tried

[90] Ibid., pp. 72–3.
[91] Minute by R. Temple, 18 April 1876, WBSA, General, Education, no. 3, April 1876.
[92] WBSA, General, Education, no. 1, Feb. 1876 and nos. B19–21, May 1878, file 27.

to forge a unity and amalgamate the two schemes. But Sarkar took a firm stance and got his IACS inaugurated formally on 15 January 1876. Temple still persisted and on 28 January called a joint meeting of the IACS and the India League. It ended in acrimony and failure. The unity talks could not have succeeded. The aims, objectives and *modus operandi* of the two schemes were entirely different. Sarkar wanted '*pure* science-learning and science-teaching, with reference to the practical applications of science so far only as are naturally and necessarily inferable while carrying on experimental investigations. The object is *not* to drill men in the arts which constitute the manual and the mechanical industries'. This was diametrically opposed to what Temple and the League had in mind. Rev. K.M. Banerjea, Chairman of the Indian League, accused Sarkar of 'soaring aloft, without looking beneath'. He wanted first 'to think of utilizing the discoveries already made before aspiring after fresh discoveries'. Another League activist, Sambhu Chandra Mukherjee, called the IACS 'unnecessary, a luxury, an anachronism and an anomaly—the scheme involved a waste. A small addition to the State grant to the State colleges, and a small grant to the Asiatic Society, might do the needful, and more efficiently'. Sarkar received a spirited defence from Father Lafont and Surendra Nath Banerjee. To quote Lafont: 'The other association wants to transform the Hindus into a number of mechanics, requiring for ever European supervision, whereas Dr Sarkar's object is to emancipate, in the long run, his countrymen from this humiliating bondage.'[93]

The association managed to survive the first onslaught. But it remained at the take-off stage for a long time. No endowed professorship could be created 'for want of adequate public support'. Almost every DPI Report laments that the IACS, though a good institution, was not receiving due public support, but it never refers to what the government proposed to do for its amelioration.[94] It was affiliated to the university only up to the FA level. Affiliation up to the graduation level came as late as at the end of 1900.[95] Nevertheless, at least a platform was there. An excellent laboratory and a reference library had been set up. The psychological fall-out was perhaps the most important. The Association symbolized the search for a distinct Indian identity in the world of science. Sarkar had undertaken an extremely difficult job. As a contemporary wrote: 'When

[93] M.L. Sarkar, Calcutta, 1877, pp. 106–23, note 89.
[94] *DPI Report*, Bengal, 1897–99, p. 38.
[95] WBSA, General, Education, nos. 47–9, Dec. 1900.

Dr Sarkar began his sowing in Bengal, the ground had lain fallow for many years and had almost become sterile . . . it was not unnatural that the learned Doctor's gospel of scientific research which offered no present tangible reward in the shape of service under government or wives with rich doweris, fell on unheeding ears.'[96] Sarkar himself grew a bit despondent as his age advanced, but his institution, taking all things into consideration, fared quite well during the first twenty-five years of its existence. He heralded a dawn of which J.C. Bose and P.C. Ray were to be the early rays.

Native Munificence

Most of the scientific societies, associations and institutions were the result of private initiative and could not have been sustained without public support and donations. The government would join only at a later phase. But the attitude and inclination of government officials very often decided the fate of a particular society or project. It has just been seen how willing the members of the Indian League were to dance to the Lieutenant-Governor's tune. Even a relatively small zamindar would offer a big donation if this helped him catch the attention of the people in power or fetch him a title like *Rai Bahadur, Khan Bahadur*, etc. In August 1856 the Commissioner of Patna, W. Taylor, drafted a scheme for an industrial school at Patna and, thanks to his official position, within six months had collected the fantastic sum of Rs 1,60,759. The trick was simple—to promise titles and bestow small favours.[97] Collecting money through fair and honest means was not easy. In September 1875 Richard Temple suggested the establishment of an agricultural college in Bihar, of course in commemoration of the Prince of Wales' visit. The zamindars of Bihar promptly promised to donate a large sum. But, interestingly enough, they did not want an agricultural college; instead they asked for a technical school. The donors had begun to assert themselves. After the initial collection they would enquire as to whether or not concrete steps had been taken. They would say—'show us a begining and we will pay up', while the officials would insist—'pay up and then we will show you a begining'.[98]

---

[96] *Indian Engineering*, XXX, 26 Oct. 1901, p. 262.

[97] Home, Public, no. 43, 23 April 1858, pt. B.

[98] S.C. Bayley, the Commissioner of Patna, reported: 'The subscribers are really hot upon this special thing of a technical or industrial school. I think I let them see that I was not, but they maintained strongly their wish to carry out the scheme.' BSAP, General, Education, nos. 11-15, May 1877, K.W.

This hide and seek game was characteristic of almost every project. The school was finally opened in March 1879, only to close down within a short period. But a fund of around two and a half lakh of rupees had already been raised, so in 1892 the school was reopened.[99]

The donors often made their donations conditional upon the grant of an equal amount as contribution from the government. This they did to ensure that the government, being the biggest beneficiary of Indian revenues, also made a contribution. Moreover, nothing moved without official patronage or at least support. In January 1864 Dadabhai Naoroji offered Rs 50,000 and along with others collected Rs 1,75000 for the establishment of a Canning Fellowship for higher studies in any branch of literature or science at the Bombay University.[100] A similar contribution was asked for from the government to which B.E. Frere, the Governor of Bombay, readily agreed. But Charles Wood, the then Secretary of State, snubbed the proposal; in the name of providing primary education to the common people, he did not want the state to pay for higher education. The proposal finally fell through as the Governor-General also felt that it would 'establish a precedent of a very inconvenient kind'. Such learning or research orientation would have certainly benefitted the already privileged sections of the society and enabled them to rival what was an exclusive European preserve. How could this be permitted?

Another major proposal for a big investment in scientific research came in 1898 from J.N. Tata, a leading industrialist of Bombay. The offer was sensational and attracted a great deal of attention at the turn of the century. Tata proposed to create a trust for the management of his income out of which Rs 30 lakhs was to be applied to the founding and maintenance of a research institute, and the balance, estimated to exceed Rs 30 lakhs, was to be held by the trustees for the benefit of Tata's descendants in perpetuity. He wanted the government to grant a similar amount to ensure the success and smooth functioning of the proposed institute. In one stroke Tata gained immense popularity, perhaps more than what M.L. Sarkar could get through his IACS movement. He symbolized the rising aspirations of the Indian bourgeoisie which had of late become conscious of the technological bearing of chemical and physical research.

This put the government in a dilemma. In public it appreciated Tata's generosity, but in private officials doubted both the necessity and result

[99] BSAP, General, Education, nos. 6–11, June 1892.
[100] MSA, Education, vol. I, 1864, file 12.

of so much expenditure on research in a colony. Moreover, Tata had lumped the proposal with his family fortunes and had asked for a legislation settling the balance of his property upon his descendants *in perpetuity.* He wanted his family to be saved from the application of the General Statute Law against perpetuities. This the government could not have conceded. The Home Secretary to GOI wrote: 'This would form a very inconvenient precedent. I believe that there are not a few of the *nouveaux riches* of India who would be liberal on these terms, but the Government of India cannot afford to purchase their "munificence" on these terms.'[101] The Secretary of State considered the proposal as 'nothing more or less than a bribe to the Government of India in order to obtain exemption from the general law as regards particular family settlement'. He added: 'I cannot see that political assistance in times of difficulty could be expected from men like Mr Tata, and in my opinion, such a departure from the principles of the general Statute Law as Mr Tata asks for could only be justified on the ground of political necessity.'[102] Sensing trouble, Tata thought it more prudent to withdraw the family clause. This took the wind out of the opposition's sails. Even Curzon confessed: 'This ready acquiscence in our views indicates a liberality and unselfishness on his (Tata's) part for which I was hardly prepared.'[103] But in the summer of 1900 Tata went to London and tried to woo the Secretary of State over the Viceroy's head. Later in a private letter to the governor of Bombay, Curzon ruefully recalled:

Having been repeatedly told by us that we could not connect the foundation of the Institute with the creation of private endowment for his family, he (Tata) went behind our back to the Secretary of State, who did not know what had passed here, and wrung from him a consent to the coupling of the two schemes. He then wrote to us and asked us how we proposed to carry out the Secretary of State's orders. The latter was as angry as we were, and he advised me to drop the whole thing. I was most reluctant to do this. . . .[104]

Both sides were suspicious of each other. Tata thought that the government was out to scuttle the move, while the latter considered the scheme too ambitious to be realized. Endowments from Tata were to yield £8000 (Rs 1,25,000) annually and an equal amount, or at least £6000 a year, was asked for from the government. Hamilton himself agreed

[101] Note by A.H.L. Fraser, 3 June 1899, Home, Education, no. 36, Dec. 1899.
[102] G. Hamilton to Curzon, 6 July 1899, Curzon Papers, IOL, Mss. Eur. F/111/158.
[103] Curzon to Hamilton, 16 Aug. 1899, ibid.
[104] Curzon to Northcote, 22 July 1901, ibid., F111/204.

that the sum was comparatively small and 'contrasts curiously with the £75,000 a year which the Prussian Government furnishes for the little University of Gottingen alone'.[105] Yet the Government of India, as usual, felt chary about such proposals. In 1864 when Naoroji had asked for a matching grant to his fund, the Governor-General had refused on the ground that it would establish 'a very inconvenient' precedent (see ref. 100). This time a member of the Viceroy's Council wrote: 'We are in no way bound to subsidize the Tata Scheme. The precedent is a dangerous one.'[106]

Meanwhile the Queen had died and enormous funds were being raised for memorial purposes. Tata wanted his scheme to be taken as an all-India memorial to the Queen. But to Curzon this proposal seemed 'to have no relation either to charity, or to suffering, or the Queen or to the 300 millions of India'.[107] Tata was no less determined. Enough pressure had been built in his favour.[108] The Indian press was almost unanimous in its support. A negative answer would have placed the entire blame on Curzon.[109] The government could hardly afford to let this happen, so Curzon now tried to clip the proposals and asked for the scheme to 'be reduced to more modest dimensions'. Finally he agreed to grant £2000 a year, but not without his usual impetuosity: 'Tata entirely owes it to me that he gets anything; and if he is not wise enough to accept it, I am ready to drop the whole thing tomorrow.'[110] Tata refused to lie low, even though this meant the non-realization of the scheme in his life-time. He died in May 1904, a partly disheartened man. He had nevertheless succeeded in creating a wider awareness and had even attracted notice in the British Parliament. In March 1903, a member of the House of Commons asked for laying before the Parliament the correspondence regarding the Tata Scheme. Afraid of further embarrassment, the Secretary of State refused to oblige to the request.[111]

[105] Hamilton to Curzon, 9 May 1901, ibid., F111/160.

[106] T. Raleigh to Curzon, 6 March 1901, ibid., F111/203.

[107] Curzon to Northcote, 19 Feb. 1901, ibid.

[108] A.M. Zaidi (ed.), *The Encyclopaedia of the Indian National Congress*, vol. IV, New Delhi, 1978, pp. 100, 415.

[109] Curzon knew this. Later is a private letter (12 July 1904) to Dorabjee Tata, he wrote: 'There seems to have been too much scheming to get the Government into an enterprise private in its origin and character and to produce a situation in which, if the project failed at any time the aspirations of its authors, it would be possible to say—it is all the fault of the Government.' Home, Education, nos. 94–106, Feb. 1905.

[110] Curzon to Northcote, 22 July 1901, Curzon Papers, IOL, Mss Eur. 111/204.

[111] Parliamentary Debates on India, 1903, p. 71, IOR, V/3/1606.

The deadlock continued and mutual suspicions ran high till Curzon's departure in 1905. Curzon could never be convinced of either the desirability or feasibility of the project. Tata's advisors, like Professor Ramsay and Padshah, were not liked by the officials; in fact they were held responsible for inflating Tata's ego and ambitions.[112] Tata himself was called 'a crafty old man'.[113] His initial atttempt to lump family fortunes with the proposed institute gave a lever to Curzon who promptly projected it as part of a merchant's innate selfishness. Another point of discord was the question of sending Parsi youths for study abroad at Tata's bequest. This did not figure in the original scheme submitted in September 1899, nor in the Clibborn-Masson Report of December 1901. This clause appeared in Ramsay's report and that too in a footnote 'added apparently by some other hand'.[114] Tata's son, Dorabjee, tried to clarify that 'they were always *meant* from the beginning'.[115] No doubt Tata had been sponsoring Parsee students since 1892 and his scheme itself to some extent emerged out of this experiment; yet the government could not have accepted a project which had an obligation designed to benefit only a particular community and had no necessary connection with the work of the institute itself. Curzon's motives can be questioned; in procedural matters, however, he stood on a safer wicket.

On the whole, Indians did come forward with donations, both large and small, to help the cause of education and, to some extent, science itself. In fact, from 1878, private aid to education in Bengal began to exceed government grants by a considerable margin.[116] Numerous instances can be cited of some *maharaja* giving money or a *diwan* instituting scholarships. But most of them carried conditions which limited their utility to a particular region, caste or creed. There were scholarships exclusively

[112] Curzon considered Ramsay himself as 'the main difficulty' in the realization of the scheme. The contempt for B.J. Padshah is clear from what H.H. Risley, the Home Secretary, wrote in a note (8.10.1904) examining the possible repercussions of Tata's death on the scheme: 'I am inclined to leave matters as they stand for the present. If we press Bombay, the result may be that Padshah will come up here (Simla) which does not seem at all desirable. He has been the evil genius of the scheme all along.' Home, Education, nos. 94–106, Feb. 1905.

[113] Curzon to Hamilton, 7 May 1903, F. 111/162, note 110.

[114] Curzon to Dorabjee Tata, 12 and 14 July 1904, Home, Education, nos. 94–106, Feb. 1905.

[115] Dorabjee Tata to Curzon, 13 July 1904, ibid. (emphasis as in original).

[116] In 1879–80 the private contributions amounted to 54 per cent of the total expenditure on educaton in Bengal and the following year it reached 56 per cent of the gross outlay. *Bengal Administration Report*, 1880–81, p. 53.

for Brahmans, Banias, Muslims, etc. Very few had the 'ethereal longing such as had consumed Dr Sarkar's soul'.[117]

Growing Demands

The people may have been slow in generating funds, but there was certainly no dearth of persons who would pin-point what had gone wrong with a particular scheme and then bring forth new proposals. In February 1876 Bholanath Das, an alumni of the Bengal Engineering College, submitted a note accounting for the the decline of his college.[118] Later, a person named Biharilal Ghosh submitted a memo on the necessity of establishing a polytechnic near Calcutta.[119] In 1886 V.R. Ghollay of the Poona College of Science gave suggestions on how to envigorate medical education in the mofussil towns. Another teacher of the same college, N.A.F. Moos, wanted men like Tait, Tyndal or Unwin to be induced to visit India for a short period.[120] This, he felt, would give a boost to science education. Numerous instances of this kind can be cited. The Indian bourgeoisie could no longer be ignored.

Even the so-called conservative farmers were amenable to change. The very first agricultural exhibition held at Alipore in January 1864 evoked excellent response, and the authorities reported: 'The purchases of machinery and of agricultural implements made at the Exhibition by the native gentlemen from the mofussil afford an undeniable proof that there is no want of disposition to give those modern appliances a fair practical trial.'[121] Once a process was proved to be advantageous the farmer was quick to adopt it.[122] The success of Ms. Thompson and Mylne's sugar-pressing mill is a case in point. It was sold by thousands, showing that Indians had no objection to machinery when it paid them to use it.[123] But large steam-power machines or implements were not wanted; only small hand or cattle-based implements found favour.[124] The main reason was the lack of funds to buy sophisticated machinery. The Behea sugar mill cost Rs 80, the winnower Rs 85, the Swedish plough for black cotton

---

[117] *Indian Engineering*, XXX, 26 Oct. 1901, p. 263.
[118] F.J.E. Spring, *Technical Education in Bengal*, Calcutta, 1886, p. 27–9.
[119] Home, Education, nos. 14–88, Oct. 1897, pt. B.
[120] *Reay Papers*, Ms. 254560(3), SOAS, London.
[121] Home, Public, nos. 32–5, 29 July 1864.
[122] *Calcutta Review*, vol. 81, no. 162, 1885, p. 428.
[123] E.C. Buck to the editor of *Implement and Machinery Review*, 14 June 1884; Revenue, Agriculture, Agriculture, nos. 4–21, Oct. 1884, file 84, K.W.
[124] E.C. Buck's Note on Agricultural Implements, 7.2.1883; Revenue, Agriculture, Agriculture, nos. 14–32, March 1883.

soil Rs 30–50, the double stilted Swedish plough for deep tillage Rs 35, the kaiser plough for light soil Rs 6, the wheat thresher Rs 42, and the jowar thresher Rs 85. But for their high price, these implements would have got a better response. In the white colonies the farmers had some capital but little experience; in India people had experience but no capital. Thiselton-Dyer regretted that India had no such men as the Earl of Leicester who, without any government assistance, had conducted experiments on a gigantic scale.[125] But was the extortionist zamindari system which the British had introduced in India geared to produce such men? The zamindars by and large had little interest in increasing the productiveness of land. On the contrary, under Section 30(b) of the Bengal Tenancy Act, the landlord could claim an increased rent on grounds of rising prices of food grains, and as the prices could be made to rise more easily by making the supply fall than by causing the demand to rise, the landlord could very easily decipher that his interest lay in seeing the supply of good grains fall by making the land produce less. The zamindars were practical people. The economics of their time taught them to behave more as tax-gatherers than as landlords. Within their own limitations, several landlords often responded favourably to various schemes, be it the acceptance of the Behea sugar mill, the well-boring scheme in NWP or education in agricultural schools. In 1882 Clibborn reported that the zamindars preferred to dig their own wells, did not want government assistance and even disliked taking *taccavi* loans for this purpose. He was subsequently proved incorrect, for when the NWP Government offered a cheap and effective well-boring equipment the response was tremendous.[126] In 1905, at the Poona College, nineteen students out of a total of twenty-nine belonged to the landowning classes.[127] And this response came in spite of the paucity of text books and the problems associated with the medium of instruction. Moreland rejected teaching in the vernacular, while Voelcker advocated it. This vaccilation prevented the percolation of agricultural education down to the lower classes. The British had all along been very particular about not disturbing the existing class relations. The upper classes were their natural allies.

The rising Indian bourgeoisie itself was not oblivious to the importance of science in agriculture. The Bengal Chamber of Commerce showed an interest in the re-organization of agricultural departments.[128] At Belgaum,

---

[125] *Jr. of the Society of Arts*, XXVIII, no. 2468, 9 March 1900, p. 341.
[126] NWP Revenue Proc., nos. 12–18, July 1885.
[127] Revenue, Agriculture, Agri., nos. 11–21, Jan. 1905.
[128] Revenue, Agriculture, Agri., no. 9, Aug. 1900, file 67.

Kirlosker was gaining fame for his chaff-cutters and other agricultural implements.[129] Tata had initiated agricultural experiments at Bangalore.[130] In 1903 the native luminaries of Bombay submitted a scheme for agricultural improvement in their presidency. They alluded to the US Department of Agiculture which comprised six divisions with hundreds of scientists working in each division and lamented that the Bombay Government employed not even one purely scientific worker. They called for the establishment of an agricultural institute with affiliated farms. But it would have initially cost Rs 5 lakhs, and this was sufficient to frighten the government. The plan died on paper. Its signatories were G.M. Gokhale, J.N. Tata, Ferozshah Mehta, W. Greaves, Marshall Reid, V.D. Thackersay, N.G. Ghorpade and Jamsetjee Jeejeebhoy. Making a polite refusal, Bombay's Revenue Secretary, R.A. Lamb, clarified: 'The Imperial Government has made substantial contributions to expenditure under medical, education, civil works and revenue settlements. . . . The Governor-in-Council therefore feels bound to restrict proposals for expenditure on agricultural improvement to the amount required to meet only immediate needs and to gain objects of which the advantage is clear.'[131]

It was easier for the government to shelve or refuse a proposal coming from an individual or a small group of persons. A big and organized forum could have exerted more pressure. The various landholders' associations were basically spineless and happy with merely toeing the official line. The scientific societies at Aligarh and Muzaffarpur did well, considering the constraints under which they worked. After their exit, M.L. Sarkar's movement did try to fill the void. A decade later the Indian National Congress (INC) was to provide the most organized and the widest possible forum to ventilate grievances and raise demands. The very third session in 1887 took up the question of technical education, and since then every subsequent session pondered over it. K.T. Telang and B.N. Seal pointed out how in the name of technical education the government was imparting merely lower forms of practical training.[132] They reeled out data in support of their criticism and looked to Germany and Japan as ideals. For D.E. Wacha, technical education signified higher education, especially of the sciences. Delivering the presidential address in 1901, he said:

---

[129] *Report on the Agricultural and Botanical Stations in the Bombay Presidency,* 1905–6, p. 50.

[130] Revenue, Agriculture, Agri., nos. 2–5, April 1905, file 33.

[131] Revenue, Agriculture, Agri., nos. 11–21, Jan. 1905.

[132] INC sessions of 1888 and 1892. A.M. Zaidi (ed.), vol. II, p. 163, note 108.

It is idle to talk of mere small industries in carpentry and brick-making and so forth. If there is to be an industrial revival of a practical characer which shall change the entire surface of this country, you will have first to lay the foundation of teaching in the Applied Sciences. You cannot have the cart before the horse. Higher education must precede industrial development.[133]

The Congress did not want the claims of higher education to be subordinated to those of primary education. It wanted the government to patronize both, equally and effectively. The inadequacies of the medical service also attracted its attention. The Indian Medical Service (IMS) came under severe criticism. In 1893 K.N. Bahadurji moved a resolution beseeching the government 'to raise a scientific medical profession in India by throwing open fields for medical and scientific work to the best talent available and indigenous talent in particular'.[134] Later N.R. Sarkar and M.N. Banerjee asked for the creation of a distinct civil medical department, separate from the military one, and 'recruited by open simultaneous competition in England and India'.[135] Discrimination against Indians engaged in scientific or technical pursuits found an echo in the Congress deliberations. Since 1878 Indians in the higher grades were getting only two-thirds of the pay of their English colleagues doing the same work. J.C. Bose and P.C. Ray suffered on this account. Bose had refused to accept a truncated salary for a long time. A.M. Bose, himself a Cambridge Wrangler, took the government to task for not only downgrading Indians but refusing even eminent British scientists like Kelvin, Lister, Ramsay, Roscoe and others who in 1897 had asked for the establishment of a central scientific laboratory in India for advanced teaching and research. It would have cost about Rs 6 lakhs, so the Government of India dutifully recorded its inability to 'entertain so costly a scheme'.[136] In 1900 N.G. Chandavarkar brought to the notice of the Congress the disparaging remarks made by the Surveyor-General, De Peree, about the native character, whereupon S.N. Banerjee moved a resolution regretting the practical exclusion of natives from higher appointments in telegraph, survey and other departments.[137] Congress was fast emerging as the most zealous vanguard of Indian interests. Whether it be education, agriculture or mining, Congress touched several

---

[133] Ibid., vol. IV, p. 251.
[134] Ninth session at Lahore in Dec. 1893. Ibid., vol. II, p. 406. This resolution was repeated at the 12th, 13th and 14th sessions in 1896, 1897 and 1898, respectively.
[135] Ibid., vol. III, p. 477, and vol. IV, pp. 144–6.
[136] 14th Session at Madras in 1898, ibid., vol. III, p. 243.
[137] 16th Session at Lahore in 1900, ibid., pp. 575–80.

problems in its wide sweep. The search for identity had found a sound and solid expression.

This search intensified during the *swadeshi* movement. Perhaps the movement itself was an outcome of this search. 'The ideas of 1905' symbolized the determination of the people in two fields: (1) the promotion of education along 'national lines and under national control' with special reference to the exact sciences and technology, and (2) the industrialization of the country and advancement of materialism.[138]

Pramatha Nath Bose, a noted geologist and science-enthusiast, was probably the first to talk about science-based industries and the need to remodel the university curriculum accordingly. In 1886 he published a pamphlet on technical and scientific education in Bengal which attracted very wide notice and is said to have inaugurated the movement for technical education in Bengal.[139] He asked for the introduction of science at the intermediate (FA) level. He found the B.Sc. course to be 'some sort of a compromise between the literary and the scientific courses' which, like many such compromises, had to a great extent proved a failure. He wanted science subjects to be taught with an eye to their application to industry. He was worried about the exploitation of Indian resources exclusively by western capital and western enterprises. So he called for some sort of higher technical education to enable Indians to also come up. This was the central idea of the Swadeshi spirit. An Indian Industries Association was formed in 1891. Its principal members were P.N. Bose himself and Trailokya Nath Mukherjee. They arranged a series of popular lectures and experimented with indigenous raw-materials but without much success.

A more ambitious scheme was launched in March 1904 with the formation of an Association for the Advancement of Scientific and Industrial Education of Indians. Its primary object was to raise funds for the purpose of enabling properly qualified students to visit Europe, America and Japan for studying science-based industries. Jogesh Chandra Ghosh was the man behind it, but he received donations from a large number of leading men in Bengal, including Europeans. Jogesh Chandra Ghosh was perhaps the only influential person who opposed the boycott of foreign goods. Naturally he could manage some official patronage.[140] In 1905, 16 students were sent, in 1906 the number rose to 40, and by 1907, 82 had been

---

[138] B.K. Sarkar, *Creative India*, Lahore, 1937, p. 625.

[139] P.N. Bose, *Essays and Lectures on the Industrial Development of India*, Calcutta, 1906, pp. 59–74.

[140] BSAP, General, Education, nos. 120–34, June 1906, K.W.

sent. An analysis of the subjects which these students took up abroad shows the industries which Indians then thought were capable of local development.[141]

Electrical Engineering	6	Sericulture	5
Applied Chemistry	5	Match-making	2
Mechanical Engineering	12	Pencil-making	6
Pharmacy	2	Soap Manufacture	2
Dyeing	4	Printing ink	
Leather Tanning	7	Manufacture	1
Agriculture	12	Button Manufacture	2
Ceramics	2	Tobacco Manufacture	1
Cotton Weaving	4	Rubber Manufacture	2
Hosiery	1	Dentistry	1
Watch-making	1	Enamelling	1
Horticulture	1	Umbrella- making	2

Meanwhile, another society called the Dawn Society, established in 1902 by Satishchandra Mukherjee, was thinking in terms of a 'national' education in literary, scientific and technical subjects, and in the wake of the political stimulus provided by the partition of Bengal, it was transformed into the National Council of Education in November 1905. It had the support of the cream of the intelligentsia and the most prominent Bengalis of the time. The Council was intended to be a National University, free of European control and aid, and aimed at a fusion between the best of the East and West. It formulated an ambitious scheme touching upon all aspects of education. But cracks soon began to appear. Influential persons like Tarakanth Palit, Neelratan Sarkar and P.N. Bose wanted the Council to confine itself only to scientific and technical education. This was not acceded to by the majority of council members, so on the very day the National Council of Education was officially registered, a second organization, its rival, was ushered into existence. This was the Society for the Promotion of Technical Education. The former set up the Bengal National College, while the latter founded the Bengal Technical Institute. Through this Institute, Benoy Sarkar called for 'mistrification' (i.e. the training of mechanics), 'factorification' and industrialization.[142] J.C. Bose, P.N. Bose, P.C. Ray and B.N. Seal served the latter while the former had the support of Rabindra Nath Thakur, Gurudas Banerjee, Ramendra

[141] J.G. Cumming, *Technical and Industrial Instruction in Bengal, 1888–1908*, Calcutta, 1908, p. 28.
[142] Sumit Sarkar, *Swadeshi Movement in Bengal*, New Delhi, 1973, p. 167.

Trivedi, Aurobindo Ghose, Benoy Sarkar and many others. The division was sharp and revealed the queer nature of Indian response. The officials watched it with unconcealed glee. As Curzon wrote: 'The government must pursue their way regardless of the Bengali chattering. So far the outcome of the agitators' efforts is to lay the foundation stone of a hall for more talk.'[143]

The activities of this era had two important features. One was that almost all the exponents of swadeshi looked to Japan as a major source of inspiration. Japan's emergence as a viable industrial power and its subsequent military victory over Russia in 1904–5 caught the imagination of another Asiatic, though enslaved, society. The lesson was that 'if the rice-eating Japs could do it why not the rice-eating Bengalis'. The success story of the Japanese was recounted at several sessions of the Indian National Congress and it again figured prominently at the first Industrial Conference.[144] For example, Puran Chand, a chemist trained in Tokyo, described how the universities 'organize the raw Japanese intellect into a tremendous force, with which the resources of the country, both of land and sea, are governed, investigated, developed and made into the national treasures'.[145]

Another characteristic was that sometimes they betrayed streaks of revivalism. The distant past comes in handy for the recovery of a lost self or the reassertion of one's identity. The search for moorings made P.N. Bose (a geologist) write *A History of Hindu Civilisation* (3 vols, Calcutta 1896); P.C. Ray (a chemist) wrote *History of Hindu Chemistry* (2 vols, Calcutta 1909), followed by Benoy Sarkar's *Hindu Achievements in Exact Science* (New York, 1918). The denunciation of Muslim rule and then the structure of the society itself was an important feature of this era. For P.N. Bose the decay of Hindu civilization began with the establishment of Muslim rule.[146] P.C. Ray blamed Manu and the brahmanas for relegating the *kalas* (arts and crafts) to the low castes.[147] Culturally, the wind was in favour of the pure and pristine glory of the *Vedas* and *Vedangas*. Some sort of mysticism had also crept in. J.C. Bose looked to *Samkhya*, and even gave Sanskrit names to the instruments he had fabricated. The instrument for recording the contractile response of the plant was christened *kunchangraph* (*kunchan* means contraction), and the appliance used for measuring suctional response was named *shoshungraph* (*shoshun* means suction).

[143] Home, Education, nos. 44–56, Nov. 1905.
[144] A.M. Zaidi, vol. IV, pp. 97, 159, note 108.
[145] *Proc. of the First Industrial Conference*, Benaras, 1905, p. 124.
[146] P.N. Bose, p. 139, note 139.
[147] P.C. Ray, *History of Hindu Chemistry*, vol. II, Calcutta, 1909, p. 195.

These terms, however, soon fell into disuse. A major work of Bose begins with a verse from the *Rigveda*: 'The real is one: wise men call it variously.'[148] Similarly, Ramendra Trivedi wanted science to explain the logic of ultimate creation. His discussion on Darwin ended with the Gita. In his posthumous publication, *Vichitra Jagat* (published in 1920), Trivedi talks of two worlds; one is *vyavaharik jagat* (visible world), the other is *pratibhasik jagat* (reflected according to the individual consciousness). The first world he relegates to the collective and demonstrable outcome of direct experience, while the other he extols. Then he talks about one more world, that of concepts (*dharana*) which can never be observed or measured.[149] The influential Brahmo leader, K.C. Sen, asked for the introduction of physical sciences on a large scale in all the schools and colleges, but added that 'no attempt should be made to exalt the physical sciences at the expense of speculative philosophy . . . no opportunity should be missed to lead the mind of the student from nature up to nature's God'. He talked about Natural Theology and wanted the professors to 'point out, in a devout spirit, the marks of Divine wisdom and mercy' while explaining the scientific facts.[150] *Tattvabodhini Patrika* invariably paid less attention to the physical sciences and more to moral and philosophical education.[151] Sustained efforts were made to counter the growing suspicion that the introduction of western science would lead to atheism or agnosticism. On 7 January 1891, M.L. Sarkar delivered a public lecture in the Town Hall of Calcutta, proclaiming that 'science leads to a firm belief in the Deity and a devout attitude of mind before the great First Cause'.[152] P.N. Bose was all for the industrial application of modern science but at the same time he did not want it to trespass certain cultural limits. Here was an intelligent conservative who wanted both the material benefits as well as the traditional values. And he was not alone. This was the spirit of his age.

Scientists Fight for Justice

Discrimination, it has been seen, was a major plank on which the British policy (perhaps the Raj itself) rested. But with the passage of time Indians were not likely to accept this without some amount of protest.

[148] J.C. Bose, *Response in the Living and Non-Living*, London, 1902, p. IV.
[149] B. Bhattacharya, pp. 230–50, note 60.
[150] G.C. Banerji (comp.), *K.C. Sen's Nine Letters on Educational Matters to Lord Northbrook in 1872*, Allahabad, 1936, pp. 37–42.
[151] Benoy Ghosh, vol. II, pp. 393–468, note 50.
[152] M.L. Sarkar, *Moral Influence of Physical Science*, Calcutta, 1892, p. 34.

The case of Bhola Nath Bose may be cited as an early example. He was one of the four native medical students, who, for the first time breaking through the trammels of caste, embarked for England in 1845 and was the first Indian to obtain the degree of M.D. from London University. The home government recommended his appointment as an assistant-surgeon in the army. But the Court of Directors took the usual stance that the time was not ripe to admit natives into covenanted services. The Government of India was advised to treat this Indian 'as if he belonged to the covenanted service'. Yet B.N. Bose found his position in every respect inferior, especially as regards pay, compared with that of European doctors of similar education and service. When petitioned, the Principal I.G. of the Medical Department refused to accept Bose's contention that the Court ever intended to assimilate his position with that of covenanted surgeons.[153] Instead his despatch was found to be very carefully worded, 'to employ those deserving young men in a manner suitable to their great expectations and advantage to the public service', thus steering clear of any covenanted-uncovenanted controversy. As a result of this petition Bose got only a raise in salary, not in grade.[154] Even the entry into a covenanted service (e.g. IMS) did not mean the end of all troubles. S.G. Chakraborty was the first Indian to top the list of successful candidates at the IMS examination in 1855. He later rose to the professorship of materia-medica and clinical medicine at the Calcutta Medical College. But in April 1867 the government, in its wisdom, decided to amalgamate this post with that of the medical store-keeper and abolished classes in clinical medicine. In view of these changes, the audit department refused to pass his bill for the pay of a 'full professorship'. Chakraborty protested, and the Lieutenant-Governor had to intervene to get his salary released.[155] The environment was certainly not conducive for higher studies, much less research.

This 'apartheid' in science made the Indians react strongly during the last quarter of the nineteenth century. P.N. Bose, J.C. Bose and P.C. Ray were the first rays of the new dawn. The ground had already been prepared by the works of Jambhekar, Sikdar, Derozians, Syed Ahmad, M.L. Sarkar and the native press. In the early 1880s P.N. Bose and K.D. Naegamvala emerged as a link between India's demand for scientific education and its fulfilment through the research works of J.C. Bose, P.C. Ray and their pupils.

[153] From B.N. Bose, Civil Surgeon, Faridpur, to the Lieutenant-Governor of Bengal, 16 July 1864, Home, Public, nos. 21–4, 10 Nov. 1864.

[154] IOR, V/6/293, Despatches to India, 25 Jan. 1865.

[155] WBSA, General, Medical, nos. 22–30, June 1867.

P.N. Bose proceeded to England on a Gilchrist Scholarship in 1874 and specialized in Geology at the University of London and the Royal School of Mines. There he took part in political meetings and often criticized the government. He would have preferred to stay on in England and pursue political work along with research. But this was not liked by the India Office which at last sought to get rid of him by giving him a graded job in the GSI. This was the first case in which the Secretary of State exercised his discretion in favour of an Indian.[156] E.C. Buck, Secretary to the Government of India, welcomed it, but Medlicott (Superintendent, GSI) was a little sceptical. 'He is a Bengali, and may be physically unfit for our work.'[157] But he later wrote:

The Bengali actually 'took me in' . . . I gave him to begin with an easy work at Nimar in which the leading features were already marked. There were, no doubt, numerous tell-tale blunders in the progress reports of his work, but his final descriptive account of the ground was so well set up that, with needful correction, I passed it for publication. There was indeed a suspiciously unnatural symmetry in his conclusions, but there was no disputing them without a re-examination of the ground which was impossible. The whole performance was undoubtedly clever, so I gave him the benefit of doubt. As an encouragement I even recommended him before the usual period for promotion to the 2nd grade. When he was afterwards moved to ground in which he had no outline to start with and the formations were new, his scientific helplessness became at once apparent.[158]

It is difficult to accept or reject Medlicott's analysis of P.N. Bose in toto, but the tenor of his writing does suggest some bias. The European surveyors used to earn extra money by working for private firms. Though there was a rule prohibiting survey officers from taking on private engagements, it was not enforced until 1895 when Bose's services were requisitioned by a Calcutta-based European firm. Though he was on leave at that time, the government did not allow him to work for the firm. The final shot came in 1903 when he was superceded by T. Holland as the Director of GSI, although Holland had been junior to him in service by about ten years. In 1876 Blanford had missed the superintendentship of the GSI narrowly on grounds of seniority (even though he had been recommended by Oldham himself).[159] The rule of seniority was then considered sacrosanct. But when the turn of an Indian came, his ten years'

[156] J.C. Bagal, *Pramatha Natha Bose*, Calcutta, 1955, pp. 32–3.
[157] Revenue, Agriculture, Surveys, nos. 44–7, May 1880, pt. B.
[158] Revenue, Agriculture, Surveys, no. 3, Aug. 1886, also Revenue, Agriculture, Surveys, nos. 4–5, Nov. 1886, file 97.
[159] Revenue, Agriculture, Surveys, no. 26, Dec. 1882, file 103.

seniority claim fell to the ground. P.N. Bose felt the injustice of being superceded so strongly that he retired from the service the same year.

There were a few other casualties also. A person named Omcharan Mukherjee applied to Medlicott for an assistantship in the GSI but without success.[160] Earlier two native assistants (appointed in 1874–75) had been dubbed as useless.[161] In July 1886, Man Mohan Lal was promised by the government an appointment as an assistant-superinten-dent in the GSI on completing his studies in England. Medlicott 'dutifully' recorded his 'disappointment'.[162]

Another glaring example is that of P.N. Datta, a Gilchrist scholar who, after a distinguished academic career at the universities of London and Edinburgh, was strongly recommended for the post of palaeontologist in the GSI by the noted British geologist, Professor Geikie. Medlicott fulminated, as expected. Though the Government of India did not totally agree with Medlicott, it advised the Secretary of State that it could be 'somewhat hazardous to appoint Mr Datta to so cardinal a post as that of palaeontologist, and therefore he should fill a less responsible position'.[163] Datta was made an assistant geologist and, being a native, was given two-thirds the salary received by others of his grade. This 'two-thirds' clause had found a place even in the 'revised service conditions of GSI' which had been issued in May 1893. By this time the clause had been dropped in other services (thanks to the fight put up by J.C. Bose and the subsequent recommendations of the Public Service Commission). The Secretary of State noticed it and asked the Government of India to give Datta the full salary of his office. This order caused a consternation among the senior officers who started finding alibies. The Governor-General finally admit-ted that this clause 'was inadvertently inserted in our office and escaped our observation'.[164]

Unlike Datta, P.N. Bose would often react sharply. But he remained basically a science-enthusiast, known more for his efforts to popularize science and for his educational activities than for geological research work. In contrast, K.D. Naegamvala, who started as a lecturer in experi-mental physics at the Elphinston College (Bombay), concentrated only on his field of interest (solar physics) and later grew confident enough to cross swords with even Kensington pundits. In 1882 he induced the chief

[160] Revenue, Agriculture, Surveys, nos. 25–33, Sept. 1880, K.W.
[161] Revenue, Agriculture, Surveys, nos. 43–5, April 1880, K.W.
[162] Revenue, Agriculture, Surveys, no. 3, Aug. 1886.
[163] Ibid., nos. 1–2.
[164] Revenue, Agriculture, Surveys, nos. 3–6, Aug. 1893, file 74.

of Bhavnagar to donate Rs 5000 for a spectroscopic laboratory. The Bombay Government gave an equal amount. For the purchase of instruments Naegamvala offered to go to England at his own expense. This gesture was appreciated and he was sent on state expense.[165]

Norman Lockyer (the Astronomer Royal) and his Committee on Solar Physics wanted certain observations to be continuously made in a tropical country like India which, unlike Europe, received plenty of sunshine. The data collected here was to be sent to London for analysis and theoretical formulations. Lockyer found in Naegamvala an intelligent and patient observer, and supported him. But the latter soon started claiming independence of action and the first crack in the relationship appeared over the possession of a particular telespectroscope. This instrument was constructed with Indian money by Colonel Strange of the Indian Surveys for use in India. In 1879 it was lent to Lockyer for his solar observations. In 1882 he got another and more powerful telespectroscope, but he still wanted to retain control over the Indian one. This was resented by Naegamvala.[166] Some correspondence took place between the India Office and the Bombay Government and Lockyer was eventually 'compelled' to hand over the instrument in 1888.

Another controversy arose as to the city in which the instrument was to be mounted. Naegamvala shifted to the Poona College of Science and wanted the instrument there as Poona offered better atmospheric conditions than Bombay. Lockyer advised that it be mounted at Dehra Dun and wanted the Indian observations to be sent to him regularly so that they could be correlated with his own, evidently to support his hypothesis on the constitution of the sun. Naegamvala was not ready to submit to this remote control style of functioning. He proposed to 'publish the observations periodically in the monthly notices of the Royal Astronomical Society and thus make them *immediately* available to *all* solar observers'. Quoting from the authority of another distinguished spectrocopist, Ft. Cortie, Naegamvala concluded that 'Mr Lockyer is not a safe guide to follow, and that the observations, which when they have accumulated for a sufficiently long period will be of considerable value and interest, ought to be carried out *independently* of South Kensington and should be published independently of Mr Lockyer or any other observer'.[167] The government was in a fix. It sought the opinion of the Meteorological Reporter, and he favoured Poona. Nevertheless, in 1892 the instrument

[165] Home, Education, nos. 40–4, June 1884.
[166] MSA, Education, vol. XII, 1885, file 166.
[167] MSA, Education, vol. XXVI, 1891, file 270. (emphasis as in original).

was shifted to Dehra Dun. Pressure from Lockyer could perhaps not be resisted and the Survey of India obviously got an upper hand. The Raja of Bhavnagar, however, again came to the rescue of Naegamvala and offered to procure a new spectroscope.[168] This ensured the continuation of his work. In 1896 he was sent to Norway to observe the solar eclipse,[169] and the next year, at his own instance, was relieved of teaching duties so that he could devote all his time to the observatory alone.[170] The study of the solar eclipse in 1898 in India was perhaps his last major undertaking. There was to be no addition to the staff or the instruments. After Dehra, Kodaikanal got official patronage and Poona was allowed to languish.[171]

Among the other important participants in India's search for identity were J.C. Bose and P.C. Ray. Both had to work against heavy odds. University life was then practically segregated into two distinct social camps. Bose took up the challenge and struggled to raise the 'professional standard and ideal above and beyond racial difference altogether'.[172] As Medlicott had opposed P.N. Bose, J.C. Bose's appointment in Presidency College was strongly objected to by its principal (Tawney) and the DPI of Bengal (A. Croft). As Bose recalled later: 'Sir Croft told me frankly that an Indian was temperamentally unfit to teach the exact method of modern science.'[173] But unlike Medlicott, Croft and Tawney somewhat relented later on, perhaps in view of the numerous honours received by Bose from Europe.

When Bose joined the service, an Indian professor's income, even in the Imperial Service, was two-thirds that of a European's. Bose found that even this two-thirds pay was to be further redued by one-half, since his appointment was only that of an officiating person. In other words, he was to get one-third of the pay normally attached to the office.[174] Bose resented

[168] MSA, Education, vol. XXVIII, 1892, file 105.

[169] Revenue, Agriculture, Meteorology, no. 1, Aug. 1896. pt. B.

[170] MSA, Education, vol. XXI, 1898, file 65.

[171] In Oct. 1902 the Royal Society Observatories Committee rather resolved 'that it is not desirable that any new course of work should be entered upon at Poona observatory pending the retirement of the Director'. Royal Society Observatories Committee Minute Book, p. 22, 30.

With Naegamvala's retirement in 1911, this observatory was abolished and the instruments were transferred to Kodaikanal. MSA, Education, vol. 51, 1911, and vol. 53, 1912, file 123.

[172] Patrick Geddes, *An Indian Pioneer of Science: the Life and Work of Sir J.C. Bose*, London, 1920, p. 35.

[173] *The Statesman*, 20 Jan. 1925.

[174] P. Geddes, p. 36, note 172, and P.C. Ray, *Autobiography of a Bengali Chemist*, Calcutta, 1958, p. 65.

this and, in protest, refused to accept a salary for three years. The rule which restricted the natives of India to two-thirds of the pay given to Europeans holding similar appointments, embodied a fixed principle of policy and finance from which the government seldom found any reason to depart. For example, in May 1897, Bose asked for the full pay of his grade, saying that he found it 'difficult to maintain himself on his small means', and that the consequent feelings of anxiety prevented him from concentrating fully on research.[175] On this petition the Finance Member of Viceroy's Council, J Westland, wrote the following note (7 September 1897):

He (Bose) is now drawing Rs 500 and it is simple nonsense on the part of a native gentleman in the service of the Government to talk under such circumstances of 'difficulty in maintaining himself on his small means'. I wonder what any of the universities in England would say to any of its staff who said, 'I am a distinguished man, and you must agree to give me, on that account, more than the allowance of my office'. I think Mr Bose has got his head a bit turned, and he can wait a bit for his distinctions and rewards.[176]

Curiously enough, the Home Member, U. Woodburn, took an entirely different stand:

The analogy of the English universities is not quite appropriate. A scientific lecturer in England would obtain private emoluments (if a man of distinction) which would make him independent of his proper salary. Mr Bose's distinction is not ordinary distinction, and as to the adequacy of his salary, I am personally aware that it has not been sufficient to meet the expenses of his experiments and tours.[177]

These two conflicting views made the Viceroy, Lord Elgin, attempt a via media by giving Bose, instead of parity in pay, a fixed scholarship of Rs 20,000 a year to be spent on scientific research.[178] This was in fact a sugar-coated consolatory gesture. Not only this, but till the Royal Society

[175] Home, Education, nos. 25–8, Nov. 1897.
[176] Ibid., p. 6.
[177] Ibid., p. 7.
[178] This was Elgin's reply to what the Secretary of State had earlier written to him: 'There is a strong feeling here that the Government should in some way mark its appreciation of Dr Bose's remarkable labours and researches in science. . . . To bring his salary up to the European standard would be an awkward precedent, but I think your ingenuity could suggest some other means of meeting his merits. I understand that he has been offered a high salary to remain and work in England. Any suggestion of the kind should emanate from India.' G. Hamilton to Elgin, 12 Nov. 1896, Elgin Papers, IOL, Mss. Eur. F. 84/14.

accorded Bose recognition, the college authorities and the government refused him any research facilities and considered his work as being purely private. Bose often felt so frustrated that he once even thought of abandoning research. Pedler always coaxed him to devote more time to teaching. But Tawney encouraged him: 'I do not like the idea of your giving up research altogether. Will not research fit in with lecture? Very few men in England achieve success as early as you have done. So you may keep your spirits up.'[179] However, Bose continued to suffer minor irritants. On 27 December 1897, Lord Rayleigh visited the Presidency College and Bose took him to the laboratory. The same day, F. Mour, the officiating principal, issued a show cause notice, asking Bose 'by what authority have you received outsiders into the laboratory, and if the visit was made by previous appointment with you, why did you not communicate Lord Rayleigh's intention and the hour arranged for his visit to me'.[180] It was Rabindranath Tagore who collected Rs 200,000 to construct a research laboratory for J.C. Bose so that he could continue his research work independently.

In 1896, in a meeting with the Lieutenant-Governor of Bengal, Bose enquired if the government could send him to England on a scientific deputation; his request was turned down. At that time the Education Board at Simla had issued a resolution expressing regret that India had never taken to scientific pursuits in spite of the efforts of the government. Bose quoted this resolution to show the contrast between profession and practice. The Lieutenant-Governor appeared irritated by such plain speaking, and changed the topic of conversation. Alfred Croft (DPI) who had by then become quite friendly with Bose restated the case and saw it through. Croft wrote about Bose: 'To help him is to promote the cause of science all over the world and this, I assume, falls properly within the functions of the Government.'[181] The only other native professor to have got this type of favour earlier was Naigamvala of Poona Observatory who had been sent to Europe in 1884 on a similar mission.[182]

Bose was a rebel in a limited sense, as was P.C. Ray, and both had to suffer. P.C. Ray, on his return from England in 1888 with a doctorate in Chemistry, had to hang around for a year and was finally offered a temporary assistant professorship. He spoke to Croft of the injustice done to

[179] C. Tawney to J.C. Bose, 12 Dec. 1895, J.C. Bose Papers, Mss. no. 41, J.C. Bose Trust, Calcutta.
[180] F. Mour to J.C. Bose, 27 Dec. 1897, ibid., Mss. no. 43.
[181] Home, Education, no. 28, Aug. 1896.
[182] Home, Education, nos. 12–13, Aug. 1884.

him. When Ray told him that if a British chemist of his qualifications had to be imported he would at once have been appointed by the Secretary of State to the Imperial Education Service with the grant of passage money, Croft almost flew into a temper and exclaimed: 'There are other walks in life open to you. Nobody compels you to accept this appointment.'[183] The temper which Croft showed was more assumed than real. A few years later he very strongly recommended Ray for promotion, but it did not find favour with the GOI.[184]

Jagdish Bose was unorthodox in another sense too. He was one of the first among modern scientists to engage in interdisciplinary research. His was a sensitive mind, capable of comprehending several problems simultaneously. On 5 March 1885, while still a student at Cambridge under Lord Rayleigh, he scribbled in his pocket-diary: 'I have been long thinking whether the vast solar energy that is wasted in the tropical regions, can in any way be utilized. Of course trees consume the solar energy. But is there no other way of directly utilizing the radiant energy of the Sun?'[185] He started as a physicist but his interest in electrical responses finally landed him in the realm of plant physiology. This was resented, particularly by Professor Sanderson of Oxford, the patriarch of this field. For this reason the paper which Bose had presented at the Royal Society on 6 June 1901 was shelved in the Society's archives. Bose sought extension of his deputation to vindicate his stand. The expert, whom the India Office consulted, refused, and it was at the personal intervention of the Secretary of State that his stay in England could be extended. Then he worked at the Royal Institution laboratory for a year and in February 1902 the Linnean Society accepted his findings with unanimous applause. Meanwhile, a physiologist who had seen Bose's experiments before the Royal Society, claimed precedence over them. A shocked Bose asked for an inquiry which finally upheld his right to absolute priority.[186] At home he had fought against administrative absurdities, abroad the academic

[183] P.C. Ray, p. 65, note 174.
[184] WBSA, General, Education, nos. 1–2, Aug. 1894.
[185] As the years passed Bose was probably much too pre-occupied to attempt this experiment. In 1972 an American scientist, Calvin, discovered that plant chlorophyll under the influence of the sun's rays can give up electrons to a semi-conductor, and he won the Nobel Prize. He calculated that a chlorophyll photo-element with an area of ten square meters could yield a kilowatt of power.
(For this information I am grateful to Dr S.D. Chatterjee, Retd. Professor of Physics, Jadavpur University. The diary is with him.)
[186] *The Modern Review*, Dec. 1915, p. 694.

challanges were no less formidable. Bose persisted and won. He tried to keep himself above politics and in the process was sometimes misunderstood by his own people. As his wife later wrote to a close friend: 'There is a feeling in India among the extremists that my husband is too fond of government patronage. You know that it is not true. He is too full of self-respect. . . . I do not care what the extremists think but I should like the nation to know he has struggled hard to keep his independence and how India has been made known to the world both by Tagore and my husband.'[187]

Identity Amidst Discrimination

As for British attitude towards the natives, one finds various shades surfacing at different levels of the official hierarchy. At the top, of course, were the loud proclamations emanating from governors and governor-generals. Addressing the annual convocation of Bombay University on 20 January 1880, Richard Temple, the Governor of Bombay, declared: 'Our object is to obtain for Natural and Physical science a larger place than heretofore in our educational system . . . this study in India has a special value. . . . It tends to correct some of the mental faults which are admitted to exist in the native mind.'[188] But theirs was, as P.C. Ray called it, 'a history of good intentions and large promises, but poor performances'.[189]

The other shade is that of the hardliners as represented by senior scientific officials like Medlicott. Their stubborn racial vanity is quite evident in what Colonel De Pree, Surveyor-General, told the Public Service Commission of 1891: 'It is suicidal for the Europeans to admit that natives can do any one thing better than themselves. They should claim to be superior in everything, and only allow a native to take a secondary or subordinate part. In my old parties I never permitted a native to touch a theodolite or an original computation, on the principle that the triangulation or scientific work was the prerogative of the highly paid Europeans.'[190]

The third distinctive feature was the ambivalent posture adopted by the

[187] Abala Bose to Patrick Geddes, 8 Jan. 1920, Geddes Papers, Mss. 10576, ff. 60–1, NLS, Edinburgh.

[188] *Jr. of the National Indian Association*, III, March 1880, p. 147.

[189] P.C. Ray, p. 63, note 174.

[190] This remark was not intended for public consumption. But somehow it leaked out and caused a good deal of furore. N.G. Chandavarkar referred to it while delivering a presidential address at the XVI session of the INC in Dec. 1900. A.M. Zaidi (ed.), vol. III, p. 580, note 108.

majority of the officials, mostly civilians, somewhere in between the above two propositions. There were men like E.C. Buck, A.O. Hume, A. Croft and Pedler. Among the scientific officials, T. Oldham (first Superintendent of the GSI) was just the opposite of Medlicott on the question of the natives' capabilities. That men like Oldham, however, were in a minority is clear from the fact that it was not their views, but those of men like Medlicott or De Pree, which shaped governmental policy and determined the course of action. The status quo was sought to be rationalized by several exhortations to the natives to soft-pedal their demands. They were advised to first concentrate upon all the lower and middle posts and only then to think of the higher ones. Campbell argued: 'I don't discourage higher aspirations, I only say, secure and possess all the ground below and thence go up higher. And after all, everyone cannot reach the very highest posts; those of the second degree are in themselves well worth securing.'[191]

The educated Indians were well aware of these crude, artificial distinctions based upon racialism. As early as in 1832 Jambhekar talked of 'the cruelty and injustice of prosecuting a whole nation from every station of trust, rank or employment, and affixing upon them the stigma of unworthiness which experience has shown to be undeserved'.[192] The Indians continued to make efforts which did take a new turn during J.C. Bose's protracted struggle against such a wide disparity in pay scales. J.C. Bose was allowed European scales of pay in 1903 only after gaining a reputation worldwide and a CIE.[193] During the 1890s the average remuneration worked out at 667 for each European and 36 for each Indian.[194] P.C. Ray had to remain content with the Provincial Service. Thus one finds men with the most distinguished attainments in the Provincial Service simply because they were Indians and men fresh from college in the Imperial Service, simply because they were Europeans.[195]

The facts also do not substantiate the basic charge that Indians were averse to change or innovation. Even the poor and illiterate peasants were found fairly intelligent and adaptive by several European observers (e.g. Voelcker). Similarly, in difficult cases, the *hakeems* and *vaidyas* would often consult allopathic practitioners.[196] While making a caste-wise

[191] George Campbell, 'The Employment of Natives of India in their own Country', *Jr. of National Indian Association*, no. 121, Jan. 1881, pp. 4–6.

[192] *The Bombay Durpan*, 13 April 1832.

[193] Home, Education, nos. 49–50, Oct. 1903.

[194] J.G. Godard, p. 267, note 12.

[195] G.K. Gokhale, 'The Employment of Indians in the Public Services', *Speeches*, Madras, 1911, p. 448.

[196] *DPI Report*, Bengal, 1865–66, p. 474. A decade later, Ravenshaw, the

analysis of the response in Bombay, Gumperz finds the proportion of Brahmanas to be fairly large in any type of education; 'they appear as strongly in technical and scientific education as in literary education'.[197] The vigorous efforts at translation by vernacular periodicals and scientific societies have already been mentioned in some detail. As Aukhoy Dutt used to say they had the common objective of 'Indianizing Western Science'. Initially the emphasis was on mathematics only, but later it diversified. For example, in 1890, from an obscure town like Arrah came a fairly comprehensive book on agriculture in Urdu, dealing with agricultural chemistry, plant physiology, soil analysis, European manure and implements, etc.[198] While making a centenary review of the scientific works of the Asiatic Society from 1784 to 1883, P.N. Bose could mention, out of 374 papers on mathematical and physical sciences, only 2 papers by two Indian authors and even in geological works, which were the most systematic under the then existing situation, only two Indians could be found who contributed 3 out of the 296 papers.[199] Once the Indians received some encouragement in geological education and operations during the 1890s, they started producing scientific papers.[200]

This is, however, not to give a clean chit to the Indians. The medical men in particular, who probably had the best possible scientific training, preferred minting money through private practice. In 1893 the Gaekwar of Baroda instituted a professorship of pharmacology at the Grant Medical College in Bombay and Dr Bahadurji was given the post on the

---

Commissioner of Orissa, reported: 'It is very gratifying to mark the advance which English medicine has made within the past ten years. At Puri, the dispensary was laid under a ban by the priests of the temple, and a threat of excommunication was held out to those who would seek to enter in: mark the great change at present, for not only do crowds of pilgrims flock to the place, and Brahmans themselves apply, and the Mahants or spiritual guides of the people have consented to become members of the charity and meet within its walls to further the usefulness of the institution.' WBSA, General, Education, nos. 1–4, Feb. 1875.

[197] E.M. Gumperz, *English Education and Social Change in Late 19th-Century Bombay*, (1858–98), Ph.D. thesis, University of California, 1965, p. 253.

[198] Syed Imdad Imam, *Keemyae Zeraet*, Arrah, 1890. Earlier he had written on horticulture, *Ketabul Asmar*, Bankipore, 1887.

[199] P.N. Bose, *Centenary Volume of the Asiatic Society*, pt. III, Calcutta, 1884, pp. 1–109.

[200] P.N. Bose contributed 16 papers; P.N. Datta published 5 articles on the lower Vindhya range and iron ores in the Central Provinces in the *Records of GSI*. N.G. Chetty wrote a manual of the Kurnool district in 1886. V.S.S. Iyer wrote a dozen

assumption that he would take up original investigations into the properties of indigenous drugs. He did not do this and instead devoted most of his time to general practice. Finally he was replaced by N.F. Surveyor.[201] From the distinguished rolls of the IMS (there were a few dozen Indians) none, except for K.R. Kirtikar and Baman Das Basu, showed any research aptitude.[202] Patience for specialization was hardly ever noticeable. This perhaps explains why there were so many LMS graduates (Licentiates) and such few MD students. Veterinary science, which was of great importance to an agricultural country like India, got the least attention. While asking for a raise in salary, N.H. Sukhia (a veterinary lecturer in Bombay) referred to his 'positive lowering in the estimation of the public as a mere horse or cattle doctor'.[203] Even in Calcutta which had the maximum exposure to metropolitan culture, the fate of science education and science popularization, much less research, fluctuated. Literary education and research continued to attract the average educated Indian. During 1856–95 an average 22 per cent of the total contribution to the literary part of the JASB regularly came from

---

papers on the geological works in Chitaldurg and Shimoga districts in the *Records of Mysore Geological Department.* S.C. Das contributed 4 articles on the characteristics of Tibet in the *JASB* during 1887–1902. A.K. Coomaraswamy presented around 25 papers in *The Geological Magazine, The Mining Journal, Ceylon Adm. Reports,* etc. on the geological structure and pecularities of Ceylon during 1900–5. See T.H.D. La Touche, *A Bibliography of Indian Geology,* Calcutta, 1917, pp. 102–4, 116–18, 196, 452–3.

R.B. Sanyal, Supdt. of the Calcutta Zoological Garden, produced some original observations in the field of zoology which were published in the *Proc. of the Zoological Society of London* during 1893–94. S.C. Mitra, 'Original Scientific Research in Bengal', *Calcutta Review,* vol. 103, 1896, pp. 360–4.

[201] Home, Medical, nos. 15–18, Dec. 1895. Dr Bahadurji was associated with the Indian National Congress and moved resolutions asking for the re-organization of the medical service, independence from army control, etc. in the INC sessions at Lahore and Poona in 1893 and 1895, respectively, A.M. Zaidi, vol. II, pp. 406, 380–1, note 108.

[202] For details, see D.G. Crawford, *Roll of the IMS,* Calcutta 1930.

[203] In the file this sentence was underlined by J.H. Steel, Principal of the Veterinary College, and in the margin he noted: 'It would seem better if Mr Sukhia's connection with a profession which he despises, cease as soon as possible.' MSA, Education vol. 16B, 1889. The situation was no better at the Bengal Veterinary College. But it was just the opposite at Lahore where the students usually came from the zamindar class, 'three-fourths being Muhammadans, thoroughly accustomed to handle animals from an early age'. Rev., Agri., Agriculture, nos. 1–8, Aug. 1906, p. 997.

Indians. But in the scientific section, only one paper was contributed during 1856–85 and during 1886–95 three Indians contributed fourteen papers,[204] which was some improvement, considering the constraints under which they laboured. Still, there were a few Indians who thought that something more could be done. *The Kayastha Samachar* cited the example of Chandrashekhar Sinha, who without any western education or western instruments, had made excellent astronomical observations, and in 1865 had composed his *Siddhanta Darpana* on a palm leaf, containing 2500 *slokas* of which 2284 were original and 216 quoted from the old Siddhantas. *Nature* (9 March 1899) hailed his labours as greater than those of Tycho, and the *Samachar* surmized that:

It is not laboratories and observatories that make the scientists, it is rather the scientist who makes them. We are aware of the extreme importance of scientific appliances. What we mean is that they cannot create the scientific spirit, without which they are entirely valueless. When our professors complain that they cannot do any original scientific work in India because there are few well-equipped laboratories here, we conclude that this is not the whole story.[205]

On the ideological plane itself, some sort of a crisis of identity had gripped the Indian mind. The lure of inching towards 'actual civilization' and the hope of producing 'a Newtonized Avicenna or a Copernicized Averroes'[206] present a curious mix of both self-criticism and a yearning for change. At a deeper psycho-social level, they indicate a crisis of identity—a crisis which had no precedence and no parallel. The realization was slow, unlike in China where it was quick and sharp. When the Christian missionaries introduced books on Western science there, the Chinese scholars objected to the word 'Western' and insisted on it being replaced by 'new'. Neither pre-colonial nor colonial India, argues Panikkar, was sensitive to this question.[207] Nevertheless, it was significant that the search had at least begun. It peeped into the distant past, looking for inspiration and sustenance. Even professional scientists like P.N. Bose and P.C. Ray found the urge irresistible. Ancient India was a safe haven, the later medieval part could have been counter-productive. Jagadish Bose val-

[204] *Annual Report of the Asiatic Society of Bengal,* A. Pedler's Presidential Address, 1896, pp. 12–13.

[205] *The Kayastha Samachar,* vol. I, no. 3, Sept. 1899, pp. 36–40.

[206] Maulavi Ubaidullah, *Essay on the possible influence of European learning on the Mahomedan mind in India,* Calcutta, 1877, p. 47.

[207] K.N. Panikkar, 'Cultural Trends in Pre-Colonial India: An Overview', *Studies in History,* II, 1980, p. 80.

iantly tried to demonstrate that the Indian ethos and the values of modern science were congruent and not poles apart. This was an extremely difficult task. Total colonization had certainly blunted the possibility of evolving perspectives rooted in indigenous intellectual and cultural heritage.

Interestingly enough, despite many tall claims about the so-called superiority of the non-tropical whites, none of the European professors sent to India annually came up to the standard of Bose or Ray. Still Bose was often asked why India could not produce a Davy or a Faraday. Original research could not flourish in the midst of discrimination. Moreover, how could modern science be woven into the fabric of Indian life, particularly when what the Raj prescribed for India was not science-as-avocation, but science-as-enterprise and, that too, under a tight metropolitan control!

CHAPTER 7

Reconstructing India

Science began in India with a contempt and repudiation of India's spiritual and cultural values. It has now ended in thinking with the spinal chord rather than with the brain, in ignoring and even ridiculing the surrounding civilization which is sick and diseased all round, and requires the balm of science to restore it to health, beauty and nobility.

Radha Kamal Mukherjee

Foreword, in K. Motwani, *Science and Society in India,*

Bombay, 1945. p. vii.

May I suggest, do not be too attracted by all the glamour of Western technology…it is wonderful, but we have in some ways industrialized too far and not made the world happier thereby. You have a chance of distilling the best out of the West and fitting it into the age-old civilization of the East. If you can improve husbandry and state of the villagers without going for too great a concentration of industry, you may in the end gain greater happiness. The key note should not be to copy and westernise, but with wisdom to fit the best of the new into the best of the old civilization.

A.C. Egerton

Transcript of All India Radio broadcast, 2 Oct.1948,

A.C. Egerton Papers, Royal Society, London

The turn of the twentieth century saw the apogee of the British Empire in India, at the same time seeds of decolonization also sprouted. The last decades of the Raj saw some flickers of 'constructive imperialism', but these came too late. By then nationalism had gathered strength, Indian leaders and the government raced to raise development issues and debate the role of science and technology therein. By 1945 many committees had been formed and reports published, and the push was on to make India a modern nation-state. At first glance, there seemed to be unity of purpose, but in reality this was not so. The British government had its own agenda and preferences while the national leaders scripted advices and raised demands. The contradictions and dilemmas of British India were not to evaporate quickly. The interlocutors had little choice but to work for both material benefits and traditional values. They wanted the best of both the worlds, and in the process strove for more autonomy and power.

Given this background, two major questions arose. First, how was the colonial government going to use scientific knowledge and techniques in the material development of the country? Second, how would the Indians articulate and press for demands for national reconstruction?

Indians craved for autonomy no doubt, but the British scientists working in Indian establishments also wanted more independence from metropolitan London control. Independence in addition, would give the Government of India authority to ensure more 'utility-oriented' work. With the creation of the Board of Scientific Advice (BSA) by governor-general Lord Curzon in 1902, two important shifts could be seen in the government's attitude. One was the idea that science in India could and should be cultivated without supervision from London; and the second was that the country's preference for 'natural history' sciences must be replaced by public sponsorship of industrial technology.[1] This was a sort of official version of the Swadeshism then raging in the country. But there was no question of any talk, much less cooperation, between the official promoters of scientific independence and the Swadeshi leaders who aimed at 'real' independence. BSA was a purely interdepartmental British affair, Indian scientists or leaders were not given any role in it. The goals of the two groups were similar, but their methods were different. Opinions remained sharply polarized in both camps: some emphasized the glory of 'pure' science, others (probably the majority) stressed the relevance of 'applied' science. The Swadeshi leaders were themselves divided.[2] Government officials were even more ambivalent.

There were some who worked for improvement. Alfred Chatterton, for example, wrote a number of pamphlets and books on the use of machines and small industries, etc. His book on lift-irrigation discussed water problems, water-lifts, cost of power, and even windmills.[3] G.A. Natesan & Co. of Madras published books on Indian agriculture, agricultural industries, art, industry, and education, etc.[4]

---

1 For details, see Roy Macleod, 'Scienctific Advice for British India: Imperial Perceptions and Administrative Goals, 1898-1923', *Modern Asian Studies*, 9, 3, 1975, pp. 343-84.

2 Dhruv Raina and S. Irfan Habib, *Domesticating Modern Science*, New Delhi, 2004, pp. 83-147.

3 A. Chatterton, *Lift-Irrigation* , Madras, 1912.

4 These were Glyn Barlow, *Industrial India*, S.R. Sayani, *Agricultural Industries*

Eventually, some hesitant steps were taken by the Indian government. In 1911, Indian Research Fund Association was created to foster medical research and public health. Three years later, Indian Science Congress Association was formed which in its very first meeting urged the government to recognize the paramount claims of science upon public funds. Finally, after almost thirty years of sustained demands from Indian leaders, the government agreed to recognize India's need to industrialize. The Indian Industrial Commission was formed in 1916 with Thomas Holland as the chairman. It is significant that Holland was also appointed president of the Indian Munition Board. For decades the colonizers had thrived on a plantation economy; the Great War convinced them of the need to foster tertiary industrialization that would provide war-related necessities.

The euphoria generated by the Industrial Commission was shortlived, and its recommendations were shelved by the British government. Its only tangible accomplishment was the creation of the Department of Industries. Also the debates which the commission provoked in India were instructive.[5] An important dissenting voice came from one of its members, Madan Mohan Malaviya, a leading nationalist. Like his cultural predecessors who had rejected rationality as a Western import, Malaviya presented a nationalist critique of British economic policies in India, and stressed that India had remained deindustrialized. The British model was inadequate. The new icons were Japan and Germany, and the new watchword was science-based technology.

In the first quarter of the twentieth century there was some optimism in the air. A publicist claimed, 'it is by no means an impossible task to make tropics healthy . . . the mosquito is no longer a nightmare, it can be got rid of'.[6] Rogers, a medical officer, felt, 'leprosy can be stamped out in the greater part of the British Empire probably within three decades'.[7] Another medical man argued,

---

in India, and E.B. Havell, *Essays on Indian Art, Industry and Education*; E. Batchelor, an ICS, wrote a tract on *The Improvement of Communication between England and India by the use of Aeroplanes*, Times Press, Bombay, 1917.

   [5] Shiv Visvanathan, *Organising for Science*, Delhi, 1985, pp. 39-96.

   [6] R.W. Boyce, *Mosquito or Man: The Conquest of the Tropical World*, London, 1910, pp. ix-x.

   [7] Leonard Rogers on the British Empire Leprosy Relief Association, 1924, Rogers Papers, PP/Rog C/13/175, Wellcome Institute for History of Medicine (WIHM), London.

Were all large villages linked up by cheap telephones with the Taluka headquarters it would doubtless be a help to general administration. Were the Taluka medical officer at the same time supplied with a motor cycle his stretch would be immensely increased and might in most cases cover all the villages in his taluka.[8]

This is an example of great foresight. But there were strong notes of despondency as well. In 1928, the Surgeon-General of Madras, J.D. Megaw reported, 'In Madras the economic problem is much more acute. The people multiply like rabbits and die like flies'.[9] What did the government think about it?

Health of the Nation

Many thought that control of diseases would promptly raise the standards of life, and would gradually induce a decline in fertility. Yet in the final analysis, the situation in India was always found too dismal to warrant such optimism. So the standard official argument was that public health does not depend only on the control of preventable diseases but also on the state of nutrition and general economic conditions. To quote Megaw again, 'a public health policy must be worked out in terms of agricultural and industrial production; it must look ahead and take into account the maintenance of a proper relation between the number of the people and the available supply of the necessities of life'.[10] This indeed was a tall order. The colonial government had neither the intention nor the wherewithal. The government responded to the growing demands for greater indigenization of the services, with a retrenchment policy.[11] Severe economy measures were further added during the Great Depression. Agriculture and public health were obviously the early casualties.

Mark Harrison rightly argues that the development of public health in India has to be seen 'in terms of a dynamic matrix of motives and sectional interests within and between European and Indian communities'.[12] But he finds the scope and effectiveness of colonial

---

[8] Note by I. Smith, Medical Adviser. Bombay dt. 13.7.27, IOR L/E/7/1150/615.

[9] J.D. Megaw to W.W. S. Carter, 29th Oct. 1928, Madras, IHD, 1.1.464, India B5 f.34, Rockefeller Archive Centre (RAC), New York.

[10] J.D. Megaw, 'Medicine and Public Health', in E. Blunt, ed., *Social Service in India*, London, 1938, p. 186.

[11] See Inchape Committee Report, 1923, which recommended a massive cut in administrative expenditure.

[12] Mark Harrison, *Public Health in British India*, Cambridge, 1994, p. 228.

medical intervention rather 'limited'. Widespread indigenous taboos and suspicions, and the reluctance of the Indian rentier class to pay, precluded the possibility of 'any vigorous programme of sanitary reform'.[13] Thus, the role of the state gets marginalized. Modern medicine entered India riding the colonial wave and, as a recent critic states it, colonial medicine did not mean altruism; it meant uncanny imperialism.[14] No wonder, it failed to make the transition from state medicine to public health.[15] Public health itself could never be defined in terms of 'human rights'; in colonial conditions, it was just the management of dangerous bodies. Notions about the 'dangers of contact' grew stronger with the broad acceptance of microbial pathology at the end of the nineteenth century. Local people were considered 'natural reservoirs' of germs. Medical theories that emerged thereupon exonerated the socio-economic conditions and the (ill) effects of imperialism, and instead blamed the victims.[16]

To ease its burden, the colonial government sought for private support. Within the Indian municipal system, it was not easy to raise finance for public health through higher taxation. In some places like Madras, the taxes were already high. In a study of the Calcutta Municipal Corporation it was found that the rentier class was extremely reluctant to pay higher taxes for public health purpose.[17] In certain areas, there was little or no surplus revenue. Under such circumstances, the health programmes of the Rockefeller Foundation (RF) came as divine intervention. It focused on the plantations in different parts of Asia and Latin America which could be made more lucrative through greater scientific input and control of diseases impairing labour productivity. In India the RFs involvement began with the visit of Dr V.G. Heiser to Madras in 1915. He saw the prevalence of hookworm disease and felt that an enormous amount of work could be done at a comparatively minor cost.[18] More work was

---

[13] Ibid., p. 234

[14] Anil Kumar, *Medicine and the Raj*, New Delhi, 1998, p. 218.

[15] David Arnold, *Colonizing the Body*, Delhi, 1993, p. 3.

[16] Warwick Anderson, 'Post-colonial Histories of Medicine', in F. Huisman and J.H. Warner, *Locating Medical History*, Baltimore, 2004, pp. 218-306. On the contrary, some scholars argue that there never was a colonial divide between colonizers and the colonized. Instead, within a generation of conquest, local collaborators had emerged who realized that their family's progress was possible through learning and imitating European ways. Sheldon Watts, *Epidemics and History*, London, 1997, p. 271.

[17] Mark Harrison, *op.cit.*, pp. 166-226.

[18] IHB, 5/2/Sp. Report, Box 49, f. 304, RAC.

done by Dr Henry Kendrick during 1921-34 who controlled hookworms by using travelling treatment units and latrine buildings. He studied the species of hookworm and the flow of groundwater in connection with board-hole latrines.[19] The most important component of this campaign was the emphasis on creating public awareness. This was done through lectures, pamphlets, and lantern shows. An exceptionally interesting mode was the use of the local story-telling method. One such pamphlet uses the popular *Baital* stories to create awareness. In one such story, the demon manifested itself in the form of millions of worms hooked to the intestines, creating sickness in the body, and the only solution lay not in religious rituals but in taking purgative medicines and adopting clean ablution methods.[20] This method proved exceptionally successful. The government soon realized that the loss in efficiency in this disease-stricken population, if expressed in terms of cash, must be enormous. Dr Clayton Lanes, who examined about 46,000 people in the tea gardens of Assam and the jails of Bengal, showed that India's entire war-debt could be paid by the wages lost by hookworm-ridden employees through sheer physical inability to work. He wrote, 'the disease is a menace and an obstacle to all that makes for civilization. As a handmaiden of poverty, a handicap of youth, an associate of crime and degeneracy, a destroyer of energy and vitality, it stands in the very forefront of diseases'.[21] The usual government excuse of 'want of funds' was considered lame. One thousand patients could be cured for a trifling sum of Rs 125, Dr Lanes argued.[22]

Public health remained limited to compilation of rather inaccurate vital statistics, vaccination, segregation, etc. Even these were so wired in government machinery and 'so hopelessly tied in bureaucratic knots' that as one foreign observer wrote, 'one doubts whether untangling can ever occur and the only hope is for a completely new system'.[23] Not even independence could meet this hope! So why blame the British? It would probably be as wrong as they blaming the poor Indians for all ills. Mahatma Gandhi was one of the few to provide self-criticism. 'Poverty is no bar to sanitation', he said, while acknowledging that, 'chronic poverty and chronic breach of the laws of sanitation are equally to blame for

[19] IHB, 6, 7, 1, 6, f.35, RAC.

[20] IHB, 5, 464, Box 49, f.306.

[21] GO No. 755, P.H. 1 June 1922, IHB, 5/3/Rep. Box 24, Hookworm Report, 1923.

[22] GO No. 755, P.H. 1 June 1922, IHB 5/3/Rep. Box 24, Hookworm Report, 1923.

[23] Diary of J.B. Grant , March 1940, R.G. 12.1 RAC.

diseases'.[24] As a recent critique notes, 'the solutions offered by Gandhi are deceptively simple in their statement, impeccable in their logic, yet frustratingly difficult to implement.'[25] Still Gandhi was persuasive. On the other hand, the British were tongue-tied, if not aloof or hostile. The real onslaught came from an American publicist, Ms Mayo. She 'meant well'[26] but lacked empathy and cut deep:

If the cellar is full of sewage, if its water supply is poisonous, if the residents are magazines of infection, and if their common habits or domestic relations scandalize me, I am but little affected by whatever virtues may offset these points, or whatever beauty the façade may show.

My belief is that the widespread, the less sparing, the most direct the criticism is, throughout the West, the better it will be for the people of India. They care what we think. They wince under our disapprobation. The hope is that power will be lashing enough to drive them to act.[27]

Possibilities for improvement were enormous. Examples and suggestions poured in, but where was the will? Pilgrims to Mecca from Egypt and Indonesia were compulsorily vaccinated against smallpox and cholera. This was considered impossible in India.[28] Rogers' suggestion to inoculate the pilgrims going to Prayag Kumbh was considered by the Pilgrim Committee as 'impracticable, inexpedient, and even dangerous'.[29] Enlightened native states such as Baroda, Mysore, and Travancore were more open to new ideas and fared better. Mysore was probably the first state in the world to open a birth-control clinic as early as 1930.[30] In 1935, the Society for the Study and Promotion of Family Hygiene was formed in Bombay. Later, the National Planning Committee of the Indian National Congress included separate studies on population and health in its programme.[31] Social workers and nationalist leaders realized the

[24] *Harijan*, 8 February 1947, quoted in Amit Misra, 'Public Health Issues and the Freedom Movement', in Deepak Kumar (ed.), *Disease and Medicine in India: A Historical Overview*, New Delhi, 2001, pp. 249–64.

[25] Amit Misra, op. cit.

[26] Ismail Mirza, *My Public Life: Recollections*, London, 1954, p.152.

[27] Miss Mayo to Miss Rathbone, n.d. (perhaps November 1927), Cornelia Sorabjee Papers, MSS. Eur. F.165/161, India Office Records, London.

[28] J.D. Graham, Health Commissioner to Government of India, to L. Rogers, 3 March 1927, Rogers Papers, PP/ROG. C10/69 , WIHM, London.

[29] Ibid., C. 10/3–4.

[30] Family Planning Association Papers, FPA, Box 419, A21/10–11, WIHM, London.

[31] J.N. Sinha, 'Science and the Indian National Congress', in Deepak Kumar, ed., *Science and Empire*, Delhi, 1991, pp. 161–81.

enormity of the task. Hence, in various religious and political congregations, publicity materials were distributed. In 1925, even a film was made on malaria by the Aurora Cinema Company of Bagbazar, Calcutta.[32]

Had the practitioners of indigenous medicine been co-opted, public health might have made more progress. After all, they remained popular even in the face of stiff competition and open hostility. Unfortunately, the modern medical men were too sure of their competence and superiority.[33] The Bhore Committee of 1944 reinforced their convictions[34] and continued to ridicule the other. As a professor of physiology at Lucknow wrote:

> The financing of Unani and Ayurvedic Institutes by Government in the hope of finding some soul of goodness in them is precisely on a par with the same Government financing archery clubs to find out the possibilities of the bow and arrow in modern warfare.[35]

Allopathy would have established itself even without government support. 'Can it be said that physics, chemistry and other sciences or the motor car, the cinema industry and the other fruits of applied science required political support to become popular here'?[36] It was difficult to oppose the West, even more difficult to ignore it. It was readily recognized that the cross-fertilization of age-old Indian civilization with post-Renaissance European culture produced 'stray, bitter-sweet fruits at first, but gradually gave rise to more satisfying and settled crop'.[37]

[32] This 5, 700-feet long film showed what a beautiful place Bengal was before the Europeans with their embankments, roads, etc. interfered with the natural flow of water which resulted in malaria. But if the Europeans could control malaria in other places, why not Bengalis? Boy Scouts were shown oiling puddles, cleaning tanks, trimming hedges, etc. Heiser Diaries, 1925–6, RG 1.2.1, RAC.

[33] 'The indigenous system of India has been a cause of disappointment to us not because they are successful rivals of modern medicine, but because we have been able to borrow or steal from them so little that is of real value.' J.W.D. Megaw, 'Confidential Note on the Working of the Punjab Medical Dept.', 6 Sept. 1928, IHD, 1.1, 464, India, Box 5, f.34, RAC.

[34] A prominent member of this committee was Sir Weldon Dalrymple Champneys (1892–1980). In a lighter vein he composed:

> My name is – er-Eustace; As you see I'am a fine specimen of the Master Race, you may think my conversation inane, But to manage these dirty Indians does not need a brain!

Champneys Papers, GC/139/H.2/10, WIHM, London.

[35] *Indian Medical Gazette*, 62, 1927, p. 223.

[36] *Srichitra Yugam* (Malayalam), 1941, pp. 47–8.

[37] *Mangalodayam*, VI, 1915, pp. 76–7.

Cash Crops to Food Crops

In 1930 Megaw referred to the education, agriculture, public health, and medical services as 'Nation Building Services'.[38] Earlier this was not the case. As a botanist had noted,

It was settled in my time that Director Agriculture was to be a man of business unacquainted with agriculture. He was to collect information and statistics to investigate markets and trade routes and complete reports on economic products but he was not to be a man who would tempt government to try agricultural experiments.[39]

We have already seen that the indigo lobby supported the creation of Pusa Agriculture Institute and the site chosen was based on its proximity with planters in north Bihar. But things were to change. Within a decade or two agricultural research was no longer conducted only along practical or empirical lines. Interdisciplinarity had emerged. Physics and chemistry pitched in terms of chemical and physiological problems while mathematics helped in the statistical examination of approximate results.[40] There did occur a change in cognitive outlook, greater accuracy, and a breaking down of arbitrary lines of demarcation which earlier existed between the mathematical sciences and biology.[41] The focus now shifted to soil-science, physiology, and genetics. Soil-science which was so far purely chemical, now involved organic chemistry, physical chemistry, bacteriology, physics, etc.

It is, however, doubtful whether these changes were noticed or nursed by the government or the public opinion. An English weekly from Poona, *The Mahratta*, wrote, 'The Pusa Institute is in the mountains, and the essays and the diagrams in the Agricultural Journal are mere hieroglyphics to the cultivators.'[42] Forecasting a 'globalisation' of food concerns and security, the *Daily Hitavadi* argued:

[38] J.D. Megaw, Note on the Formation of a Public Health Board, 8 Aug.1930, IHD, 1.1, 464A, Box5, f.35, RAC

[39] C.B. Clarke to I.H. Burkill, 26 March 1903, Letters to Burkill, Kew Garden Archives

[40] See for example, P.C. Mahalanobis, 'The Probable Error in Field Experiments in Agriculture, *Agricultural Journal of India*, XX, 1925, pp. 96–116.

[41] 'Modern Agricultural Research', *Indian Scientific Agriculturist*, II.12, Dec.1921, pp. 1–2.

[42] *The Mahratta*, 1 Dec. 1907, *Native Newspapers Report* (*NNR*), Bombay Dec. 1907, p. 1722; also *The Belgaum Samachar*, 16 Sept. 1907, *NNR*, Bombay, Sept. 1907, p. 1315.

India's economic future rests not upon industries so much as upon agriculture. Inspite of any amount of Swadeshi and boycott, Indians can never seriously challenge the industrial supremacy of Germany, England, etc. with their immense resources in capital, trained skill, etc. A successful competition in the industrial sphere being thus out of question, it is in the matter of raw produce, therefore, that India should look for her economic supremacy. [43]

The question which now arises is how to excel in raw production, how to achieve this? According to *Deshabhimani*, 'The best means of achieving success in these matters will be to afford facilities for individual private enterprise'. If these succeed then they will be imitated by others.'[44] For long, the government had been organizing agricultural exhibitions and fairs to inspire the cultivators. In 1926 a huge agricultural exhibition was held at Poona. But most of the exhibits were hardly suitable for Indian conditions.[45] The efficacy of such exhibitions and the diffusion of scientific methods of cultivation actually depended on the financial resources of the agriculturists. There existed (and still exists) a vicious circle between poverty and low production. The government and the local landlords were interested more in surplus from the land than in investment. The emerging middle class had little or no interest. As an influential doctor and educationist, Nilratan Sarkar, regretted, 'The *bhadraloks* of Bengal should always remain *bhadraloks*. They should not have anything to do with ploughshares.'[46]

During the inter-war period agricultural experts did ask for improvement of the plant by means of selection or hybridization, control of water supply, and improvement in the physical conditions of the soil.[47] There was a possibility of establishing new agricultural institutions, for example, a central entomological research institute.[48] A Royal Commission on Agriculture was constituted in 1926 which recommended the establishment of an Imperial Council of Agricultural Research (ICAR). This Council was to promote and coordinate agricultural and veterinary

[43] *Daily Hitavadi*, 9 July 1910, *NNR*, Bengal, no. 29 of 1910, pp. 840–1.

[44] *Deshabhimani*, Cuddalore, 15 July 1911, *NNR*, Madras, no. 30 of 1911, pp. 1084–5.

[45] 'Such shows held at an emormous cost coming ultimately out of the tattered pockets of the ryots are much like putting the cart before horse' *Kesari*, 26 Oct. 1926, *NNR*, Bombay, no. 44, 1926, p. 1201.

[46] Minutes of the Senate of Calcutta University, No. 10, 1938, p. 7.

[47] H. Martin Leake, *The Foundations of Indian Agriculture*, Cambridge, 1923, pp. 186–91.

[48] Revenue, & Agriculture, Agriculture, Proc. 22–23, Aug. 1922, pl. A, NAI.

research throughout India. The model selected was the Australian Council for Scientific and Industrial Research and not the Indian Research Fund Association (IRFA) which looked after medical researches.[49] Unlike medical research which was concentrated at very few centres, there were many agricultural research stations and there was lack of cooperation between different provincial agricultural departments. ICAR was supposed to rectify it. Unfortunately the Great Depression had begun and like others agricultural plans went into a limbo. It was under the stress of the Second World War that emphasis shifted from the cash crops to food crops. Bad harvests as well as the war exigencies forced the government to launch a 'Grow More Food' campaign.[50] The concept of planning was now taken seriously and as a government official noted:

Planning will be a mere shell if we do not put first thing first in post-war reconstruction. Food and clothing are the first material needs of every man, woman and child. Both come from the soil. . . . Nation building on the foundation of an ignorant and debilated people has little hope of fully succeeding. . . . The Indian farmer needs first to be educated to appreciate the advantages of the contemplated changes in his way of life arising out of the plans.[51]

Industrialize or Perish

An interesting characteristic of the period was the cautious yet firm approach towards industrialization. They believed that in industrialization lay salvation, but it was also considered necessary to avoid the pitfalls of blind imitation and crude industrialization. Efforts were to be made not

[49] Education, Health & Land, Agriculture, Proc. Nos. 1–3, March 1929, pt. A, NAI.

[50] S.M. Mishra, 'Agriculture, Environment and National Reconstruction: A Study of the Debates in the Central Legislature of India: 1937-1947, *Proceeding of the Indian History Congress*, Bareilly Session, Dec. 2004.

[51] Roger Thomas, Adviser to Government of Sind on Agriculture and Post-War Reconstruction, *Notes on the Planning for Agriculture*, Simla, 1944.

He printed this Note privately because his views did not find favour with the ICAR members. Of one such experience he writes, 'I had a stern fight at the July meeting of the Indian Central Cotton Committee over cotton floor prices... I pleaded for the man behind the plough. My greatest satisfaction was to be told by a rabid nationalist member that he knew no Indian who could have better pleaded the cause of the unprivileged and added that if there were more Europeans of my way of thinking in India there would be no need for the Quit India slogan!

Letter dt. 6.8.1944, Roger Thomas: Papers, IOR MSS. Eur. F/235/48.

to lose human, nay Indian face. The colonizers discussed about moral regeneration for a long time. This, the nationalists viewed as propagandist in nature. Instead, they dwelt upon a 'synthetical' economic and industrial regeneration. This regeneration was not to be achieved at the cost of peasants and artisans. Whether it was the *Dawn Society Magazine* of Calcutta or the *Kayastha Samachar* (later the *Hindustan Review of Allahabad*), or the *Swadeshmitran* of Madras, the tenor was the same—industrialization was in the national interest and should be conducted on national terms. Benoy Kumar Sarkar, an important interlocutor of the period, used interesting terms like 'mistrification' and 'factorification'[52] (*mistri* refers to technicians). The importance of artisans and technicians was thus brought into focus. The demand for chemical industries was ably advocated and pushed by scientists like P.C. Ray. All this had been preceded by a vociferous demand for techno-scientific education. There was to be no diminution in that. Rather the new argument was that science should be taught in a scientific way and not by the literary method.[53] The overall picture that emerges is of an all-embracing 'socio-cultural transformation'. *Suraj* (good rule, which many genuinely believed the British provided) was to be replaced by *Swaraj* (self-rule) which, coupled with *Swadeshi* (self-reliance), constituted a weltanschauung powerful enough to transcend the barrier imposed by colonial rule. The process of decolonization was thus much more than a political process. As Habib puts it, it contained within it the development of an alternative developmental philosophy.[54]

This industrialization route was not linear. There were so many disjunctions. The government explored some industrial possibilities. In 1905, for example, a steel plant in Kashmir to meet the railway requirements was considered. The cost was found to be 50 per cent more than in England and so was promptly abandoned.[55] Due to the increasing nationalist demand for greater Indianization of the different services and also in view of the incessant bickering among the various departments,

[52] B.K. Sarkar, *Education for Industrialisation*, Calcutta, 1946, p. 3.

[53] R.D. Patel, *The Claims of Science in National Life*, Surat, 1921, p. ix.

[54] S. Irfan Habib, 'Science, Technical Education and Industrialisation: Contours of a Bhadralok Debate, 1890-1915', in Roy Macleod and Deepak Kumar, (ed.), *Technology and the Raj*, Delhi, 1995, pp. 235–49.

[55] Letter from J.W. Haibard (Royal Indian Engineering College, Coopers Hill) to Maj. A. Hamilton, 21 Nov. 1905, L.W. Dane Papers, MSS. Eur. D.1103, box 1, IOR.

Thomas Holland planned a reorganization of the scientific services in India. This was opposed by many who preferred to enjoy more autonomy under provincial or local administrations. Even a memorandum was submitted and Holland had to withdraw.[56] The demand for Indianization had irked many Indian Medical Services officials as well. D.G. Megaw (DG, IMS) tried his best to stem the tide but it was not possible. The Acts of 1919 and 1935 had conceded to the Indians considerable share in administration. Yet the colonial juggernaut continued to roll, albeit haltingly. From the Indian side, there was not let up. Pressure continued which not only harped on the government's reluctance but also threw open internal inconsistencies. Where to go, how to go; and to this debate every thinking Indian contributed. The notions of science and its terminologies entered so deep in the political and cultural lexicon of the country that no politician or social reformer could afford to ignore them. Writing on nationalism, Rabindranath Tagore (perhaps the most influential literary figure of his times) called the government 'an applied science . . . it is like a hydraulic press, whose pressure is impersonal, and on that account completely effective'. And 'power' appeared to him as 'a scientific product made in the political laboratory of the Nation'.[57] Criticizing the different royal commissions and other government efforts, an economist from Benaras wrote 'mechanically':

It seems that these engines were not complete units at all; they were only parts of a machine-unit, and what is more unfortunate, they were indifferently constructed by different batches of designers and craftsmen who were, moreover, never given a chance to cooperate. The products of this motley crowd of workmen were clumsy in shape or form, and, designed as they were under widely different conditions and at different times, were turned out more or less according to the whims and fancies of a particular batch of designers and not according to the requirements of other component parts of the machine. The result was that, try as our rulers might, these parts could never be fitted together to form a complete machine-unit. [58]

What then was the alternative?

[56] R.S. Hole (FRI, Dehradun) to D. Prain (Kew), dt. 10.1.1920, Kew Misc. Reports, Indian Advisory Committee, 1898–1920, Royal Botanic Garden, Kew.

[57] Rabindranath Tagore, *Nationalism*, London, 1921, pp. 11–17.

[58] H.R. Soni, *Indian Industry and its Problems*, London, 1933, p. v.

Extraordinary Dissent: Gandhi

Amidst the growing demands for self-rule, democracy, industrialization, and development, Gandhi emerged as an extraordinary dissenter. Gandhi condemned the West for precisely those virtues in which it took pride—modernization and industrialization. Gandhi seldom used the terms 'science' or 'technology'. His concern was with civilization and mechanization, and on these topics he talked and wrote profusely. He considered machinery 'the chief symbol of modern civilization'; 'It represents a great sin', he wrote. 'It is machinery (and Manchester) that has improverished India. . . . I can not recall a single good point in connection with machinery. Books can be written to demonstrate its evils.' [59] Yet, many times Gandhi would say he was not opposed to machinery *per se*:

How can I be when I know that even this body is a most delicate piece of machinery? The spinning wheel itself is a machine. What I object to is the craze for machinery...today machinery helps a few to ride on the backs of millions. The impetus behind it all is not philanthropy to save labours but greed. It is against this constitution of things that I am fighting with all my might.' [60]

So machinery as an ally of capitalism was Gandhi's anathema. How about machinery under state control as in a Marxist system? Here also he could foresee state emerging as capitalist and equally oppressive. Both means and ends had to be 'pure' for him. The supreme consideration for him is man. The machine should not atrophy the limbs of man. But he would make some 'intelligent exceptions', for example, the Singer sewing machine. [61] He would not mind villagers plying their implements and tools with the help of electricity. But then the village communities or the state would own power houses, just as they have their grazing pastures. [62] Later, in 1946, he even agreed that 'some key industries are necessary'. He would not enumerate them but would prefer it to be under state ownership if the state professed non-violence. [63] Gandhi was worried less

---

[59] A.J. Parel (ed.), *Gandhi: Hind Swaraj and other Writings*, Cambridge, 1997, pp. 109–10.

[60] *Young India*, Nov. 13, 1924, pp. 377–8.

[61] Gandhi noticed 'a romance' in the birth of this machine. Singer saw his wife labouring over the tedious process of sewing and seaming with her own hands, and simply out of his love for her he devised the sewing machine, in order to save her from unnecessary labour. Here motive was love, not greed. Ibid.

[62] *Harijan*, 22 June 1935.

[63] *Harijan*, 1 Sept. 1946.

about production and more about distribution. He was not for mass production but production by the masses and that too with instruments that the masses could afford to possess and repair.[64] A simple cottage machine like charkha, the spinning-wheel solved the problem of distribution. There would be no concentration of wealth and power. As Gandhi once beautifully explained:

No amount of human ingenuity can manage to distribute water over the whole land as a shower of rain can. No irrigation department, no rules of precedent, no inspection and no watercess. Everything is done with an ease and gentleness that by their very perfection evade notice. The spinning-wheel, too, has got the same power of distributing work and wealth in millions of houses in the simplest way imaginable.[65]

Gandhi was equally worried about the growing number of idle hands. Greater use of heavy machinery would render more and more people idle. Charkha was Gandhi's prescription for the idle moments. A simple yet very potent device, it made sound economics as well. P.C. Ray, a great chemist, educationist, and entrepreneur (1861–1914), wrote 'Economics of Charkha' and citing from Buchanan's survey of 1800, he explained how the rural women generated extra income through spinning and weaving in spare time.[66] As an activist during the famine relief work at Khulna (1921), Ray found that the impact of crop failure would have been much less severe if the *ryot* (peasant) had a supplementary means of livelihood to fall back upon. When charkha and cotton were given to them, they looked upon it as a godsend. A little different experiment was made during floods in Rajshahi district. Some thousand acres of land had remained untilled and the soil needed to be loosened before monsoon but the cattle were lost because of the floods. So in consultation with P.C. Ray, it was decided to use motor tractor. This machine with only two persons did in one day as much ploughing as fifty yoke of oxen, driven by fifty cultivators. Now came up the question of tractor ploughing under normal circumstances. It was felt that to introduce the tractor permanently might mean an uprooting of village population and it might involve driving them into the towns. Ray and his fellow-workers could not face the thought of 'these same country people' being driven to the town while the tractors displaced their labour. [67]

[64] *Harijan*, 2 Nov. 1934, pp. 301–2.
[65] *Young India*, 27 Dec. 1923.
[66] *Young India*, 3 Aug. 1922.
[67] *Young India*, 20 Nov. 1924.

Several eminent contemporaries, however, remained skeptical about the real value of Gandhian economics and his charkha. Rabindranath Tagore, the laureate whom Gandhi adored as Gurudev, was not very convinced about the efficacy of the method of boycott and non-cooperation. In a brilliant essay published in *Modern Review*, he hinted that charkha was being accepted more out of blind faith in Gandhi rather than reasoned necessity. Gandhi published a strong rejoinder in *Young India*. In the 'The Great Sentinel', he first thanked the poet for his call to be rational and not to follow an idea of a leader blindly. Then he clarified that he had no blind followers and that his arguments were rational. He reiterated his conviction thus:

India is a house on fire, because its manhood is being daily scorched, it is dying of hunger because it has no work to buy food with. Khulna is starving not because the people cannot work. Our non-cooperation is neither with the English nor with the West. . . . Our non-cooperation is with the system the English have established, with the material civilization and its attendant greed and exploitation of the weak. . . . Indian nationalism is not exclusive, nor aggressive, nor destructive.[68]

In reply to Tagore, Gandhi talked of developing around the charkha a programme of anti-malaria campaign, improved sanitation, conservation, and breeding of cattle. His critique was thus no mere denunciation of the West. Rather it involved the development of a whole network of what Gandhi called 'ameliorative activity'.[69] Tagore's difference of opinion with Gandhi was, as a recent work argues, probably a matter of emphasis. Tagore too ended as a critic of the modern West.[70] Both registered their protest against the dominant civilization in their own distinctive ways. This was the culmination of the cautious approach which the late nineteenth-century cultural nationalists of India had adopted towards the Western civilization and industrialization. They were aware of the benefits of the latter but wanted these in national interest and on national terms. Indian values have been a constant refrain right from the writings of Mahendra Lal Sircar to Rabindranath Tagore.

One of the very few occasions on which Gandhi touched the issue of science and people directly was while he was talking to some students at Trivandrum in 1925 and when he visited the Indian Institute of Science

[68] *Young India,* 18 Oct. 1921.
[69] Gandhians find in it a concept of 'technology practice'. See Sunil Sahashrabuddhe, 'Gandhi and the Challenge of Modern Science', *Gandhi Marg,* V, Sept. 1983, pp. 330–7.
[70] Ashis Nandy, *The Illegitimacy of Nationalism,* Delhi, 1994, pp. 2–3.

at Bangalore in 1927. He declared that he was not an opponent or a foe of science. He recalled how he once wanted to become a doctor but could not stand the idea of vivisection and retreated. He wanted to put certain limitations 'upon scientific research and upon the uses of science'.[71] These limitations were those of humanity and morality. So cruelty to animals in the name of medical research was considered undesirable. Secondly, science should be pursued for the sake of knowledge, accurate thought and action. He cited how J.C. Bose and P.C. Ray pursued science for the sake of science and not for money or fame. J.C. Bose had told him that all his researches were devoted 'to enable us to come nearer our Maker'. Then Gandhi asked the important question:

How will you infect the people of the villages with your scientific knowledge? Are you then learning science in terms of the villages and will you be so handy and so practical that the knowledge that you derive in a college so magnificently put—and I believe equally magnificently equipped—you will be able to use for the benefit of the villagers?[72]

He posed a similar question to the students of the Indian Institute of Science in 1927. The money that the Tatas or the Mysore Darbar gave them ultimately came from the villages and the rule of 'no taxation without representation' applied to the villagers too. So what was the Institute doing for them?[73]

There were no easy answers. The system prepared them for an urban or industrial India or for a slot abroad. An Indian student specializing in wood chemistry in America once asked Gandhi: 'Would you approve of my going into industrial enterprise, say pulp or paper manufacture? Do you stand for the progress of science? I mean such progress which brings blessings to mankind. For example, the work of Pasteur of France or that of Dr. Benting of Toronto?' Gandhi had no objection to an industrial enterprise so long as it remained humanitarian. After all it added to the productive capacity of the country. Similarly with all admiration for the 'scientific spirit of the West', he would again plead for putting limitations

[71] Raghavan Iyer (ed.), *The Moral and Political Writings of Mahatma Gandhi*, vol. I, Oxford, 1986, pp. 310-15.

[72] *The Hindu*, 19 March 1925, quoted in Raghvan Iyer (ed.), op. cit., pp. 310-15.

[73] 'All research will be useless if it is not allied to internal research which can link your hearts with those of the millions. Unless all the discoveries that you make have the welfare of the poor as the end in view, all your workshop will be really no better than Satan's workshops', *Young India*, 21 July 1927.

'upon the present methods of pursuing knowledge'.[74] One may not agree with Gandhi's great emphasis on non-violence and anti-vivisection, but he definitely provided a notion of accountability in his addresses at Trivandrum and Bangalore with which it is almost impossible to differ.

P.C. Ray appreciated Gandhi more sympathetically than the other scientists of the day. He had Western education. Later he established the first great chemical industry in the country and yet in full consciousness of the relevance of modern machinery, he also adopted and pleaded for charkha. He knew that a little self-contained village was gone for ever, never to return. Ray would gladly accept the new modes of communication. As he put it, 'the Ganges cannot be forced back to Gangotri'.[75] Yet the relevance of Gandhi's emphasis on moral values and social control, which the march of Western civilization had gradually eroded, was not lost on him. Even in old age he worked on Gandhian projects and it was not without reason that Gandhi called him 'a scientist of scientists'. Another Indian scientist, J.C Bose, might have agreed with Gandhi philosophically but at a more mundane level he seems to have kept aloof. His sociologist friend and biographer Patrick Geddes, severely criticized the Mahatma for reverting to the spinning-wheel at a moment when the fundamental poverty of the masses called for the most resourceful application of the machine both to agricultural and industrial life.[76] Meghnad Saha, a pioneer astrophysicist and well known for his deep social commitments, wanted India to choose 'the cold logic of technology' and not the vague utopia of Gandhian economy.[77] In the very first issue of his celebrated journal, *Science and Culture*, while appreciating Gandhi's genuine sympathy with 'victims of an aggressive and selfish industrialism', he firmly refuted the claim that 'better and

---

[74] Raghvan Iyer, ed., op.cit., pp. 317–18.

[75] P.C. Ray, 'The Call of the Charkha', *Young India,* 27 Dec. 1923.

[76] J. Tyrwhitt, ed., *Patrick Geddes in India,* London, 1947, pp. 8–9.

[77] In a lecture delivered at Santiniketan on 13 Nov. 1938, Meghnad Saha said: Many of our well-wishers insist that we should remain basically agricultural. I think this smacks of conspiracy. If all of us return to village it becomes easier for a handful of capitalists to exploit us. In the West all the key industries like power-production, machine-making, road-construction are controlled by the state. The same policy should be adopted here. Our country can be prosperous like Europe or America if only we can develop industry and raise capital through state supervision. See Santimoy Chatterjee, ed., *Meghnad Rachna Sankalan,* Calcutta, 1986, pp. 113–16. To such expectations, Gandhi had already argued, 'to industrialise India in the same sense as Europe is to attempt the impossible', *Young India,* 6 Aug. 1925.

happier conditions of life can be created by discarding modern scientific technique and reverting back to the spinning-wheel, the loin cloth, and bullock cart'.[78]

Some of Gandhi's political contemporaries were so vexed that they just did not know where to place him—a mad man or a saint.[79] Irwin, the Viceroy, wrote in a private letter that 'far from being a saint, Gandhi is the most astute Machiavellian little bania that ever was bred. It is only we people who are in close touch with him and have been for long years who can understand what a slippery little devil he is'.[80] A British mining prospector records: 'Gandhi was making the future so uncertain that nobody dared to start anything. It was not any good prospecting with Gandhi on the spot.'[81] Through his ideas and actions Gandhi had caused a turmoil from which there was hardly any escape.

Much of this confusion arose because Gandhi was not a philosopher or a theorist in the conventional sense. He rather embodied a curious amalgamation of utopian vision and ground realism. He was neither a traditionalist nor a modernist, neither a technophile nor a technophobe. He provided a critique of both. Perhaps this explains why postmodernists find in him a delightful subject! He talked of morality in an age when immorality was fast becoming a way of life. When the European travellers and imperial ideologues stressed the need for a moral regeneration in India, it was perceived as a colonial propaganda. In a remarkable inversion of the same logic, Gandhi called the West immoral, warned against its imitation, and then imposed his notion of morality on everyone who cared (or dared) to come closer to him. His extreme emphasis on 'moral fibre' often took him to limits which he could have better avoided. For example, he argued that our ancestors knew how to invent machinery but they 'after due deliberations . . . set a limit to our indulgences' and decided 'that we should only do what we could with our hands and feet'.[82] A similar flight of imagination in a speech in London in 1931 made him make the following claim:

---

[78] *Science and Culture*, Vol. 1, June 1935, pp. 3–4. This editorial was repeated in June 1942.

[79] Gandhi replied: 'I am certainly mad in the sense every honest man should be'. See *Young India*, 19 March 1925, p. 98.

[80] The Viceroy to H. Butler, 9 January 1932 and 22 January 1932, IOR Mss. Eur. F11654–55.

[81] Recording of E.J.Beer IOR Mss.Eur.T.175.

[82] M.K. Gandhi, *Hind Swaraj*, Ahmedabad, 1939, p. 55.

Without fear of my figures being challenged successfully, I can say that today India is more illiterate than it was fifty or hundred year ago, because the British administrators when they came to India, instead of taking hold of things as they were, began to root them out. They scratched the soil and began to look at the root, and left the root like that, and the beautiful tree perished.[83]

Philip Hartog (former vice chancellor, Dacca University) took up the challenge and asked for the figures. Despite promises Gandhi could not supply them. They simply did not exist.[84] Gandhi could have avoided this unhistorical way of idealizing and romanticizing the past. But perhaps the necessities or requirements of a political discourse made him do so.

This, however, should not deflect us from the importance of his deep social commitment. As a politician responsible for bringing the masses into the politics of national movement, he could not have left the masses out in any programme of national emancipation or reconstruction. So not technology *per se* but the impact of technology was his great concern. He realized that modern technology has its own dynamics and that it would create new hierarchies in an already caste-ridden society and would further marginalize the poor. As Ashis Nandy argues, he rejected technicism, not technology.[85] He rejected scientism as well. He denied science a monopoly of knowledge and scientists a monopoly of science.[86] But it is a matter of conjecture whether he would have written about the scientists in the same way as he did about lawyers or doctors. A professional class of scientists was yet to emerge then. But he would have definitely seen the difference between the profession of lawyers and scientists. Gandhi would not have called for the destruction of large laboratories as some scholars would like us to believe,[87] but he would

[83] Hartog Papers, IOR, Mss. Eur. D551/30. See also Dharampal, *The Beautiful Tree*, Delhi, 1983, pp. i-ii.

[84] Philip Hartog to M.K. Gandhi, 27 October 1931, Hartog Papers, IOR, Mss.Eur. D551/30.

[85] Ashis Nandy, *Traditions, Tyranny, and Utopias,* Delhi, 1987, pp. 136–7.

[86] Bhikhu Parekh, *Colonialism, Tradition, and Reform: An Analysis of Gandhi's Political Discourse,* New Delhi, 1989, p.103.

[87] Shiv Visvanathan, 'Reinventing Gandhi', International Unit on Militarization and Demilitarization in Asia (IUMDA), Vol. 5, 1992, pp. 34–63.

Gandhi has been an object of academic appropriation. During the heydays of Nehruvian ideas (1950s–1980s), it was argued that modern science, technology, and civilization are very much compatible with the main ideas of Gandhism and that 'Swaraj gets enriched only in close association with modern science, technology, and civilization.' See R.K. Patil, 'Self-sufficiency and Modern Industrialization',

have definitely asked for moral responsibility and social accountability. The notion of social accountability remains the most important contribution of the Gandhian ideology. And in reality it did receive attention, at least at the policy level, in the nascent scientific establishments of the country. For example, presiding over the meeting of the Board of Scientific and Industrial Research in 1946, C. Rajagopalachari invoked Gandhi and asked the Board to keep in mind the unalterable facts of our rural economy.[88]

Dualism Ends

By 1930, the Great Depression, the Civil Disobedience Movement, etc. had depressed the colonial government but the nationalists' morale was upbeat. On the scientific firmament new stars had emerged; S. Ramanujan (1887–1920), S.N. Bose (1894–1974), C.V. Raman (1888–1970) and M.N. Saha (1893–1955). Ramanujan, it is said, knew infinity; he was *svayambhu.*[89] Satyendra Bose collaborated with Einstein[90]; Raman got Nobel Prize for his 'Effect'[91] while Saha pioneered new enquiries in astrophysics.[92] Politically, the long-drawn freedom struggle had begun to yield significant results; provincial autonomy was introduced in 1935 and debates had begun on planning for national prosperity. *Reconstructing India* by M. Visvesvaraya (1860–1962) was probably the first attempt to make the Indians plan-conscious.[93] In 1934 he wrote *Planned Economy for India.* His optimism is unmistakable. He wrote,

---

Gandhi Marg, Vol. 5, January 1961, pp. 46–52' and N. Malla, 'Swaraj, Science, and Civilization', *Gandhi Marg,* Vol. 3, September 1981, pp. 345–53.

[88] *Proceedings of the Board of Scientific and Industrial Research,* Delhi, 16 September 1946.

[89] Robert Kanigel, *The Man Who Knew Infinity: A Life of the Genius Ramanujan,* Calcutta, 1993, pp. 353–9.

[90] J. Mehra, 'Satyendra Nath Bose', *Biographical Memoirs of Fellows of the Royal Society,* 21, 1975, pp. 117–54.

[91] G. Venkataraman, *Journey into Light: Life and Science of C.V. Raman,* New Delhi, 1994; Rajinder Singh, *C.V. Raman's Science, Philosophy and Religion,* Bangalore, 2005.

[92] D.S. Kothari, 'Meghnad Saha', *Biographical Memoirs of Fellows of the Royal Society,* 5, 1960, pp. 217–36.

[93] M. Visvesvaraya, *Reconstructing India,* London, 1920. This work inspires the title and tenor of this chapter.

The people have most of the facilities required for a big step forward. Enormous numbers of trained men and huge masses of uneducated population are waiting by the roadside to be picked up, drilled and put to work to increase production and service.[94]

Unlike P.C. Ray who saw the remedy in *Swadeshi* and charkha, Visvesvaraya asked for rapid industrialization through Indian capital and enterprise. Saha, on the other hand, believed that the changes were to be induced through a mixed economy in which both the state and private enterprise would have significant roles to play.[95] Saha was little suspicious of the intentions of the Indian capitalists and would therefore assign the state a greater role.[96] So when Visvesvaraya suggested that applied research be preferred as far as possible, Saha found it 'rather hazardous to insist on the mere industrial aspect'.[97]

In 1934 G.D. Birla, a major industrialist and follower of Gandhi, pleaded for planning.[98] But it was Saha who convinced the national political leadership of its necessity. In 1937, he persuaded the Congress president, Subhas Chandra Bose, to constitute a National Planning Committee (NPC). This materialized at the end of 1938 under the chairmanship of a promising politician, Jawaharlal Nehru. As many as twenty-nine expert subcommittees were formed to address different areas of national reconstruction, including agriculture, industries, population, labour, irrigation, energy, communication, afforestation, health, housing, and education.[99] Nehru was amazed at both the utility and the vastness of the planning process. He wrote:

One thing led to another and it was impossible to isolate anything or to progress in one direction without corresponding progress in another. The more we thought of

[94] M. Visvesvaraya, *Planned Economy for India*, Bangalore, 1934, pp. vi–vii.

[95] Dinesh Abrol, 'Colonised Minds or Progressive Nationalist Scientists: The Science and Culture Group', in R. MacLeod and D. Kumar (eds.), *Technology and the Raj*, New Delhi, 1995, pp. 265–88.

[96] *Science and Culture*, IV, 1938, p. 367.

[97] Ibid., V, 1940, p. 572.

[98] G.D. Birla, *Indian Prosperity: A Plea for Planning*, annual speech delivered on 1 April 1934 at FICCI, New Delhi, later published by Leader Press, Allahabad, 1950.

[99] The Second World War disrupted the work of the NPC which could be resumed at the end of the war. Before dissolving itself in 1949, the NPC published 27 volumes of reports outlining a Ten Year Plan to be implemented by a government of free India. K.T. Shah, *Report: National Planning Committee*, Bombay, 1949.

this planning business, the vaster it grew in its sweep and range till it seemed to embrace almost every activity.[100]

With nationalist feelings bursting at the seams, the NPC provided an opportunity to recast the old arguments in terms of national regeneration, self-sufficiency, and all-around progress.[101] But it also laid bare inherent tensions and contradictions. Gandhians like J.C. Kumarappa and S.N. Agarwal preferred traditional technology and village industries, and attacked the NPC.[102] Even within the Congress, younger leaders leaned towards modern science, technology, and heavy industrialization. Some wanted to follow the socialist path, others favoured capitalist models such as that envisaged in Purshotamadas Thakurdas's *Plan for Economic Development of India, 1944*, popularly known as the Bombay Plan. The indigenous business class was no less divided. Gandhi's politics were convenient, but his economics were not. Although socialism remained an ideal for many (including Nehru and Saha), a version of democratic socialism with a mixed economy was accepted by the NPC as the basis for future development.[103] Saha resented this dilution, and even called the Congress leaders 'puppets in the hands of big industrialists'. He and Nehru gradually drifted apart.[104]

Saha would insist on 'scientific method' in every aspect of national life. A contemporary chemist S.S. Bhatnagar (1894–1955) and a young physicist H.J. Bhabha (1909–66) preferred building centres of excellence in frontier areas of scientific research. Bhatnagar was to build a chain of laboratories under the aegis of the Council of Scientific and Industrial Research (CSIR, 1942). Bhabha played a greater role in independent India pioneering nuclear technology. But both Bhabha and Saha were theorists who had an impulse to test ideas. Bhabha saw the source of power in the controlled release of fission while Saha saw power in the huge rivers of India. Bhabha argued for quick import of models which could be adapted and reproduced in India, and thereby gain time and

[100] Jawaharlal Nehru, *The Discovery of India*, Calcutta, 1946, p. 396.

[101] J.N. Sinha, 'Technology for National Reconstruction: The National Planning Committee, 1938-49', in R. MacLeod and D. Kumar (eds.), *op.cit.*, pp. 250–64.

[102] J.C. Kumarappa, *Why the Village Movement?*, Wardha, 1936; S.N. Agarwal, *The Gandhian Plan for Economic Development for India*, Bombay, 1944.

[103] This was done to accommodate the views of the Indian industrialists as outlined in Purshotamadas Thakurdas, *Plan for Economic Development for India*, Bombay, 1944.

[104] R.S. Anderson, *Building Scientific Institutions in India: Saha and Bhabha*, McGill, 1975, p. 29.

leverage. Saha believed in the development of India though a wholly independent science and technology firmly embedded in a socialist economic pattern.[105] Both were substantially different. Later Nehru leaned more towards Bhatnagar and Bhabha, which led to government-controlled industrial and defence research. By 1945 the political leadership had made a conscious decision to modernize and the dualism of the previous decades came to an end.

A Belated Realization

The British government awoke to these new nationalist demands, rather late. The Board of Scientific Advice was allowed to languish and disappear. The much-publicized Holland Commission Report was shelved. Meanwhile, in 1928 the Government of India preempted the question of industrial research by linking it to agriculture. The Great War and later the Great Depression were cited as reasons for the Indian government's inactivity, while in other countries these acted as catalysts.[106] Britain had established Department of Scientific and Industrial Research (DSIR) in the middle of the war (1916). Similar organizations were established in Australia and New Zealand, but not in India. In 1933, Richard Gregory, editor of *Nature*, visited Indian universities and appealed to the then Secretary of State for India, Samuel Hoare, to create an equivalent of the DSIR for the development of natural resources and new industries. Hoare supported the idea, as did the provincial governments of Bombay, Madras, Bengal, Bihar, and Orissa, but the Government of India did not.[107] The finance member of the Government of India, George Schuster, did feel the winds of change and was probably the first official to talk about planning India's development. But his dangerous 'Keynesian' ideas were neither appreciated nor shared by his

---

[105] Ibid., pp. 101–2.

[106] What to say of initiating new industrial researches as recommended by the Holland Commission, the Government of India in 1931 proposed drastic reductions in the grants to even well-established areas of research like geology, botany, agriculture, etc. This was strongly resented by several fellows of the Royal Society who in a resolution condemned such action 'as risking a permanent loss to India, to the whole Empire, and to the world at large', *Minutes of the Royal Society*, Vol. XIII, 1931, p. 230.

[107] V.V. Krishna, 'Organisation of Industrial Research: The Early History of CSIR 1934-47', in R. MacLeod and D. Kumar, *op.cit.*, pp. 289–323.

colleagues.[108] Except for an innocuous Industrial Intelligence and Research Bureau set up in 1934 to gather and disseminate industrial intelligence information, the Government of India dithered. It witnessed in silence the stormy sessions of the Indian National Congress, the rumblings of Indian industrialists, and the establishment of the NPC.

The onset of the Second World War changed the scenario fundamentally. The Secretary of State for India suddenly realized that it was possible to outmanoeuver the Congress by attempting, 'regardless of conventional financial restraints,' a 'complete overhaul of India's national life,' with the British playing the postwar role of a 'bold, far-sighted and benevolent despot . . . in a series of five-year plans, to raise India's millions to a new level of physical well-being and efficiency.'[109] This was not a mere reassertion of the 'white man's burden'. It had within it a plausible answer to Gandhi's call to the British to 'Quit India'. The same year, after the failure of his political mission to transfer power to the Indians, Stafford Cripps wrote 'A Social and Economic Policy for India' in which he proposed giving more attention to education, population control, agricultural productivity, factory legislation, and industrialization. He also recommended the use of 'modern techniques of economic planning and the modern device of the Public Corporation.'[110] This was also what nationalists like Saha were demanding, yet their motives were different. From the British point of view; it was a reassertion of constructive imperialism. But the likes of Saha would not accept an 'empire-driven development'. They would not settle for anything less than *Purna Swarajya* (total self-rule) and complete control over the agenda and its implementation.

Schuster and Cripps were not alone in trying a different strategy. Two successive viceroys, Linlithgow and Wavell, were 'sincerely concerned to see that science is properly used to give India a chance of developing on right lines.'[111] Linlithgow had even written a tract on Indian

[108] S. Bhattacharya and B. Zachariah, 'A Great Destiny: The British Colonial State and the Advertisement of Post-War Reconstruction in India, 1942-45', *South Asia Research*, 19, 1999, pp. 71–100.

[109] L. Amery to Linlithgow, 27 May 1942, quoted in ibid.

[110] Ibid. This strategic realization soon percolated down the hierarchy. For example, Roger Thomas (an agricultural expert and adviser on postwar reconstruction in Sindh) described planning "as the essence of post-war development. . . . Rather than a reversal of *laissez faire* ... to avoid the pitfalls of the earlier mercantile doctrine . . .' Roger Thomas Papers, MSS Eur. F.235/46, IOR, London.

[111] A.V. Hill (Secretary Royal Society, London) to C.J. Mackenzie (National

peasants.[112] When the Second World War began (in September 1939 for England), a Board of Scientific and Industrial Research was established under S. S. Bhatnagar with a charter similar to that of the British DSIR. By the end of 1940, it had about eighty researchers working on the purification of Baluchistan sulphur and the development of vegetable oil blends, dyes, and emulsifiers. Impressed by its utility and success, the Government of India elevated this board to the status of a council (i.e., CSIR) in September 1942. Although their mandates were similar, in constitution and work the British DSIR and the Indian CSIR differed. The former was answerable to the British Parliament through the Privy Council and was free from bureaucratic hassles, while the latter's agenda was set by the Department of Supply and Munitions. Still, the establishment of the CSIR in the middle of the war was a remarkable development. The chemist in Bhatnagar envisioned a 'spectrum of research with pure and applied research at either end'.[113] Even with a limited budget, and in the midst of political turmoil, the CSIR began work in radio research, statistical standards, building materials, and electrochemistry. But, unlike the DSIR or similar organizations in Canada and Australia, the Indian CSIR was not encouraged to contribute substantially and directly to the war effort. It remained a distant, poor cousin. All that it could boast of at the end of the war were thirty-two processes, none of which could be classified as a major breakthrough.[114]

Hill's 1944 Visit: A Turning Point

This phase of transition (1937–47) was, however, significant in view of the ever-increasing attention being paid to development and the role of science and technology as catalytic agents. It was an intellectually stimulating period in which age-old problems were revisited and a new vision emerged. The efforts of M. N. Saha and others helped, including the British scientists A. V. Hill (1886–1977) and J.D. Bernal (1901–71).

Research Council, Canada), 14 Nov. 1944, A.V. Hill Papers, MDA/A7, preserved at Royal Society, London.

[112] Linlithgow, *The Indian Peasant,* Faber & Faber, London, 1932. Later, in the midst of a famine in Bengal, experts discounted the possibilities of 'nation building on the foundations of an ignorant, illiterate and debilated people.' Roger Thomas, *op.cit.,* Simla, 1944.

[113] S.S. Bhatnagar, 'Indian Scientists and the Present War', *Science and Culture,* VI, 1940, p. 195.

[114] V.V. Krishna, *op.cit.,* p. 315.

In early 1944, Hill (biological secretary of the Royal Society) visited India on an official mission to advise and report on the state of scientific research.[115] Hill had an excellent personal rapport with Bhatnagar and he corresponded with more than fifty other Indian scientists.[116] Hill came to India with no political views, but quickly put economic problems high on the agenda. He regretted that, since 1931, several scientific operations had been starved by 'false economy'.[117] He spoke of a quadrilateral dilemma, that is, population, health, food, and natural resources.[118] To him, the fundamental problems of India were 'not really physical, chemical or technological, but a complex of biological one[s] referring to population, health, nutrition, and agriculture all acting and reacting with another.'[119] To represent this, he formulated H = f(X, Y, Z, W), where H is total human welfare, X is population, Y is health, Z is food, and W is other natural resources.[120] In all his lectures in and on India Hill dwelt upon the need to control population. Several times he repeated, 'You can not keep cats without drowning the kittens:' which, put in terms of *Homo sapiens* instead of *Felis cattus,* simply meant, 'You cannot have a higher standard of life without limiting reproduction.'[121] Hill was not the first to stress population control.[122] But the political and industrial leadership of India remained complacent. To quote Hill:

When I asked one of the authors of the Bombay Plan why population was practically not mentioned in their report, he replied that his colleagues could not agree and so had decided to leave the population problem to God. I asked him why they did not leave industry and housing to God, too. . . . If public health and food are to be

---

[115] A.V. Hill, *Scientific Research in India,* Royal Society, London, 1944.

[116] A list of his correspondents in India is given in A.V. Hill Papers, AVHL II, 5/115, Churchill College. Cambridge.

[117] A.V. Hill to F.W. Oliver, 5 June 1945, Hill Papers, MDA, A7.9. Royal Society, London.

[118] Hill's speech before the East India Association, 4 July 1944, *Asiatic Review,* XI, Oct. 1944, pp. 351–6.

[119] Messel Lecture before the Society of Chemical Industry. 1944, later published in A. V; Hill, *The Ethical Dilemma of Science and Other Essays,* New York, 1960, p. 375.

[120] A.V. Hill to John Mathai. 5 Feb. 1954, Hill Papers, Cambridge, AVHL II, 4/79.

[121] *Ibid.,* 6/4/548.

[122] Eugenic ideas had already found some acceptance in India. N.S. Phadke pleaded for it in his *Sex Problem in India,* Bombay, 1927, pp. 328–41. See also R. P. Paranjpye, *Rationalism in Practice,* Calcutta, 1935, pp. 20–3, and Marie Stopes Papers, PP/MCS/A.313–15, WIHM, London.

planned, so inevitably must population be, or all our efforts will be brought to nought.[123]

Hill's diagnosis was correct, but. the remedies that he suggested were mixed in nature and value. He wanted government to recognize that science was not just a hand-maiden but 'an equal partner in statecraft'. In England, the Medical Research Council (1914), the DSIR (1916), and the Agricultural Research Council (1931) were not under respective user ministries, but under the Lord President of the Council. In India, meteorology was under the Postal Department, the Geographical Survey was under Labour, the Survey of India was under the Education, Health and Land Department, and the CSIR was under the Department of Commerce. If scientific research in India was to make a concerted and coordinated contribution to national development, it had to be brought under 'some more systematic plan' or 'under one umbrella'. So Hill recommended the creation of a Central Organization for Scientific Research with six research boards, covering medicine, agriculture, industry, engineering, war research, and geological and botanical surveys. He felt that, under a central organization, the different subjects would be treated with 'some degree of uniform encouragement'. The user-departments or ministries could have separate development or improvement councils that would be entrusted with the task of translating pure research into production.[124]

Hill made two additional points. Health and population problems were to be taken seriously, and agriculture and industry were to be brought together. As a biologist, he emphasized the research component in medical education, and asked for an all-India medical centre.[125] He regretted that biophysics was completely neglected in the country, except at the Bose Institute in Calcutta. When Bhabha was planning for the founding of the Tata Institute for Fundamental Research in 1944, Hill advised him to 'take biophysics under its wing'.[126] Agricultural progress, he argued, would depend largely on mechanization, land utilization, fertilizers,

[123] AVHL.II 4/42, Cambridge.
[124] A.V. Hill, *Scientific Research in India, op.cit.*, pp. 40–4. Hill would not prescribe similar centralization for England 'because already there is so much of scientific work going on in different departments that to centralise would do harm rather than good.' A.V. Hill to Bhatnagar, 21 June 1948, Hill Papers, AVHL II, Cambridge.
[125] The All India Institute of Medical Sciences is now known more as a referral hospital than a research centre.
[126] '. . . many of the most important application[s] of physics will be in biology.' Hill to Bhabha, 22 June 1944, Hill Papers, London, MDA, A4.6.

irrigation, transport, roads, food processing, and a great variety of other technoscientific factors. Probably he wanted to reduce India's traditionally heavy dependence on agriculture, but would not say so openly.[127] He preferred a safer, middle course. As he put it, 'the factor of safety in India is far too low for luxuries like bloody revolutions or for monkeying about with machinery already groaning under a heavy overload.'[128] This was a subtle denunciation of both the communists and the Gandhians. Hill had a sense of mission, probably *a la* Kipling. To a British metallurgist, he wrote:

India is at the parting of the ways, and that either she may go ahead to an efficient and prosperous economy or may sink back into inefficiency, disorder and disaster. We can of course, like Pilate, wash our hands of it and clear out, but that scarcely relieves us of the responsibility for having got them so far and then having left before the job is finished.[129]

Another interesting facet of Hill's interest in India is his understanding of the Indian scientists. Of C.V. Raman, India's best-known scientist, Hill wrote, 'he is queer fish . . . Nobel Prize more or less turned his head! Meghnad Saha is also a rather an odd fellow but much more reasonable than Raman.'[130] 'Birbal Sahni has done good work in Palaeo-botany. Unfortunately he too is a politician and is unreasonable and stiff-necked.'[131] Bhatnagar was his best bet, 'a fighting Punjabi, extremely energetic, dashing hither and thither like mercury but always to good purpose, a most loyal and friendly fellow'.[132]

Other scientists with whom Hill developed lasting friendships were H.J. Bhabha, S.S. Sokhey (Director, Haffkine Institute in Bombay), S.L. Bhatia (Indian Medical Service), J.C. Ghosh (Director, Indian Institute

---

[127] In contrast, more than five decades before Hill, a perceptive judge had asked, 'Why are we to suppose that the inhabitants of India. ingenious, quick, receptive, skilful of hand, and authentic in taste, are to confine themselves, as in the past, almost exclusively to agriculture, and will not some day be in a position to take a hint from England, which for its small population, imports more than two millions' worth of food per week?' Justice Cunningham, 'The Public Health in India,' *Journal of the Society of Arts,* XXXVI, 1888, pp. 241–65.

[128] A.V. Hill papers, AVHL II 4/42, Cambridge.

[129] A.V. Hill to A. MacCance, 16 May 1944, Hill Papers, London, MDA, A5.12, Royal Society, London.

[130] Hill to the Vice-Chancellor, Cambridge University, 26 March 1946, Hill Papers, London, MDA, A5.5.

[131] A.V. Hill to F.W. Oliver, FRS, 22 Dec. 1944, Hill Papers, MDA, A.7.

[132] A.V. Hill to Ralph Glyn, M.P., 18 Oct. 1944, Hill Papers, MDA, A5.14.

of Science), and J.N. Mukherjee (Director, Indian Agricultural Research Institute). To Bhabha he advised, 'I very much hope that you will get Chandrasekhar (a renowned astrophysicist). . . . There are not many of his quality in India that you can afford to give him away to the Americans.'[133] How prophetic! Chandrasekhar later decided to settle in the United States.

Another influential British scientist who took a keen interest in India was J.D. Bernal. Bernal firmly believed that 'science does not exist in a social and economic vacuum' and he debunked the idea of pure science as a 'convenient fiction' that enabled 'the wealthy to subsidize science without fearing to endanger their interests and scientists to avoid having to ask awkward questions as to the effects of their work in building up the black hell of industrial Britain.'[134] Bernal wanted India to make the most scientific use of its resources. For example, he believed water was 'even more important than power in the economy of India'. So he advocated sand reservoirs (as in Soviet Asia) rather than open tanks and enclosed agriculture (as in the United States) for northern parts of India. Control of soil erosion was vital, as was microbiology for tapping biological resources. Bernal was also one of the earliest to stress the value of solar energy.[135]

Like Hill, Bernal was a keen observer of people. He was present at the opening of both the National Chemical Laboratory in Poona and the National Physical Laboratory in Delhi. Of the Delhi ceremony he recounted later:

It was astonishing the respect which the authorities in India give to science, it is difficult to imagine an opening of any building in England which would be attended by the King, the Prime Minister and two other ministers but that was the case there, because the retiring Governor-General (Mr. Rajgopalachari) was making his last public appearance on this occasion. . . . As the speeches went on in their inanity they gradually got me down, finally I had to listen to Patel (Home Minister) himself talking about the restoration of law and order which seemed hardly worthwhile. As a result I threw away the speech I had prepared and even proceeded to give quite a different one, telling them that they would never get anywhere unless they did the job with popular support and *put some more money into the universities and the*

---

[133] A.V. Hill to H.J. Bhabha, 22 June 1944, Hill Papers, MDA, A4.6.

[134] J.D. Bernal, 'Science and Liberty', *Science and Society*, I, VI, 1938, pp. 348–50.

[135] 'Notes on the Utilisation of Research on Short and Long-term Developments in India', J.D. Bernal Papers, MSS Add. 8287, box 52.4.55, Cambridge University Library, Cambridge.

teachings and saw that the plans which were made were actually carried out instead of remaining on paper.[136]

This period of transition had an even more unfortunate dimension. It was the communal one. In 1932 Megaw (IMS) wrote, 'we are trying to keep communal considerations out of the picture but the tendency at the moment is very much in the direction of insisting on proportional representation of the Indian communities and ofcourse the introduction of this policy would be fatal to the scientific departments'.[137] Twelve year later another IMS surgeon noted, 'as Hindus and Muslims tended to vote only for someone of their own religion, the Government of Bengal was largely in the hands of Muslims. . . . There was political pressure to give Muslims equal number of places in administration, teaching, etc. so that very often an inferior Muslim was given a job in preference to a more competent and more intelligent Hindu.'[138] But then there were officials who took delight in this piquant situation. An ICS stationed at Chittagong during the Japanese threats recorded gleefully.

I never had the slightest political trouble; all the Congressites were taking contracts. All the stories of India not supporting the war seemed nonsense to me.

[136] *Ibid.,* Box 19.L1.33. The inauguration in Poona was no less interesting. To quote Bernal again,

Bhatnagar was running the whole show. He had even provided complete speeches for everyone to make. I saw the one that had been given to Irene [Curie]. It was mostly an account of the wonderful energy, enterprise, foresight and public spirit of Sir Bhatnagar himself. However, Irene was sufficiently high-spirited to alter it all and hardly to mention the admirable gentleman in her speech. After about two hours speeches the doors were opened with some difficulty with a golden key by the beloved leader (Pt. Nehru) and we all trooped in. The effect however was somewhat disconcerting because behind the facade there was very little indeed; most of the building had not been built at all. There was only one storey and that contained no apparatus that actually worked.

[137] J.D. Megaw (D.G., IMS) to W.S. Carter (Rockefeller Foundation, USA), 27 July 1932, RAC, IHB, 1.1, 464 India Box 6, f.37.

[138] E.J. Somerset Papers, CSAS, Cambridge. Around same time a Rockefeller official also noted,

Bengal unfortunately has a Muslim Minister of Public Health whose IQ almost is 50. Despite written agreement between the Government of Bengal and the Government of India that appointments should be made through the Technical Advisory Committee instead of through the Civil Service Commission the Minister held up the Committee's selection in the first instance on the basis that they should be through the Service Commission. After this ignorance has been corrected the next issue was communal ratio.

Diary of J.B. Grant, Feb. 3–28, 1944, RG 12.1, RAC.

We had lot of trouble in eastern part of Bengal; the Hindus there who had given us all the terrorists in the old days got themselves carved up by the Muslims and Gandhi went there too.[139]

Reflections: Disunity in the Development Discourse

The twentieth century, like the eighteenth century, was an era of transition for India. Swadeshi and *swaraj* reverberated in the air. These were more than political slogans, they symbolized an intense yearning for change.[140] The direction of change, however, remained unclear. Many critiques appeared, and many options were debated. A renowned art critic, Ananda Coomaraswamy, criticized his contemporary Swadeshi protagonists for having ignored the skilled artisans and village craftsmen.[141] Officials like Alfred Chatterton (Director of industries, Madras) differed and asked for industrial development under government patronage.[142] Towards this end, the Holland Commission attempted some sort of 'a national planning in a bureaucratic guise.' But nothing could impress the British government, which conveniently and regularly took refuge in the situation created by war and depression.

By the 1930s, nationalism had gathered strength, and thanks to the movement's demand for total indigenization, the British government could conveniently leave the core sectors of agriculture, health, and education in Indian hands. Indian leaders and government now vied to raise development issues. Dissenters were gradually marginalized. To the government, Gandhi was important politically, but not otherwise. His vision of a new India was not fully shared even by his own followers!

The period 1937–47 deserves more attention. Besides NPC publications, the Bombay Plan, and Hill's report, other significant studies emerged. Health and industrial research, for example, were explored in committees under the chairmanship of Joseph Bhore and Shanmukham

---

[139] M.M. Stuart Papers, Acc.9260/8, National Library of Scotland, Edinburgh.

[140] As a recent work argues, 'It would be erroneous to conceive Swadeshi's nativism as an atavistic upsurge of a reified tradition in the face of modernization. Rather, nationalism's nativist particularism must be situated within a broader understanding of the perceived decentering dynamic of capitalist expansion.' Manu Goswami, 'From Swadeshi to Swaraj: Nation, Economy, Territory in Colonial South Asia, 1870-1907', *Comparative Studies in Society and History*, 40, 4, 1998, pp. 609–37.

[141] Ananda Coomaraswamy, *Art and Swadeshi*, Madras, 1910.

[142] For details see Nasir Tyabji, *Colonialism and Chemical Technology*, Delhi, 1995.

Chetty, respectively.[143] All these committees had at least one thing in common: they asked for greater institutionalization. Everyone recognized the importance of coordination, and even for this they asked for a coordinating institution![144] But the proliferation of institutions did not necessarily result in quality research, nor could it bring together academic science and industrial needs. Indians were encountering a situation similar to Britain's, where the work of the Board of Education, the University Grants Commission, and the DSIR reflected widely varying pressures.[145] All of India's major political leaders, and even the colonial government, had joined in the development discourse chorus.[146] The new watch-words were planning, industrialization, institutionalization, and coordination.

While there seemed to be unity of purpose among the sectors taking part in the debate, in reality this was not so. The discourse was run by 'experts': middle-class professionals who stood for the state and, through the state, for the nation. Politicians and bureaucrats added their own flavour. And all this was done in the name of the 'masses' who 'entered the picture only as the somewhat abstract ultimate, beneficiary.'[147] In the name of the masses, the authors and defenders of the Bombay Plan asked for a shift from an 'over-agriculturalised' economy to an industrial economy.[148] At the plan's core was the notion that industrial production was more capitalistic than agricultural production. Even those who argued for industrialization felt the pressure of class interests:

[143] Report of the Health Survey and Development Committee, 4 vols, Delhi, 1946; Report of the Industrial Research Planning Committee, Delhi, 1945.

[144] The Chetty Committee recommended a National Research Council (consisting of representatives of science, industry, labour, and administration) to coordinate all research activities and to function as a National Trust for Patents. Ibid.

[145] Roy MacLeod and E.K. Andrews, 'The Origins of DSIR: Reflections on Ideas and Men, 1915-1916', Public Administration, Spring 1970, pp. 23-48.

[146] Bidyut Chakraborty, 'Jawaharlal Nehru and Planning, 1938-41: India at Crossroads', Modern Asian Studies, 16, 2, 1992, pp.275-87; I. Talbot, 'Planning for Pakistan: The Planning Committee of the All India Muslim League 1943-46, Ibid., 28, 4, 1994, pp. 877–86; Vinod Vyasulu, 'Nehru and the Visvesvaraya Legacy', Economic & Political Weekly, 29 July 1989, pp. 1700–4.

[147] Benjamin Zachariah, 'The Development of Professor Mahalanobis', Economy and Society, 26, 3, 1997, pp. 434–44. I am grateful to Sanjoy Bhattacharya for this subtle review article.

[148] An interesting debate for and against the Bombay Plan can be seen in P.C. Malhotra and A. N. Agarwala, 'Agriculture in the Industrialists' Plan', Indian Journal of Economics, XXV, 99, 1945, pp. 502–10.

The real bottleneck is the attitude of capitalists, and civil servants in league with them, who oppose all industrial production in excess of the proved absorptive capacity of the market and threaten that such production is unprofitable. They forget that planned development ultimately aims at giving plenty to all and not profits to few.[149]

A more viable proposition was planned development using the socialist (Soviet) pattern. Nehru's heart lay here, but the pragmatist in him led to compromises at every step. Lest right-wing leaders feel alienated, Nehru spoke of independence and a democratic structure first, to be followed by socialism and planning.[150] After independence, he perfected the art of mixed economy and mixed politics.[151] He tried to combine Gandhi and Visvesvaraya, and finally could do justice to neither.

Another casualty of the era of mixed policies was the traditional distinction between pure and applied science. Bhatnagar, Bhabha, and many others looked for a composite structure that combined the two. But to purists like Meghnad Saha, pure science was the seed of applied science, and 'to neglect pure science would be like spending a large amount on manuring and ploughing the land and then omit the sowing of any kind.'[152] Satyen Bose and C.V. Raman had similar views. Academic science still held a greater appeal to the Indian mind.

However, the notion that science and technology were two sides of the same coin had two interesting results for an undeveloped country. First, it meant assigning a far greater authority and responsibility to the state.[153] Second, in the name of coordinating the two, the tendency in

[149] J.C. Ghosh, 'Industrial Planning for India', *Science and Culture*, XII, 1947, pp. 345–7.

[150] S. Gopal, (ed.), *Selected Works of Jawaharlal Nehru*, vol.IX, New Delhi, 1976, pp. 377–99.

[151] In a chance meeting in Beijing in 1954, Nehru confided to J.D. Bernal,

'Most of my Ministers are reactionary and scoundrels but as long as they are my Ministers I can keep some check on them. If I were to resign they would be the Government and they would unloose the forces that I have tried ever since I came to power to hold in check. . . I have to work with the people who are actually influential with the country. They may not be the kind of people I like but it is the best I can do.'

To this Bernal added, 'He treated the rule of a country of hundreds of millions as if it were the management of a college in Cambridge'. *Bernal Papers*, MSS Add. 8287, Box 48, B.3.349.

[152] *Science and Culture*, IX, 1943, p. 571.

[153] This was not something new. The colonial requirements had introduced the concept of state scientist. The postcolonial state strengthened it further.

practice was to centralize. Hill himself argued for centralization (which he would not prescribe for Britain), and this suited Bhatnagar. Centralization and concentration of power was to become the hallmark of the scientific establishment in post independent India. Bhatnagar acquired the reputation of being an 'empire-builder'.[154] He built a chain of eleven national laboratories from 1947-1954 and twenty more would follow. Was this done at the cost of the university system? Even Hill was alarmed to see 'the great developments in Government research laboratories in India,' and warned Bhatnagar 'lest by getting all the best people away from the universities you may dry up the source of scientific talent, or at least training, for the next crop of scientists.'[155] How right he was!

The development discourse in India was thus both intense and instructive. Its reconstruction pleas and plans were neither inane nor mere pious hopes: they were sincere attempts to attain a well-deliberated goal. In recent years, the concept of planning for development has been discredited, and the state now appears more 'predatory' than developmental.[156] In today's India one sorely misses the quality and intensity of the early to mid-twentieth century development debates.

[154] G.F. Heany Papers, Memories (TS), box VII, f.340, Centre for South Asian Studies, Cambridge.

[155] A.V. Hill to S.S. Bhatnagar, 11 May 1951.

To this Bhatnagar replied, 'The universities in this country have not suffered for want of government help but the public interest in the universities has declined largely because the universities are having vice-chancellors not on the consideration of their attainments but of their political affinity.'

S.S. Bhatnagar to A.V. Hill, 18 May 1951, Hill Papers, AVHL II, Churchill College, Cambridge.

[156] S. Chakravarti, 'Predatory State: The Black Hole of Social Science, ' *The Times of India,* 22 Sept. 1999.

CHAPTER 8
Conclusions

Policy in its genuine signification being the art of ordering all things for the benefit of the citizens of the state . . . is the common sense of Government, or rather common sense as applied to Government; is everywhere requisite serving in some nations to restrain, in others to excite, in all to methodize, and direct the Endeavours of a Nation.

> Robert Kyd in a 'Memorandum on Agriculture Productions,
> Commerce, Population and Manufactures', Calcutta, 1786

The above statement, coming from an army officer and an amateur botanist stationed in a colonial outpost, is significant. He talks of policy as the common sense of government', and is concerned with 'ordering all things' for the benefit of the state and its 'citizens'. The range of his memorandum includes almost everything. This is quite characteristic of the early explorers who came as an integral part of the colonial process. A recent work on French colonialism and science in eighteenth-century Haiti finds science and scientists marching virtually as a 'productive force', 'at the vanguard, not the rear guard of colonialism'.[1] I have described a similar situation in Chapter II. The colonizers were fully aware of the importance of science as a very effective instrument of colonization and control. Their concept of science was closely related to the needs of the empire.

The thing that strikes most in any debate on colonial science is the role of the state. As the colonial state owed its origin primarily to mercantilist activities, the notion of 'science for profit' makes an early appearance. Yet, in the early stage, the colonial scientists (those days mostly surgeon-naturalists) had more freedom and flexibility. There were tremendous difficulties but also enormous opportunities to discover and sight new things. Support from metropolitan scientists added to their confidence and their agenda was not entirely derivative. They did enjoy a certain

---

[1] Unlike India, in French Haiti, science served slavery, 'unfreedom' and human oppression, fames E. Me Clellan III, *Colonialism and Science: Saint Domingue in the Old Regime*, Baltimore, 1992, pp. 7-9, 289-93.

amount of autonomy and they too influenced metropolitan discourses (for example, on the deposition of coal-seams, nature of cholera, etc.). A recent work has shown that the idea of environmental conservation came from the colonies, and colonial planters, botanists and foresters contributed a great deal to the initiation and maturation of conservation debates in the metropolitan circles.[2] Moreover, the very concept of a state scientist emerged in the colonies, and this shows how aware the trading companies who ran the colonial business were of the importance of scientific explorations. A knowledge of the local terrain, local resources, customs and traditions was vital for the founding of a colonial state. The process of acquiring this knowledge was not an easy one; almost insurmountable physical and conceptual problems came in the way. Colonial science did not come in a neat distinct package; the claims of universality and the utility of science messed up its identity.

The early colonizers in India realized that they had to tread cautiously. The state they sought to establish had to borrow, and yet be substantially different, from the pre-colonial structures of power. In order to legitimize their own rule, they first had to delegitimize several pre-colonial structures and texts. For this, the condemnation of the immediate past was considered necessary. Indians were declared unscientific, superstitious and resistant to change; India was identified with dirt and disease. Travellers, scholars and officials of both the Orientalist and Anglicist variety subscribed to this view. William Jones, the foremost Orientalist, declared that in scientific accomplishments the Asiatics were 'mere children' when compared with the Europeans.[3] Thus was established a paternalistic Raj which would be caring and dismissive at the same time. It was to be based on claims to not only superior musketery, but to a superior knowledge as well. This sense of superiority came from western discourse on rationality and progress, and was promptly used to denounce whatever scientific knowledge (e.g. in astronomy and medicine) the Indians could boast of. Yet this denunciation was not total. Several early colonial scholars showed respect for certain indigenous scientific traditions and techniques. They wanted western knowledge to permeate slowly and cause gradual displacement. Even a 'piggy ride' on the traditional

---

[2] Richard Grove, *Conservation and Colonial Expansion: A Study of the Evolution of Environmental Attitudes and Conservation Policies on St. Helena, Mauritius, and in India, 1660–1860*, Ph.D. thesis, University of Cambridge, 1988.

[3] Quoted in Michal Adas, *Machines as the Measure of Men*, Delhi, 1990, p. 107. Even a hundred and fifty years later, Indians were described as 'half-devil and half-child'. J. Thomson, *The other Side of the Medal*, London, 1925, p. 29.

caste structure was advocated even though it meant some dilution of colonial authority. Brahmins, for example, were recruited for popularizing vaccination. But these largely remained strategies for penetration and control. Though appropriation of whatever was found 'useful' in the native repertoire was actively encouraged, a syncretic acculturation formed no part of the colonial agenda.

Things, however, changed a bit by the middle of the nineteenth century. The state had become extremely powerful, and race had become a trope of ultimate, irreducible difference.[4] India came to be viewed through the 'cultural blindness' of Mill and Macaulay. The favourite rhetoric was: 'India is too old to be rational . . . only the litmus test of British colonialism will usher the sub-continent into rationalism and modernity.'[5] In short, the colonial arteries had hardened.

Colonial scientists were now expected to work for the consolidation of the empire—consolidation through organized economic exploitation. Exploratory activities continued, but they were now organized under institutional umbrellas. Several survey organizations were established. The government had, of course, its own motives, but the institutionalization of geological and botanical works proved to be of historic importance—a milestone in both the history of science and the history of imperialism.

In many cases the provincial governments were responsible for the administration of scientific organizations and institutions, and on-the-spot decisions were often taken as and when required. But the final authority rested in London, and metropolitan interests, pressures and pulls weighed heavily on colonial administrators. Excessive administrative control exercised at different levels, in turn, ensured that colonial scientists would always dance to the official tune. This bred dissatisfaction, and often led to demoralization, among the scientists. Some found themselves saddled more with administrative responsibilities than with research work, while others resented their hopeless dependence on bureaucracy for every minor favour. Sometimes they found their role chafing. A teacher wrote: 'The chill wind of Indian officialdom makes one's vanity wilt soon. One is merely senior or junior, Oxford and public school or not. The rest is of no importance.'[6] Oldham, Watt and Ross expressed their disappointment on several occasions. But they could never

[4] H L. Gates Jr. (ed.), *Race, Writing and Difference*, Chicago, 1986, p. 5.

[5] Sara Suleri, *The Rhetoric of English India*, Chicago, 1992, p. 34.

[6] Letter from C.J. Sisson, Elphinstone College, Bombay, 21 Nov. 1902, Ms. Sar Coll. 34, Edinburgh University Library.

transcend the economic and political considerations on which the Raj itself rested. Colonial pragmatism demanded the complete subjugation of personal view-points to economic interests. In 1902 a young and promising botanist was advised thus:

> . . . for the present it will be good policy on your part to leave the counting of organs of *Ranunculus arvensis* (and such like amusements) to Scotch Professors of Botany and ply Government with roseate agricultural reports. You will have observed that in the Forest Department, the man who can sell a great amount of teak gets to the top, in the geology the man who can find coal (or even gold)— when you have got to the top, you can indulge in counting organs of *Ranunculus arvensis* or any other scientific trifling that amuses you.[7]

It is not that the Raj bureaucracy deserves opprobrium all the time. Several district level officials reported and sometimes published interesting accounts of scientific relevance. In 1878, for example, a magistrate of Monghyr district published a detailed account of the 'natural history' of the area and was reviewed in *Nature*.[8] When such officials graduated to senior positions, they showed respect for 'local knowledge' and tended to favour local talents. The civilian officials naturally felt inclined to study the social ecology of the society they were sent to govern. Local knowledge was vital for them but not for the technical experts who were products of a 'technical discourse' of science, and who used certain universal parameters to which the local variations had either to conform or get rejected. This perhaps explains the difference between the attitudes of a person like E.C. Buck and someone like Medlicott. The colonial system in any case excelled in hierarchization and marginalization.

The British educational experiments in India have been severely criticized. Education was no doubt an important segment of the whole colonial enterprise and was definitely meant to strengthen it. Viswanathan calls it a 'mask of conquest',[9] and Goonatilake considers it a tool for 'cultural blanketing'.[10] Are these sweeping judgments? S. Ambirajan raises the important question as to whether the system was planned and erected for just this aim or whether there were other forces that brought

[7] C.B. Clarke (Royal Botanic Garden, Kew) to I.H. Burkill, 9 Dec. 1902, Kew Archive, Letters to Burkill, f. 35.

[8] Edward Lockwood, *Natural History, Sport and Travel*, London, 1878; reviewed in *Nature*, XIX, 13 Feb. 1879, p. 337.

[9] Gauri Viswanathan, *Masks of Conquest: Literary Study and British Rule in India*, New York, 1989.

[10] Susanta Goonatilake, *Crippled Minds: An Exploration into Colonial Culture*, New Delhi, 1982.

about the same results. He believes that chance, more than foresight, determined the future. 'There is a bureaucratic momentum', he argues, 'which propels institutions along a path, though not necessarily the one charted by the initiators'.[11] 'Chance' and 'bureaucratic momentum' are valid arguments if we do not lose sight of the fact that it was a colonial bureaucracy. This bureaucracy ensured the primacy of colonial requirements. Engineering colleges existed for the Public Works Department and were called 'civil' engineering colleges.[12] The nature and pattern of engineering education in India differed from that of Britain. Whereas in England it evolved from below and gradually became a part of the University curriculum, in India it was organized from above. Though it was organized from above in France also, the motive and situation differed greatly. In Europe, engineering education was developed in order to facilitate the process of industrialization. In India there was no such imperative. Here hope were pinned not on 'material' but on 'moral' uplift. In fact, the whole aim of colonial education was 'moral development' and 'character formation'. The 'native' character was considered defective, immoral and superstitious. The 'new' education armed with western rationality was supposed to correct it. But the PWD-oriented education could not have achieved this.

The latter half of the nineteenth century was a period of consolidation and institution-building. These institutions not only 'imported' knowledge; they imparted and, to some extent, generated knowledge. But could they diffuse new knowledge and to what extent? Telegraphs and railways were the high-technology areas in those days. Telegraph operations remained a purely governmental exercise, while the railways, raised on guaranteed profits, depended on wholesale import from Britain. Even the repair-cum-manufacturing establishments, like the Jamalpur workshop, proved to be islands in themselves. No technological spin-off could emerge, much less galvanize, the neighbourhood of a railway colony. Mechanical engineering came later and remained a poor distant cousin of engineering in the public works department. Irrigation and later hydraulic engineering definitely benefited from the large irrigation works. The Roorkee Engineering College was closely linked to Cautley's Ganges canal. Whether the generation or refinement of irrigation technology at

[11] S. Ambirajan, 'The Content of Science and Technology Education in South India', in Roy MacLeod and Deepak Kumar (eds), *Technology and the Raj* (forthcoming).

[12] For details, see Arun Kumar, *Engineering Education and Public works Department*, Ph.D. thesis, University of Delhi, 1989, pp. 1–75.

Roorkee or Guindy reduced or increased the economic dependency of India is arguable and a matter of several statistical debates. These enterprises were basically technology projects with specific aims, and not technology systems with a wider canvas and greater results. A geographical relocation of technology (as in the case of railways) was possible and was achieved, but a cultural diffusion of technology is so different and much more complex. Moreover, the professional colleges were so controlled that they could not induce change at a perceptible or a faster pace. The medium of instruction was also a factor to be considered. The Japanese had insisted on their own language. The result was that modern knowledge and the scientific spirit could percolate down to the masses. In India, colonial education widened the gulf and accentuated the age-old divide. Even in government institutions, growth was kept under a self-regulatory check. The Tokyo Engineering College was established in 1873, much later than the Engineering College at Roorkee, and by 1903 it had a staff of 24 professors, 24 assistant-professors and 22 lecturers. The Massachussets Institute of Technology was established in 1865 and by 1906 it had 306 teachers.[13] On the other hand, Roorkee, even after hundred years of existence (i.e. in 1947), had only 3 professors, 6 assistant-professors and 12 lecturers. The inference is simple. As Headrick points out, colonial rulers educated their subjects only up to a point. Beyond that point, they withheld the culture of technology.[14]

The logic of the metropolis–colony relationship was thus not in favour of the latter getting anything like a higher form of techno-scientific education. Unlike Australia and Canada, what Victorian India received was a low form of scientific and technical education administered under 'controlled' conditions. Fundamental research work were simply out of the question. Till 1890, research remained an exclusive governmental exercise, and was carried out mostly through survey and military organizations. It is true that the surveyors were sometimes more than surveyors and contributed significantly to scientific research, but the fact remains that economic considerations weighed heavily on the choice and character of the research works undertaken. The colonial scientists themselves preferred to depend on the metropolis. MacLeod has suggested 'that the idea of a fixed metropolis, radiating light from a single point source, is inadequate. There is instead a moving metropolis—a function of empire,

[13] D.E. Alexander, *The Development of Engineering Education in the United States*, Ph.D. thesis, Washington State University, 1977, p. 96.

[14] D.R. Headrick, *The Tentacles of Progress: Technology Transfer in the Age of Imperialism 1850–1940*, New York, 1988, p. 345.

selecting, cultivating intellectual and economic frontiers'.[15] No doubt a colony sometimes offered more than mere data. The establishment of the Board of Scientific Advice in India raised hopes that a similar type of centralized official body would be formed in England also. The periphery thus inspired the metropolis. A 'peripheral' scientist, J.C. Bose, made original contributions and made his way into the metropolitan scientific circles. In some cases, the tables were turned and the metropolis found itself deeply indebted to colonial science. But it would not frankly admit this, perhaps due to political and racial arrogance. With India the relationship was a little more complex. After all it was not a white-settler colony. In between the two there existed a big cultural gap. In 1881 the French held a meeting of their Association for the Advancement of Science at Algiers (their colony).[16] Their British counterpart (BAAS) never thought of holding a scientific meet in India. They held one in Canada,[17] and even in South Africa, but never in India.[18]

In the application of western science, some scholars have spotted a 'planned' and 'conscious' policy, and have hailed this as the 'constructive side of imperialism'.[19] To them, science formed an important and integral part of the colonial development policy. Here the key-word is 'development', but development of whom? It is doubtful to what extent the experimental farms, their emphasis on cash crops, and the state subsidies to cotton, tea and indigo research were 'constructive'. Tropical medicine did make some progress and the microscope replaced the sword to some extent, yet it remained confined to the army and touched only the fringe of the local population.[20] The curative drug industry was to benefit more in the years to come at the cost of social-preventive medicines. 'Conscious' efforts were made in certain directions (thanks to the individuals who

[15] Roy M. MacLeod, 'On Visiting the Moving Metropolis: Reflection on the Architecture of Imperial Science', *Historical Records of Australian Science*, V, 3, 1982, n.p.

[16] *Nature*, XXIV, 12 May 1881, p. 31.

[17] To quote Sir Lyon Playfair: 'The inhabitants of Canada received us with open arms, and the science of the Dominion and that of the United Kingdom were welded.' *Nature*, XXXII, 10 Sept. 1885, p. 438.

[18] *Nature*, LXXIII, 23 Nov. 1905, p. 77.

[19] C. Forman, 'Science for Empire: Britain's Development of the Empire through Scientific Research 1895–1940', Ph.D. thesis, University of Wisconsin, 1941, p. 1; M. Worboys, *Science and British Colonial Imperialism, 1895–1940*, D.Phil. thesis, University of Sussex, 1979, p. 2, 44.

[20] Warwick Anderson, 'Laboratory Medicine as Colonial Discourse', *Critical Inquiry*, 18, Spring 1992, pp. 506–29.

pushed the Government of India to sponsor their work), but 'planning' was never done. Science planning was not possible in those days. This, however, does not mean that there was no binding thread. The fact that colonial scientific activities lacked coherence does not reduce them to a farce. The story is certainly not that of accidents or unforseen developments. The evolution and progress of colonial science does reveal a pattern which can be discerned. It had an ideology, a string of institutions and a set of committed people to serve its ends.

Colonialism involved not only exploration and classification but also coding and decoding cultures. Its cultural projects showed a deeper penetration and a greater resilience than its economic forays. Pre-British India had never seen an ouslaught of this magnitude. Islam was not a civilizational threat in the sense the Raj was. With the help of schools, universities, text books, museums, exhibitions, newspapers, etc., the local discourse was influenced and colonized. Modernity was presented as a colonial import and not something intrinsic to man's rational nature. The new rule, with its sharp tools, dissected and bared 'differences'; it was not inclined to synthesize. This is what Macaulay did. Justifiably proud of his heritage and power, he could conceive of creating nothing other than 'a mimic man'—'Indian in blood and colour, but English in tastes, in opinions, in moral and in intellect'. Yet this colonial mimesis had its own tensions. The new mimic man would be *almost the same but not quite.*[21] To the colonizer, he would be a collaborator and a threat at the same time. He might well be equally uncomfortable with his own traditions. He had to deal with a double text—one internal, the other alien; and had to carry on a two-pronged fight—one against the traditional order, the other against colonial hegemony. The pedagogic weapon he wielded had no primordial affinity with western learning nor was it rooted in indigenous traditions. This resulted in an uneasy mix, an unequal war and a split soul.

Colonialism usually stalls the possibilities of 'exchange' and prefers a one-way traffic. One may talk of 'transfer'—transfer of knowledge, systems or technologies, but it was a transfer 'restricted' or 'guided' to achieve certain determined objectives. The way this transfer was conceived and executed left much to be desired. Teachers and professionals were sent from Europe and it was thought they would provide the magic touch. But infact they were 'inorganic' intellectuals who reproduced the same breed in India.[22] Transfer of knowledge through an 'alien' medium

[21] Homi Bhabha, 'Of Mimicry and Man: The Ambivalence of Colonial Dicourse', 28 Oct., Spring 1984, pp. 126–33.
[22] Not 'organic' in the Gramschian sense.

had its own problems. Scientific societies undertook the task of translating into the vernacular in a big way. But translations were never adequate, at least conceptually. Translations were neither preceded nor succeeded by original works. There were sharp cleavages in the style of thinking, speaking and writing. Education lagged behind and research work was government-oriented. Yet, notwithstanding the 'limitations', both the colonizers and the colonized did enjoy certain moments of autonomy. It is in these moments that Thomas Oldham produced a stratigraphical map; Ramchandra attempted a mathematical project rooted in indigenous tradition; Ronald Ross stumbled upon the malaria vector; and J.C. Bose criss-crossed the domains of physics and botany with *Samkhya* as his intuitive guide! In these 'hybrid' moments a colonial scientist produced some original work, and a native wrote not a copy of the colonial original, but a qualitatively different work in itself.

An important aspect of this cultural encounter was an almost obsessive concern of the contestants with what they thought of each other. The British were worried about what the natives thought of them. So an insane or degenerate European had to be removed from the native gaze, and the scientific pretensions of even an ill-educated colonizer had to be protected. The Raj after all rested on prestige. They realized that the new interactions might prove counter-productive. The very First Dicennial Missionary Conference in 1872 noted: 'The higher education of heathen minds sets them the more against us; they use their education as a club to break our heads.'[23]

Similarly, the Indians, especially the neo-elites, were worried about what the British thought of them. So efforts were made right from the days of Raja Rammohun to project certain Indian traditions and ideas as fully compatible and not opposed to modern science. The early orientalists found it easier to propagate the new astronomical truths by citing the *Siddhantas*. Wilkinson rightly exclaimed: 'What can be more flattering to the vanity of the Hindu nation, than to see their own great and revered masters quoted by us with respect, to prove and illustrate the truths we propound.'[24] Master Ramchandra looked to a twelfth-century text (Bhaskara's *Bija Ganita*) as the starting point for his work on differential calculus.[25] Bankim Chandra found Darwinism closer to the Hindu

[23] D.L. Gosling, *Science and Religion in India*, Madras, 1976, p. 13.

[24] J. Wilkinson, 'On the Use of the Siddhantas in the Work of Native Education', *JASB*, III, 1834, pp. 504–19.

[25] Dhruv Raina, 'Mathematical Foundations of a Cultural Project or Ramchandra's Treatise Through the Unsentimentalized Light of Mathematics', *Historia Mathematica*, 19, 1992, pp. 371–84.

concept of Trinity and declared the philosophy of *Karma* and *Maya* as 'far more consistent with science'.[26] J.C. Bose gave Sanskrit names to his instruments. Numerous popular articles traced the seeds of modern advancements to ancient texts. A Tamil journal, for example, claimed:

Western science has brought telegraph, railways and small industry. The westerners think that they have made a unique contribution in above areas. If somebody reads *Surya-Siddhanta* they would be able to find how the qualities and usages of steam had been known to Hindus. Half-baked English-educated Indians make some noise without realizing what they have in Indian tradition. Ancient Indians knew about space, telescopes, watches, chemical warfare, and also travelled in air. Now all that is gone.[27]

Were such arguments exercises in revivalism or revitalization, cultural self-defence or self-assertion? It was perhaps a combination of both but one does notice a fairly strong obsession with the distant past. A relatively weak process of secularization was also at work. In some instances, the Hindu scriptures were presented as representing *jnana* (knowledge) and not *vijnana* (science). Syed Ahmed presented the *Quran* as a book meant for moral guidance and not one in which to seek scientific knowledge.[28] The positivists and the Brahmos emphasized the importance of reason and observation, though their reason was not without God and was mixed with a heavy dose of moral and spiritual teaching. In any case, modern science was not seen as an alien import. Darwinism did not cause a ripple in India. The new paradigms in science were quickly accepted. But what garnered more attention were the 'practical skills' of Europe. These were to be learned and absorbed in India's inherited culture in a gradual, organic fashion',[29] a task in which the native interlocutors of techno-science largely failed. The fault perhaps lay in the transfer mechanisms and not in the recipients' capacity to assimilate. The application of western science in a non-western set-up was a new experiment and was riddled with numerous difficulties.[30] The urban elite avoided the rural hinterland. Agricultural exhibitions and fairs could not become vectors of rapid change. Museums, popularly called *jadoo-ghars* (magic houses), attracted a large number of middle and lower classes, but could do little more than inspire awe and admiration. It is not that the so-called

[26] *Bankim Rachanavali*, II, Calcutta, 1965, p. 280.

[27] P.N. Chokkalingam, 'Science and Technology in Ancient India', *Vivek Bhanu*, Aug.–Sept. 1906, pp. 313–19.

[28] Pervez Hoodbhoy, *Islam and Science*, London, 1991, pp. 56–7.

[29] Tapan Raychaudhury, *Europe Reconsidered*, Delhi, 1988, p. 92.

[30] Gyan Prakash, 'Science Gone Native in Colonial India', *Representations*, 40, Fall 1992, pp. 153–78.

conservative peasants were not amenable to change. They had no objection to the new tools provided these brought profits and were within their means. The problem was not cultural stagnation or social conservatism of the Indians; rather it was finding economically viable and appropriate technological solutions. The colonial government at first ignored science education and later tried to use science as a crutch which could at best provide a halting gait.

Such was the milieu in which independent India woke up. Science and technology were taken seriously. These were to make a new India. There was enormous enthusiasm which was used by some to serve their own professional or even selfish interests. Several brilliant scientists took up administrative responsibilities. Committees became more important than classrooms, J.B.S. Haldane who made India his home, was worried about organizing science in India:

. . . the first thing to do is to utilise the scientists whom we have got. That is to say they should be given time for research and teaching and as far as possible not asked to do anything else. . . . If I am asked to spend a day listening to platitudes by politicians and administrators, this means insult to me.[31]

But Haldane had few takers. The post-colonial state was no less powerful than the colonial state. The concepts of state science and state scientists now had a permanent place both in the Indian system and psyche.

Academically, quality was there and many Indian scientists were simply world-class. But the environment was deficient, the attitude was lax, and the economy poor. As a keen American visitor noted:

Although the Indians of all classes look to American and western civilization, they do not wish this influence to come to them as a product of a master race. They resent the 'master race' discrimination . . . the difference in their scientific and technical outlook is one primarily of environment rather than of genetic constitution. . . . The greatest fault I should feel among scientists is a certain disinclination to use their hands. This is a fact I have also found among scientists in South America. The lack of the appreciation of the dignity of manual labour is also a handicap to their technological advancement. One of the American vice-consuls in India told me that the difference between Indians and Americans could be shown by the fact that when an American got a Ph.D. degree that was the start for his going to work more intensively, but when an Indian got a Ph.D. that was likely to be the end of his work—he had reached his goal![32]

[31] J.B.S. Haldane Papers, 20626, n.d. p.39, Scottish National Library, Edinburgh.
[32] Confidential Note by Prof. A.F. Blakeslee (a noted botanist) on the Indian

Forty-seven years later, President of the Indian Science Congress lamented:

When the plans of our future were being drawn up, science *per se* had little part to play; its role was almost entirely usurped by technology. And when people like Shanti Swarup Bhatnagar started setting up national institutes totally independent of universities, the obvious negation of Nehru's grand vision of science and technology had probably not been foreseen. History of science in India would record that intentions were for excellence, but steps taken have not been conducive to achieve the same.[33]

No wonder, technology has overtaken contemporary India; its brilliant practitioners function as 'techno-coolies'; the state obsessed with security concerns pegs up technology demonstrations mostly of military nature; and with educational institutions on decline, the research for cutting-edge knowledge remains elusive.

Science Congress, dt. 29 April 1947. Rockefeller Archive Centre 1.1, 464, Box 1, f.5.

[33] P.N. Srivastava, Presidential Address: Science in India, Excellence and Accountability, Indian Science Congress, 81st session, Jaipur, 1994, pp. 12–3.

Bibliography

I enjoyed working most on the archival sources, though some friends warned me that these tools are conventional, official, ancient and so forth. It was rewarding to go through numerous, often contradictory, notes written by lower-rung officials filed between dusty covers. They give an idea of how a decision was arrived at, the tensions involved and the perceived threats. The files preserved in the Indian archives lay bare the inner thoughts of the official mind and show what went into the making of a particular decision. They have this advantage over those preserved at the India Office Library and Records in London. It may be erroneous to believe that the official sources give only a particular picture. Through them it is possible to know about the 'other' side as well. This, however, is not to understate the importance of 'local' sources, especially those written in the Indian languages. During the second half of the nineteenth century numerous pamphlets, articles and tracts were written in the Indian languages, especially in Bengali. In them one gets sharp critiques of official policies and actions. Similarly, in private papers, several official participants appear critical, outspoken and forthright—virtues they could not afford while writing official notes or reports. The letters written by Roxburgh, Wallich, Watt and Ross to their peers in London and the replies they received make exceptionally interesting reading and reveal what is not normally available in official documents or contemporary publications. They often contradict what one finds in official records and give new insights. Moreover, the way they have been preserved at several British institutions can be a lesson. In contrast, at the J.C. Bose Trust in Calcutta, what I was shown in the name of J.C. Bose papers were some press clippings and printed materials!

The National Archive of India is perhaps the largest repository of documents on the theme. For any account of the Company period, proceedings of the Home Department, Public Branch, are indispensable. The details of all survey operations, educational matters, telegraphs, etc. are to be found there. After takeover by the Crown, the Education Branch was opened, and sometime later the Medical Branch. The next important set of data are from the proceedings of the Revenue and Agriculture

Department, especially those of the Agriculture Branch from 1871 to 1907. In the last quarter of the nineteenth-century new and more specialized branches like the Surveys, Meteorology, Forests, etc. came into being. Perhaps the only example of a branch devoted to science and its applications is the Industry and Science Branch (1872–77) from the West Bengal State Archive.

Apart from drab files and spicy correspondences, the Raj produced tons of printed official reports. These are of two kinds. Some were published annually or in a series, for example, the DPI Report, Bengal (1856–96), Records of the Geological Survey of India 1891 (1868–96) and Report on Sanitary Measures (1867–1906). There were several one-time reports addressed to a specific problem like the Report of the Indian Education Commission (1882), Report of the Central Indigenous Drugs Committee (1901), etc.

Among the primary sources, much more important than the official reports are the contemporary tracts, books and journals. These provide not only first-hand information but analysis, very often critical and piercing, and sometimes laudatory and one-sided. In these publications the past comes alive in its myriad shades and colours. They can be put under two categories—those written by Europeans and those written by Indians. In the first category there are numerous travelogues and books of a general nature. There are many books on specific problems as well. Education, for example, has been discussed most extensively. Several books were published on surveys, botanical and mineral resources, agricultural practices and health problems. Some tracts make extremely interesting reading. W. Adam, a registrar at Bombay High Court, worked and wrote on, *Solar Heat: A Substitute for Fuel in Tropical Countries* (Bombay, 1878). Indian writings, on the other hand, are mostly of a responsive nature. There appeared numerous tracts, pamphlets and articles written by Indians, though not enough books. Vernacular publications picked up as the years rolled by. They deserve more attention than they have received so far.

PRIMARY SOURCES

I. Archival Documents

National Archives of India, New Delhi
Proc. of the Home Department, Public Branch, 1786–1905.
  Education Branch, 1857–1905.
  Revenue Branch, 1839–46.

Medical Branch, 1876–1909.
Sanitary Branch, 1895–97.
Patents Branch, 1876–87.

Proc. of the Finance Department, Accounts and Finance Branch, 1880–93.
Statistics and Commerce Branch, 1880–93.
Salaries & Establishment Branch, 1880–98.

Proc. of the Revenue Agriculture Department, Agriculture and Horticulture Branch, 1871–82.
Agriculture Branch, 1882–1907.
Fibres and Silk Branch, 1881–86.
Forests Branch, 1883–1907.
C.B. and C.D. Branch, 1881–86.
H.B. and A.S. Branch, 1890–93.
C.V. Administration Branch, 1900–8.
General Branch, 1881–1906.
Meteorology Branch, 1896–1902.
Minerals Branch, 1877–1905.
Practical Arts and Museum, 1900–2.
Surveys Branch, 1871–1905.

West Bengal State Archives, Calcutta
Proc. of the General Department, Education Branch, 1865–1907.
Medical Branch, 1861–1907.
Industry & Science Branch, 1872–77.
Miscellaneous Branch, 1890–95.

Proc. of the Revenue Department, Agriculture Branch, 1864–1906.
Forest Branch, 1869–1906.

Bihar State Archives, Patna
Proc. of the General Department, Education Branch, 1859–1908.
Proc. of the Revenue Department, Agriculture Branch, 1861–1908.
Proc. of the Municipal Department, Medical Branch, 1875.

Maharashtra State Archives, Bombay
Education Volumes 1 to 53, 1864–1912.

Agriculture & Horticulture Society, Calcutta
Records of the Agriculture and Horticulture Society of India, 1827–99.

India Office Records, London
Despatches from the Secretary of State to the Government of India,
    1858–76; 1879–97.
Despatches to Madras Government, 1807–58.
Despatches to Bengal Government, 1837–59.
Board of Control Letters on Telegraph, 1857.
Parliamentary Debates on Indian Affairs, 1893–1906.

Royal Society, London
Indian Government Advisory Committee Minutes, 1903–20.
Indian Observatories Committee Minutes, 1886, 1897–1912.
Imperial Institute Committee Minutes, 1887.

Royal Botanic Garden, Kew
Misc. Reports, Indian Botanic Survey, 1884–1920.
Misc. Reports, Indian Agriculture, 1869–1921.
Misc. Reports, Reporter of Economic Products, 1859–1900.
Misc. Reports, Imperial Institute.
Misc. Reports, Indian Advisory Committee, 1898–1920.

II. Private Papers

India Office Library, London
Anderson Papers, IOL, Photo. Eur. 85.
Banks' Memorial on Tea, 27 Dec. 1788, IOL, Mss. Eur. D. 993.
Buchanan Papers, IOL, Mss. Eur. D. 541.
Carey Papers, IOL, Mss. Eur. B. 230.
Curzon Papers, IOL, Mss. Eur.F. 111/158–230.
Elgin Papers, IOL, Mss. Eur. F. 84/74.
G.G. Spilsbury Papers, IOL, Mss. Eur. D. 909.
Greville Papers, IOL, Mss. Eur. E. 309/2.
George Watt Papers (1886–1930), IOL, Mss. Eur E. 337.
Lamb Papers, IOL, Mss. Eur. D. 893.
Morley Papers, IOL, Mss. Eur. D. 573/43.
Podgson Papers (1869–81), IOL, Mss. Eur. B. 251.
R.E. Fife Papers (1854–1911), IOL, Mss. Eur. D. 1015.
Richard Temple Papers, IOL, Mss. Eur. F. 86.
Robert Kyd Papers, IOL, Mss. Eur. F. 95.
Roxburgh Papers, IOL, Mss. Eur. D. 809.
Waugh Papers, IOL, Mss. Eur. F. 181.

British Museum and Library, London
A.O. Hume Papers, B.M., Natural History Mss. HUM, 1967–81.
Banks to Roxburgh, B.M., Add. Mss. 33980.
Campbell Bannerman Papers, B.M., Add. Mss. 41218.
J.D. Hooker to Lord Ripon, B.M., Add. Mss. 43531.
L. Davis to Banks, B.M., Add. Mss. 8968.
Major-General Hardwick Papers (1755–1835), B.M., Add. Mss. 12615,
     Add. 9869–70, Add. 9910–12.
Peel Papers, B.M., Add. Mss. 40572.
Roger de Candolle Papers, B.M., Natural History, B.M.S. CAN.
Roxburgh to Banks, B.M., Add. Mss. 33979.
Roxburgh to A.R. Lambert, B.M., Add. Mss. 28545.

Royal Society, London
Herschel Papers, HS. 10, 263–4.

Royal Geographical Society, London
C.R. Markham Papers.
J.T. Walker Papers (1876–81).
Major T.B. Jervis Papers.

Royal Botanic Garden, Kew
Kew Papers, Indian Letters (Bengal), 1853–1900.
     Indian Letters (Madras, Bombay, Bengal), 1901–14.
     Indian Letters (Calcutta Botanic Garden), 1901–14.

London School of Hygiene and Tropical Medicine
Alcock Papers.
Ross Papers.

Imperial College of Science and Technology, London
T.H. Huxley Papers.

Welcome Institute of History of Medicine, London
Leonard Rogers Papers, GOR/A.55/1–110.

Institute of Geological Sciences, London
T. Holland to J.J.H. Tea, G.S.M. 2/284.

School of Oriental and African Studies, London
Reay Papers, Ms. 25460(3).

National Archives of India, New Delhi
Lansdowne Papers (1888–94), Micro. 1050–1661.

J.C. Bose Trust, Calcutta
J.C. Bose Papers.

Bangiya Sahitya Parishad, Calcutta
R.C. Dutt Papers, Newspaper Cuttings in 10 vols, 1869–1900; Speeches
    and Papers on Indian Questions, 1897–1902, 4 vols.

III. Official Reports

Annual Report of the Chemical Analyser to the Bombay Government,
    1878–1902.
Annual Report of the Veterinary College, Bombay, 1886–1906.
Bengal Administration Report, 1875–86.
DPI Report, Bengal, 1856–96.
General Report on the Topographical Survey of India, 1860–71.
General Report on the Operations of the Survey of India, 1878–82.
Memoirs of the Geological Survey of India, 1856–1931.
Papers Regarding Culture and Trade of Sugar in India, 1822.
Papers Relating to Indigo Cultivation in Bengal, Calcutta, 1860.
Papers Regarding Tea Industry in Bengal, Calcutta, 1873.
Papers on the Establishment of Universities in India, Calcutta, 1856.
Papers on Technical Education in India, 1886–1904, Calcutta, 1906.
Quinquennial Review of Education in Bengal, 1892–96, 1897–1901,
    1902–6.
Records of the Geological Survey of India, 1868–96.
Record of the Progress of Mining Instruction in Bengal, 1905–9.
Report of the Committee for Investigating the Coal and Mineral Resour-
    ces of India, Calcutta, 1838.
Report of the Medical College of Bengal, 1849–50.
Report on Sanitary Measures, vols I–XXV, 1867–1906.
Report of the Plague Committee, 1897–98.
Report of the Bombay Plague Research Laboratory, 1896–1902, Bombay,
    1903.
Report of the Leprosy Commission in India, 1890–91, Calcutta, 1893.

Report of the Proceedings of the Central Indigenous Drugs Committee of India, Calcutta, 1901.

Report to the Government of India on Indigo Research Work performed at the University of Leeds, 1905–10.

Report on the Agricultural and Botanical Stations in the Bombay Presidency, 1905–6.

Report of the Indian Education Commission of 1882.

Report on Practical and Technical Education by E.C. Buck, Calcutta, 1901.

Report on Technical Education in Primary and Secondary Schools by A.W. Thomson, Bombay, 1901.

Report on the Observations of Total Solar Eclipse at Sahdool in Central India, Dehra Dun, 1898.

Report of the Royal Indian Engineering College Committee, Simla, 1903.

Report of the Indian Survey Committee, Calcutta, 1905.

Report of the Committee to enquire into the position of Natural Science in the Educational System of Great Britain, Delhi, 1918.

Report of the Indian Industrial Commission, 1916–18, Calcutta, 1918.

Revenue Administration Report, Central Provinces, 1903–4.

Scientific Memoirs by the Indian Medical Service Officers, 12 parts, 1885–1901.

Selections from the Records of the Bengal Government, XIV, Calcutta, 1854.

Selections from the Records of the Bengal Government, XXV, Calcutta, 1857.

Selections from the Records of the Bengal Government, XXVIII, Calutta, 1858.

Selections from the Records of the North West Provinces, Calcutta, 1858.

Selections from the Records of the Government of India, LIV, Calcutta, 1867.

Selections from the Records of the Government of India, LXIV, Calcutta, 1868.

Selections from the Records of the Government of India, LXXVI, Calcutta, 1870.

Selections from the Records of the Government of India, no. 377, Calcutta, 1900.

Selections from the Educational Records of the Government of India, vol. I (Educational Reports, 1859–71), New Delhi, 1960, vol. II

(Development of University Education, 1860–87), New Delhi, 1963, vol. IV (Technical Education in India, 1886–1907), New Delhi, 1968.

IV. Contemporary Tracts and Publications

A Brief Account of the Solar System in Hindi for the Use of Native Schools, Agra, 1840.
Account of the Operations of the Great Trigonometrical Survey of India, II, Dehra Dun, 1879.
Adam, W., *Report on Vernacular Education*, Calcutta, 1868.
Adam, W., *Solar Heat: A Substitute for Fuel in Tropical Countries*, Bombay, 1878.
Annual Report of the Indian Association for the Cultivation of Science, 1888–1905.
Ascoli, F.D., *A Memoir upon the Map of Bengal*, Calcutta, 1914.
Bacon, Thomas, *First Impressions and Studies from Nature in Hindustan*, 2 vols, London, 1837.
Ball, Valentine, *A Manual of the Geology of India*, pt. III, Calcutta, 1881.
Balfour, E. (ed.), *Cyclopaedia of India, Commercial, Industrial and Scientific*, Madras, 1871–73.
Ballantyne, J.R., *A Discourse on Translation*, Mirzapur, 1855.
Basu, B.D., *Education in India Under East India Company*, Calcutta, n.d.
Bhatta, Onkar, *Bhugolsar* (in Hindi), Agra, 1841.
Bhatwedekar, K.S., *Pictorial Translation of C.E. Blunt's Book on Astronomy into Marathi*, Bombay, 1861.
Birch, W.B., *Report on the Production and Manufacture of Sugar*, Calcutta, 1880.
Black, C.E.D., *A Memoir on the Indian Surveys, 1875–90*, London, 1891.
Blanford, H.F., *Instructions to Meteorological Observers in India*, pt. 2, Calcutta, 1876–77.
————, *Rudiments of Physical Geography for the Use of Indian Schools and a Glossary of Technical Terms Employed*, London, 1881.
Blanford, W.T., *Observations on the Geology and Zoology of Abyssinia*, London, 1870.
————, *The Fauna of British India*, London, 1895.
Bose, J.C., *Response in the Living and Non-Living*, London, 1902.
Bose, P.N., *Rudiments of Physical Geography* (in Bengali), Calcutta, 1890.
————, *A History of Hindu Civilization*, III, Calcutta, 1896.
————, *Essays, Lectures on the Industrial Development of India*, Calcutta, 1906.

Buchanan, F., *Journey from Madras through the Countries of Mysore, Canara and Malabar*, 2 vols, London, 1807.

Buck, E., *Note on Tobacco Culture and Curing*, Calcutta, 1878.

Carey, W.H., *The Good Old Days of John Company*, 3 vols, Simla, 1882–87.

Carter, H.J., *Geological Papers on Western India*, Bombay, 1857.

Centenary Review of the Asiatic Society of Bengal, pt. III, Calcutta, 1885.

Centenary vol. of the Indian Museum, 1814–1914, Calcutta, 1914.

Centenary vol. of Calcutta Medical College, Calcutta, 1935.

Colebrooke, H.T., *Miscellaneous Essays*, I, London, 1837.

Convocation Addresses, University of Calcutta, 1858–90, I–II, Calcutta, 1914.

Croft, A., *A Review of Education in India in 1886*, Calcutta, 1886.

Cumming, J.G., *Technical and Industrial Education in Bengal 1888-1908*, Calcutta, 1908.

Curzon, *Speeches*, IV, Calcutta, 1906.

De Campignuelles, S.J. Rev., *Observations Taken at Dumraon, Behar, during the Eclipse of 22nd January 1898*, London, 1899.

Dey, K.L. and Mais, W., *The Indigenous Drugs of India*, Calcutta, 1896.

Dubuat, C., *Principles of Hydraulics with their Practical Application to the Indian Method of Irrigation*, Madras, 1822.

Fayrer, Joseph, *The Thantophidia of India*, London, 1872.

Francotte, E. Rev., *Meteorological Observations at St. Xavier's College (1868–1913)*, Calcutta, 1918.

Gammie, G.A., *A Note on the Plants Used during Famines in Bombay*, Bombay, 1902.

Ghose, N.M., *Higher Education in Bengal*, Calcutta, 1901.

Godard, J.G., *Racial Supremacy: Being Studies in Imperialism*, Edinburgh, 1905.

Gokhale, G.K., *Speeches*, Madras, 1911.

Goldhingham J., *Astronomical Observations at Madras*, III, 1825.

Gupta, Russicklal, *Hindu Practice of Medicine*, Calcutta, 1892.

Hart, E., *The Medical Profession in India*, Calcutta, 1894.

Harrison, J., *The Origin and Progress of Bengal Medical College*, Calcutta, n.d.

Hedin, S., *Scientific Results of a Journey in Central Asia, 1899–1902*, IV. Stockholm, 1907.

Heyne, B., *Tracts on India*, London, 1814.

Hooker, J.D., *Himalayan Journals*, 2 vols, London, 1854.

Howell, Arthur, *Education in British India*, Calcutta, 1872.

Huxley, T.H., *Science and Education*, London, 1893.

India Reform Tract, VI, London, 1853.

Indian Register of Medical Science, vol. I, Calcutta, 1848.

Imam, Syed Imdad, *Ketabul Asmar,* Bankipore, 1887 (in Urdu).

Imam, Syed Imdad, *Keemayae Zeraet,* Arrah, 1890 (in Urdu).

Jameson, W., *Report upon the Botanical Gardens of the North West Provinces,* Roorkee, 1855.

Johnston, Rev., *The Higher Education and the Education of the Masses in India,* Calcutta, 1882.

Lees, W.N., *Tea Cultivation and Other Agricultural Experiments in India,* Calcutta, 1863.

Lockwood, E., *Natural History, Sport and Travel,* London, 1878.

Long, J., *Introduction to Adam's Report,* Calcutta, 1868.

Macgeorge, G.W., *Ways and Works in India,* Westminster, 1915.

MacKenna, J., *Agriculture in India,* Calcutta, 1915.

MacLeod, K., *History of the Medical Schools of Bengal,* Calcutta, 1872.

Markham, C.R., *A Memoir on the Indian Surveys,* London, 1871.

————, *Peruvian Bark: A Popular Account of the Introduction of Cinchona Cultivation into British India,* London, 1880.

Medical Selections, I, Calcutta, 1833.

Merival, H., *Colonization and Colonies,* London, 1861.

Mitra, Rajendaral, *Scheme for the Rendering of European Scientific Terms into the Vernaculars of India,* Calcutta, 1877.

Money, J.W.B., *Java: or How to Manage a Colony: Showing a Practical Solution to the Questions now Affecting British India,* London, 1861.

Morehead, Charles, *Clinical Researches on Disease in India,* 2 vols, London, 1856–58.

Mouat, F.J., *Proposed Plan of the University of Calcutta,* Calcutta, 1845.

Mukherjee, N.G., *Handbook of Sericulture,* Calcutta, 1899.

————, *Improvement of the Sugar Growing Industry of India,* Calcutta, 1901.

Mukherjee, Asutosh, *History of Indian Museum,* Calcutta, 1914.

Murdoch, J., *Educational Reform,* Madras, 1893.

Nowrojee, J. and Merwanjee, H., *Journal of a Residence of Two Years and a Half in Great Britain,* London, 1841.

Oldham, T., *Coal Resources and Production in India: Being Return Called for by the Secretary of State,* Calcutta, 1864.

Padartha Vidya Sara: Elements of Natural Philosophy and Natural History in a Series of Familiar Dialogues (in Hindi), Calcutta, 1843.

Piddington, H., *On the Scientific Principle of Agriculture,* Calcutta, 1839.

Pratt, Hodgson, *University Education in England for Natives of India*, London, 1860.

Proceedings of the First Industrial Conference, Benaras, 1905.

Ranade, M.G., *Essays on Indian Economics*, Madras, 1906.

Ray, P.C., *History of Hindu Chemistry*, 2 vols, Calcutta.

——, *Autobiography of a Bengali Chemist*, Calcutta, 1958.

Rennell, J., *A Bengal Atlas*, Calcutta, 1781.

——, *Memoir of a Map of Hindustan*, London, 1788.

Reports of the Association for the Advancement of Scientific and Industrial Education of Indians, Calcutta, 1905–8.

Ribbentrop, B., *Forestry in British India*, Calcutta, 1900.

Royle, J.F., *Essays on the Productive Resources of India*, London, 1840.

——, *The Fibrous Plants of India*, London, 1855.

Samvad: Collections from Bengali Periodicals on History, Biography, Natural History, etc., Calcutta, 1853.

Sarkar, M.L., *On the Desirability of a National Institution for the Cultivation of the Sciences by the Natives of India*, Calcutta, 1869.

——, *David Hare and the Obligations of the Hindu Community to Scientific Education*, Calcutta, 1876.

——, *The Projected Science Association for the Natives of India*, Calcutta, 1875.

——, *The Indian Association for Cultivation of Science*, Calcutta, 1877.

——, *Moral Influence of Physical Science*, Calcutta, 1892.

Sathianadhan, S., *History of Education in Madras Presidency*, Madras, 1894.

Shah, M.K., *The Principles of Agriculture for India: Compiled from Various Sources for the Use of Practical Farmers, Land Owners and Agricultural Students*, Ahmedabad, 1888.

Sly, F.G., *Report on the Development of Land Records and Agriculture*, 1900–1.

Smith, David, *Report on the Coal and Iron Districts of Bengal*, Calcutta, 1866.

Spring, F.J.E., *Technical Education in Bengal*, Calcutta, 1886.

Tackeray, E.T., *Biographical Notices of Officers of the Royal Bengal Engineers*, London, 1900.

Taylor, W., *Brief Sketch of the Behar Industrial Institution*, Calcutta, 1857.

Temple, R., *India in 1880*, London, 1881.

The Tea Cyclopaedia, Calcutta, 1881.

Tomory, A., *Technical Education in Europe, A Possibility for Bengal*, Calcutta, 1892.

Townsend, C.C., *The Mineral Wealth of India*, Bombay, 1891.
Tracts on Indian Education, Calcutta, n.d.
Trevelyan, G.O., *The Competitionwallah*, London, 1864.
Voelcker, J.A., *Report on the Improvement of Indian Agriculture*, 2nd ed., Calcutta, 1897.
Walker, J.T., *Account of the Operations of the Great Trigonometrical Survey of India*, I, Dehra Dun, 1870.
Waring, E.J., *Remarks on the Uses of Some of the Bazaar Medicines and Common Medicinal Plants of India*, London, 1897.
Watson, J.F., *Prize Essay on the Cultivation and Manufacture of Tea in India*, Calcutta, 1871.
————, *On the Establishment of the Indian Museum*, London, 1875.
Watt, George, *Dictionary of the Economic Products of India*, 6 vols, Calcutta, 1889–93.
————, *Pests and Blights of the Tea Plant*, Calcutta, 1898.
————, *Memorandum on the Organisation of Indian Museum*, Calcutta, 1900.
Wisset, R., *On the Cultivation and Preparation of Hemp*, London, 1804.

V. Contemporary Journals

The Agricultural Gazette of India, Bombay, 1871–75.
The Agricultural Journal of India, I, Calcutta, 1906.
Annals of Royal Botanic Garden of Calcutta, V, 1895.
Asiatic Researches, 1788–1829.
The Asiatic Journal, 1816–45.
The Athenaeum, 1855–1905.
The Bombay Quarterly Magazine, I, 1851.
The Calcutta Journal of Natural History, 1841–48.
The Calcutta Monthly Journal, 1797–1841.
Calcutta Review, 1845–1915.
Canadian Record of Science, 1884–93.
The East India and Colonial Magazine, 1832–38.
East and West Simla, 190–5.
Gleanings in Science, 1829–31.
The India Review and Journal of Foreign Science and Arts, 1837–41.
The Indian Agricultural Gazette, I–IV, 1885–88.
The Indian Agriculturist, 1875–89.
The Indian Annals of Medical Science, 1854–63.
Indian Annals and Magazine of Natural Science, I, Bombay, 1887.
Indian Engineering, 1881–1901.

The Indian Forester, Dehra Dun, 1875–1904.
The Indian Medical Gazette, 1866–1906.
The Indian Textile Journal, 1891–1904.
Journal of the Asiatic Society of Bengal, 1832–43.
Journal of the Bombay Branch of Royal Asiatic Society, Centenary vol., 1905.
Journal of Calcutta Medical and Physical Society, I, 1837.
Journal of National Indian Association, III, 1880–81.
Journal of the Royal Society of Arts, 1875–1911.
The Kayastha Samachar, 1899–1904.
Nature, 1870–1910.
Proceedings of the Asiatic Society of Bengal, 1873–1900.
Science and Culture, 1935–41.
Transactions of the Agricultural and Horticultural Society of India, 1820–28.
Transactions of the Medical and Physical Society of Calcutta, 1826–63.
Transactions of the Mining and Geological Institute of India, 1907.

SECONDARY SOURCES

I. Articles

Alam, 'Imperialism and Science', *Social Scientist,* VI(5), Dec. 1977, pp. 3–15.

Ali, M.A., 'The Eighteenth Century—An Interpretation', *Indian Historical Review,* 1–2, 1979, pp. 175–86.

Anderson, W., 'Where Every Prospect Pleases and only Man is Vile! Laboratory medicine as Colonial Discourse', *Critical Inquiry,* 18, Spring 1992, pp. 506–29.

Ansari, S.M.R., 'The Establishment of Observatories and the Socio–Economic Conditions of Scientific Work in Nineteenth Century India', *Indian Jr. of History of Science,* XIII(1), 1978, pp. 63–71.

Archer, M., 'India and Natural History: The Role of the East India Company, 1785–1858', *History Today,* IX, 1959, pp. 736–44.

Arnold, 'The Indian Ocean as a Disease Zone 1500–1950', *South Asia,* XIV(2), Dec. 1991, pp. 1–21.

Askari, S.H., 'Medicines and Hospitals in Muslim India', *Jr. of Bihar Research Society,* XLIII, 1957, pp. 7–21.

Basalla, G., 'The Spread of Western Science', *Science,* 156, Mar. 1967, pp. 611–22.

Bhabha, H.K., 'The Other Question—The Stereotype and Colonial Discourse', *Screen,* 24 Nov. 1983, pp. 18–35.

————, 'Of Mimicry and Man ; The Ambivalence of Colonial Discourse', 28 Oct. Spring 1984, pp.126–33.

————, 'Signs Taken for Wonders: Questions of Ambivalence and Authority Under a Tree Outside Delhi, May 1817', *Critical Inquiry*, 12, Autumn 1985, pp. 144–65.

Bhattacharya, S., 'Laissez-faire in India', *IESHR*, II(1), 1965, pp. 1–22.

————, 'Cultural and Social Constraints on Technological Innovation: Some Case Studies', *IESHR*, III(3), 1966, pp. 240–67.

————, 'Colonial Power and Micro-Social Interactions: Nineteenth-Century India', *Economic & Political Weekly*, 1–8 June 1991, pp. 1399–403.

Breckenridge, C.A., 'The Aesthetics and Politics of Colonial Collecting: India at World Fairs', *Comparative Studies in Society and History*, 32(2), 1989, pp. 195–216.

Brown, P., 'This Thing of Darkness I Acknowledge Mine: The Tempest and the Discourse of Colonialism', in Dollimore, S. (ed.), *Political Shakespeare*, Manchester, 1985, pp. 48–71.

Chandra, B., 'Reinterpretation of 19th Century Indian Economic History', *IESHR*, V, 1968, pp. 35–75.

Chaudhary, B.B., 'The Growth of Commercial, Agriculture and Its Impact on the Peasant Economy', *IESHR*, VII(1), Mar. 1970, pp. 25–60; VII(2), June 1970, pp. 211–51.

Cohn, B.S., 'The Command of Language and the Language of Command', in Guha, R. (ed.), *Subaltern Studies*, IV, Delhi, 1985, pp. 276–329.

Fanon, F., 'Medicine and Colonialism', in Ehrenreich, J. (ed.), *The Cultural Crisis of Modern Medicine*, New York, 1978, pp. 229–51.

Ghosh, G.C., 'The Utilitarianism of Dalhousie and the Material Improvement of India', *Modern Asian Studies*, XII(1), 1978, pp. 97–110.

Gilman, S.L., 'Degeneracy and Race in the Nineteenth Century: The Impact of Clinical Medicine', *The Journal of Ethnic Studies*, 10(4), 1983, pp. 27–50.

Gorman, M., 'Sir W.B. O'Shaughnessy: Pioneer Chemical Educator in India', *Ambix*, XXX(2), July 1983, pp. 107–16.

————, 'Introduction of Western Science in Colonial India: Role of Calcutta Medical College', *Proc. of American Philosophical Society*, 132(3), Sept. 1988, pp. 276–98.

Grout, A., 'Geology and India 1775–1805: An Episode in Colonial Science', *South Asia Research*, 10 May, 1990, pp. 1–18.

Habib, I., 'Technology and Barriers to Social Change in Mughal India', *Indian Historical Review*, 1–2, 1979, pp. 152–74.

————, 'Changes in Technology in Medieval India', *Studies in History*, II(1), 1980, pp. 15–40.

————, 'The Technology and Economy of Mughal India', *IESHR*, XVII(1), 1980, pp. 1–34.

————, 'Studying a Colonial Economy without Perceiving Colonialism', *Social Scientist*, 139, Dec. 1984, pp. 3–27.

Habib, S.I. and Raina, D., 'Copernicus, Columbus, Colonialism and the Role of Science in Nineteenth Century India', *Social Scientist*, 13(3–4), 1989, pp. 51–66.

————, 'The Introduction of Scientific Rationality into India: A Study of Master Ramchandra—Urdu Journalist, Mathematician and Educationalist', *Annals of Science*, 46, 1989, pp. 597–610.

Harrison, M., 'Tropical Medicine in Nineteenth Century India', *British Jr. of History of Science*, 25, 1992, pp. 299–318.

Inkster, I., 'Scientific Enterprise and Colonial Model: Observations on Australian Experience in Historical Context', *Social Studies of Science*, XV, 1985, pp. 677–704.

————, 'Prometheus Bound: Technology and Industrialization in Japan, China and India prior to 1914: A Political Economy Approach', *Annals of Science*, 45, 1988, pp. 399–426.

Jan Mohamed, A.R., 'The Economy of Manichean Allegory: The Function of Racial Difference in Colonialist Literature', *Critical Inquiry*, 12, Autumn 1985, pp. 59–87.

Jones, G.G., 'The State and Economic Development in India, 1890–1917: The Case of Oil', *Modern Asian Studies*, 13(3), July 1979, pp. 353–75.

Krishna, V.V., 'The Emergence of the Indian Scientific Community', *Sociological Bulletin*, 40(1–2), 1991, pp. 89–107.

Kumar, D., 'Patterns of Colonial Science in India', *Indian Journal of History of Science*, 15(1), May 1980, pp. 105–13.

————, 'Racial Discrimination and Science in Nineteenth Century India', *IESHR*, XIX, 1983, pp. 63–82.

————, 'Science, Resources and the Raj: A Case Study of Geological Works in Nineteenth-Century India', *Indian Historical Review*, X (1–2), 1984, pp. 66–89.

————, 'Science in Higher Education: A Study in Victorian India', *Indian Journal of History of Science*, 19(3), 1984, pp. 253–60.

Larwood, H.J.C., 'Western Science in India before 1850', *Journal of Royal Asiatic Society*, 1962, pp. 62–76.

MacLeod R., 'Scientific Advice for British India: Imperial Perceptions and Administrative Goals, 1898–1923', *Modern Asian Studies*, IX(3), 1975, pp. 343–84.

————, 'On Visiting the Moving Metropolis: Reflections on the Architecture of Imperial Science', *Historical Records of Australian Science*, V(3), 1982, n.p.

MacLeod, R. and Dionne, R., 'Science Policy in British India', *Proceedings of the Sixth European Conference on Modern South Asian Studies*, Sevres, 1978, pp. 55–68.

Mohamed, A.R.J., 'The Economy of Maniehean Allegory: The Function of Racial Differences in Colonial Literature', *Critical Inquiry*, 12, Autumn 1985, pp 59–87

Mohanty, S.P., 'Us and Them: On the Philosophical Bases of Political Criticism', *Yale Journal of Criticism*, II(2), 1989, pp. 1–31.

Morris, D.M., 'Towards a Reinterpretation of Nineteenth Century Indian Economic History', *IESHR*, V, 1968, pp. 1–15.

Panikkar, K.N., 'Cultural Trends in Pre-Colonial India: An Overview', *Studies in History*, II(2), 1980, pp. 63–80.

————, 'Indigenous Medicine and Cultural Hegemony: A Study of the Revitalisation Movement in Kerala', *Studies in History*, 8(2), 1992, pp. 283–307.

Parry, B., 'Problems in Current Theories of Colonial Discourse', *Oxford Literary Review*, 7(1–2), 1987, pp. 27–58.

Prakash, G., 'Science "Gone Native" in Colonial India', *Representations*, 40, Fall 1992, pp. 153–78.

Raina, D., 'Mathematical Foundations of a Cultural Project or Ramchandra's Treatise Through the Unsentimentalised Light of Mathematics', *Historia Mathematica*, 19, 1992, pp. 371–84.

Raina, D. and Habib, S.I., 'Ramchandra's Treatise through "The Haze of the Golden Sunset": An Aborted Pedagogy', *Social Studies of Science*, 20, 1990, pp. 455–72.

————, 'The Unfolding of an Engagement: "The Dawn" on Science, Technical Education and Industrialization: India, 1896–1912', *Studies in History*, 9(1), 1993, pp. 87–117.

Rashed, R., 'Science as a Western Phenomenon', *Fundamenta Scientiae*, I, 1980, pp. 7–21.

Ridley M. and Newton, P., 'Biology Under the Raj', *New Scientist*, 22 Sept. 1983, pp. 857–64.

Roach, J.P.C., 'Victorian Universities and the National Intelligentsia', *Victorian Studies*, II, 1959, pp. 131–50.

Sangwan, S., 'Plant Colonialism', *Proc. of the Indian History Congress,* Burdwan Session, 1983, pp. 414–24.

——, 'European Impressions and Interpretation of Indian Science and Technology 1650–1850', *Social Science Probings,* II(3), Sept. 1985, pp. 353–77.

——, 'Indian Response to European Science and Technology 1750–1850', *British Journal of History of Science,* 21, 1988, pp. 211–32.

——, 'Science Education in India under Colonial Constraints 1792–1857', *Oxford Review of Education,* 16(1), 1990, pp. 81–95.

Sen, S.N., 'Introduction of Western Science in India', *Indian Journal of History of Science,* Nov. 1966, pp. 112–22.

Simmons, C.P., 'Indigenous Enterprise in the Indian Coal Mining Industry, 1835–1939', *IESHR,* XIII(2), 1976, pp. 189–218.

Sirkin, G. and Sirkin, N.R., 'The Battle of Indian Education', *Victorian Studies,* 14(4), 1971, pp. 407–28.

Stokes, E.T., 'Bureaucracy and Ideology: Britain and India in the Nineteenth century', *Transactions of Royal Historical Society,* 30, 1980, pp. 131–56.

Todd, J., 'Science at the Periphery: An Interpretation of Australian Scientific and Technological Dependency and Development prior to 1914', *Annals of Science,* 50, 1993, pp. 33–58.

Tucker, R., 'Forest Management and Imperial Politics: Thana District, Bombay, 1823–87', *IESHR,* XVI(3), 1979, pp. 273–300.

Vicziany, M., 'Imperialism, Botany and Statistics in Early Nineteenth-Century India: The Surveys of Francis Buchanan 1762–1829', *Modern Asian Studies,* 20(4), 1986, pp. 625–60.

Wurgaft, L.D., 'Another Look at Prospero and Caliban: Magic and Magical Thinking in British India', *The Psychohistory Review,* VI(1), 197, pp. 2–26.

II. Ph.D. Thesis

Alexander, D.E., '*The Development of Engineering Education in the United States,* Ph.D. thesis, Washington State University, 1977.

Basalla G., *Science and Government in England (1800–70),* Ph.D. thesis, Harvard University, 1963.

De Vecchi, V.M.G., *Science and Government in Nineteenth Century Canada,* Ph.D. thesis, University of Toronto, 1978.

Forman, Charles, *Science for Empire: Britain's Development of the Empire through Scientific Researches, 1895–1940,* Ph.D. thesis, University of Wisconsin, 1941.

Ghose, S.K., *The Introduction and Development of Electric Telegraph in India*, Ph.D. thesis, Jadavpur University, 1974.

Griffin, H.M., *T B Macaulay and the Anglicist-Orientalist Controversy in Indian Education*, Ph.D. thesis, Pennsylvania University, 1972.

Grove, Richard, *Conservation and Colonial Expansion: A Study of the Evolution of Environmental Attitudes and Conservation Policies on St. Helena, Mauritius, and in India, 1660–1860*, Ph.D. thesis, University of Cambridge, 1988.

Gumperz, E.M., *English Education and Social Change in late 19th Century Bombay (1858–98)*, Ph.D. thesis, University of California, 1965.

Hoare, Michael, *Science and Scientific Associations in Eastern Australia, 1820–90*, Ph.D. thesis, Australian National University, 1974.

Hume, J.C. Jr., *Medicine in the Punjab, 1849, 1911*, Ph.D. thesis, Duke University, 1977.

Kelham, B.B., *Science Education in Scotland and Ireland, 1750–1900*, Ph.D. thesis, University of Manchester, 1968.

Kumar, Anil, *Development of Medical Science in India, 1835–1911*, Ph.D. thesis, University of Delhi, 1991.

Kumar, Arun, *Engineering Education and Public Works Department*, Ph.D. thesis, University of Delhi, 1989.

Mangamma, J., *Technical, Industrial and Agricultural Education in Madras Presidency (1854–1921)*, Ph.D. thesis, University of Delhi, 1971.

Mitra, Udayan, *Social Factors in the Rise and Development of the Bengali Middle Class*, Ph.D. thesis, Rabindra Bharati University, 1982.

O'Keefe, T.J., *British Attitudes Towards India and the Dependent Empire, 1857–74*, University of Notre, Dame, Indiana, 1968.

Worboys, M., *Science and British Colonial Imperialism, 1890–1940*, University of Sussex, D.Phil. thesis, 1979.

III. Books

Adas, M., *Machines as the Measure of Men, Science, Technology and Ideologies of Western Dominance*, Ithaca, 1989.

Ahmed, A., *In Theory: Classes, Nations and Literatures*, Delhi, 1992

Alatas, S.H., *The Myth of the Lazy Native: A Study of the Image of the Malays, Filipinos and Javanese from the 16th to the 20th Century and its function in the Ideology of Colonial Capitalism*, London, 1977.

Alter, P., *The Reluctant Patron: Science and the State in Britain 1850–1920*, Oxford, 1987.

Alvares, C., *Homo Faber: Technology and Culture in India, China and the West, 1500 to the Present Day*, New Delhi, 1979.

Ambirajan, S., *Classical Political Economy and British Policy in India*, Cambridge, 1978.
Arnold, D. (ed.), *Imperial Medicine and Indigenous Societies*, Manchester, 1988.
———, *Colonizing the Body*, Delhi, 1993.
Asad, T. (ed.), *Anthropology and the Colonial Encounter*, Ithaca, 1973.
Ashby, E., *Universities: British, Indian, African: A Study in the Ecology of Higher Education*, London, 1966.
Bala, P., *Imperialism and Medicine in Bengal*, New Delhi, 1991.
Ballhatchet, K., *Race, Sex and Class under the Raj*, London, 1980.
Basalla, G., *The Evolution of Technology*, Cambridge, 1988.
Bagal, J.C., *Pramatha Nath Bose*, Calcutta, 1955.
Bagchi, A.K., *Private Investment in India*, 1900–30, Cambridge, 1972.
———, *The Political Economy of Underdevelopment*, Cambridge, 1982.
Barker, F. et al. (eds), *The Politics of Theory*, Colchester, 1983.
Basu, A., *Essays in the History of Indian Education*, New Delhi, 1982.
Berman, M., *Social Change and Scientific Organisation: The Royal Institution, 1799–1844*, Ithaca, 1978.
Bernal, J.D., *The Social Function of Science*, London, 1939.
———, *Science and Industry in the Nineteenth Century*, London, 1953.
———, *Science in History*, IV, Harmondsworth, 1969.
Bernstein, H.T., *Steamboats on the Ganges*, Calcutta, 1960.
Bhattacharya, B., *Banga Sahitye Vijnan* (in Bengali), Calcutta, 1960.
Bhattacharya, S. and Redondi, P.C. (eds), *Techniques to Technology*, New Delhi, 1990.
Biswas, K., *150th Anniversary of the Royal Botanic Garden*, Calcutta, 1942.
Bose, D.M., Subbarayappa, B.V. and Sen, S.N. (eds), *A Concise History of Science in India*, New Delhi, 1971.
Brockway, L.H., *Science and Colonial Expansion: The Role of the British Botanic Gardens*, New York, 1979.
Burkill, I.H., *Chapters on the History of Botany in India*, Delhi, 1965.
Cardwell, D.S.L., *The Organisation of Science in England*, London, 1980.
Chakravarty, S., *The Raj Syndrome: A Study in the Imperial Perceptions*, New Delhi, 1991.
Charlesworth, N., *British Rule and the Indian Economy 1800–1914*, London, 1982.
———, *Peasants and Imperial Rule: Agriculture and Agrarian Society in Bombay Presidency, 1850–1935*, Cambridge, 1985.
Chattopadhyay, D., *History of Science and Technology in Ancient India: The Beginnings*, Calcutta, 1986.

Chattopadhyay, G. (ed.), *Awakening in Bengal,* I, Calcutta, 1985.

Chattopadhyay, S.K. (ed.), *Ramendra Rachanavali Samgraha* (in Bengali), Calcutta, 1958.

Cipolla, C., *Guns and Sails in the Early Phase of European Expansion, 1400–1700,* London, 1965.

Crawford, D.G., *A History of the Indian Medical Service,* 2 vols, London, 1914.

———, *Roll of the Indian Medical Service,* Calcutta, 1930.

Curtin, P.D., *Europe's Encounter with the Tropical World in the Nineteenth Century,* Cambridge, 1989.

Desmond, R., *The Indian Museum, 1801–79,* London, 1982.

———, *The European Discovery of the Indian Flora,* Oxford, 1992.

Dewey, C. and Hopkins, A.G. (eds), *Imperial Impact: Studies in Economic History of Africa and India,* London, 1978.

Dharampal, *Indian Science and Technology in the Eighteenth Century: Some Contemporary European Accounts,* Delhi, 1971.

Ellsworth, E.W., *Science and Social Science Research in British India, 1788–1880,* Westport, 1991.

Ernst, W., *Mad Tales from the Raj: The European Insane in British India, 1800–58,* London, 1991.

Falola, T. (ed.), *Britain and Nigeria: Exploitation or Development,* London, 1987.

Gaeffke, P. and Utz, D.A. (eds), *Science and Technology in South Asia,* Philadelphia, 1985.

Gates, H.L. Jr. (ed.), *Race, Writing and Difference,* Chicago, 1986.

Geddes, P., *Universities in Europe and India,* Madras, 1915.

———, *An Indian Pioneer of Science: The Life and Work of Sir J C Bose,* London, 1920.

Ghose, B., *Samayikepatre Banglar Samajchitra 1840–1905* (in Bengali), 5 vols, Calcutta, 1962–68.

Gosling, D.L., *Science and Religion in India,* Madras, 1976.

Goonatilake, S., *Crippled Minds: An Exploration into Colonial Culture,* New Delhi, 1982.

Greenberger, A.J., *The British Image of India: A Study in the Literature of Imperialism, 1880–1960,* New York, 1969.

Griffiths, P., *The History of Indian Tea Industry,* London, 1967.

Guha, R., *The Unquiet Woods: An Ecological Change and Peasant Resistance in the Himalayas,* Delhi, 1989.

Harnetty, P., *Imperialism and Free Trade,* Manchester, 1972.

Headrick, D.R., *The Tools of Empire: Technology and European Imperialism in the Nineteenth Century,* New York, 1981.

————, *The Tentacles of Progress: Technology Transfer in the Age of Imperialism, 1850–1940*, New York, 1988.

Hehir, P., *Hygiene and Diseases of India*, Madras, 1913.

Hirschmann, E., *White Mutiny*, New Delhi, 1980.

Hobsbawm, E. and Ranger, T. (eds), *The Invention of Tradition*, Cambridge, 1983.

Home, R.W. and Kohlstedt, S.G. (eds), *International Science and National Scientific Identity: Australia between Britain and America*, Dordrecht, 1991.

Hoodbhoy, P., *Islam and Science*, London, 1991.

Hunt, L. (ed.), *The New Cultural History*, Berkeley, 1989.

Hutchins, F.G., *The Illusion of Permanence: British Imperialism in India*, Princeton, 1967.

Howarth, O.J.R., *The British Association for the Advancement of Science: A Retrospect, 1831–1931*, London, 1931.

Inkster, I and Morrell, J. (eds), *Metropolis and Province: Science in British Culture 1780–1850*, Philadelphia, 1983.

Iyer, R. (ed.), *The Glass Curtain between Asia and Europe*, London, 1965.

Jain, P.K., *The Indian Agricultural Service 1906–24*, New Delhi, 1978.

Jambhekar, G.G. (ed.), *Memoirs and Writings of Bal Gangadhar Shastri Jambhekar*, I, Poona, 1960.

Kotb, Y.S., *Science and Science Education in Egypt*, New York, 1951.

Kumar, Deepak (ed.), *Science and Empire: Essays in the Indian Context*, Delhi, 1991.

Kumar, Dharma (ed.), *The Cambridge Economic History of India*, II, Delhi, 1984.

Kumar, K., *Political Agenda of Education: A Study of Colonialist and Nationalist Ideas*, New Delhi, 1991.

La Touche, T.H.D., *A Bibliography of Indian Geology*, Calcutta, 1917.

Lemaine, G. (ed.), *Perspectives on the Emergence of New Disciplines*, Paris, 1976.

Leslie, C. (ed.), *Asian Medical Systems*, California, 1977.

Lewis, A.R. and McGann, T.F. (eds), *The New World Looks at History*, Austin, 1963.

Lockyer, W.L., *Life and Work of Sir Norman Lockyer*, London, 1928.

Louis, W.R. (ed.), *Imperialism*, New York, 1976.

Mackenzie, J. (ed.), *Imperialism and the Natural World*, Manchester, 1990.

MacLeod, R. (ed.), *Government and Expertise*, Cambridge, 1988.

MacLeod, R. and Lewis, M. (eds), *Disease, Medicine and Empire*, London, 1988.

Mannoni, O., *Prospero and Caliban: The Psychology of Colonisation*, New York, 1956.

Mansel, D. (ed.), *A Selection from the Writings of Joseph Needham*, Lewes, 1990.

McClellan, J.E., *Colonialism and Science: Saint Domingue in the Old Regime*, Baltimore, 1992.

McGuire, J., *The Making of a Colonial Mind: A Quantitative Study of Bhadralok in Calcutta, 1857–85*, Calcutta, 1983.

Meade, T. and Walker, M. (eds), *Science, Medicine and Cultural Imperialism*, New York, 1991.

Meadows, A.J., *Science and Controversy: A Biography of Sir Lockyer*, London, 1972.

Megroz, R.L., *Ronald Ross, Discoverer and Creator*, London, 1931.

Mendelssohn, K., *Science and Western Domination*, London, 1976.

Meulenbeld, G.J. and Wujastyk, D. (eds), *Studies on Indian Medical History, 1834–1947*, Groningen, 1987.

Misra, B.B., *The Indian Middle Classes: Their Growth in Modern Times*, Delhi, 1978.

Mitchell, T., *Colonizing Egypt*, Berkeley, 1991.

Mommsen, W.J. and Osterhammel, J. (eds), *Imperialism and After: Continuities and Discontinuities*, London, 1986.

Morehouse, W. (ed.), *Science and the Human Conditions in India and Pakistan*, New York, 1968.

Mukhopadhyay, G., *The Surgical Instruments of the Hindus*, 2 vols, Calcutta, 1913–14.

———, *History of Indian Medicine*, 3 vols, Calcutta, 1923–29.

Nandy, A., *Alternative Sciences*, New Delhi, 1980.

———, *The Intimate Enemy: Loss and Recovery of Self Under Colonialism*, Delhi, 1983.

———, *Science, Hegemony and Violence*, Delhi, 1990.

Nakayama, S., *Characteristics of Scientific Development in Japan*, New Delhi, 1977.

Needham, J., *The Grand Titration*, London, 1969.

Neelmeghan, A., *Development of Medical Societies and Medical Periodicals in India, 1780–1920*, Calcutta, 1963.

Novarro, V., *Imperialism, Health and Medicine*, New York, 1981.

Parekh, B., *Colonialism, Tradition and Reform*, New Delhi, 1989.

Parry, B., *Delusions and Discoveries: Studies on India in the British Imagination, 1880–1930*, Berkeley, 1972.

Petitjean, P. et al. (ed.), *Science and Empires*, Dordrecht, 1992.

Phillimore, R.H., *Historical Records of the Survey of India*, III, Dehra Dun, 1954.

Phillips, C.H. (ed.), *The Correspondence of Lord William Bentinck*, I, Oxford, 1977.

Pyenson, L., *Cultural Imperialism and Exact Sciences: German Expansion Overseas, 1900–30*, New York, 1985.

————, *The Empire of Reason: Exact Science in Indonesia, 1850–1950*, Leiden, 1989.

Civilizing Mission: Exact Sciences and French Overseas Expansion, 1830–1940, Baltimore, 1993.

Qaisar, A.J., *The Indian Response to European Technology and Culture, 1498–1707*, Delhi, 1982.

Rahman, A. (ed.), *Science and Technology in Medieval India: A Bibliography of Source Materials in Sanskrit, Arabic and Persian*, New Delhi, 1982.

Ramasubban, R., *Public Health and Medical Research in India*, Stockholm, 1982.

Ratnam, R., *Agricultural Development in Madras State prior to 1900*, Madras, 1966.

Ray, R.K., *Industrialisation in India*, Delhi, 1979.

Raychaudhury, T., *Europe Reconsidered Perceptions of the West in Nineteenth Century Bengal*, Delhi, 1988.

Raychaudhury, T. and Habib, I. (eds), *The Cambridge Economic History of India*, I, Delhi, 1984.

Rawat, I.S., *Indian Explorers of the Nineteenth Century*, Delhi, 1973.

Reingold, N. and Rothenberg, M. (eds), *Scientific Colonialism*, Washington, 1986.

Rogers, L., *Happy Toil: Fifty Five Years of Tropical Medicine*, London, 1950.

Ross, R., *Memoirs*, London, 1923.

————, *Ross-Manson Letters*, London, 1929.

Roy, A. and Bagchi, S.K. (eds), *Technology in Ancient and Medieval India*, Delhi, 1986.

Rowland, J., *The Mosquito Man: Sir Ronald Ross*, London, 1958.

Said, E.W., *Orientalism*, New York, 1978.

Culture and Imperialism, London, 1993.

Saldana, J.J. (ed.), *Cross-Cultural Diffusion of Science: Latin America*, Mexico, 1987.

Saletore, B.A. (ed.), *Fort-William-India House Correspondence, 1782–85*, IX, Delhi, 1959.

Sangwan, S., *Science, Technology and Colonisation: An Indian Experience, 1757–1857,* Delhi, 1991.

Sanyal, R., *Voluntary Associations and the Urban Public Life in Bengal, 1815–76,* Calcutta, 1980.

Sarkar, B.K., *Creative India,* Lahore, 1937.

————, *Education for Industrialisation,* Calcutta, 1946.

————, *India in Exact Science: Old and New,* Calcutta, 1947.

Sarkar, S., *The Swadeshi Movement in Bengal,* New Delhi, 1977.

Scott, H.H., *A History of Tropical Medicine,* 2 vols, London, 1937.

Sen, S.N., *Scientific and Technical Education in India,* New Delhi, 1991.

Stearns, R.P., *Science in the British Colonies of America,* Urbana, 1970.

Stone, J., *Canal Irrigation in British India: Perspectives on Technological Change in a Peasant Economy,* Cambridge, 1984.

Strachey, J., *The End of Empire,* Delhi, 1959.

Subbarayappa, B.V. (ed.), *Proceedings of the Indo-Soviet Seminar on Scientific and Technological Exchanges between India and Soviet Central Asia in Medieval Period,* New Delhi, 1985.

Suleri, S., *The Rhetoric of English India,* Chicago, 1992.

Thornton, A.P., *Doctrines of Imperialism,* New York, 1965.

Thorner, D., *Investment in Empire: British Railway and Steam Shipping Enterprise in India, 1825–49,* Philadelphia, 1950.

Tirmizi, S.A.I., *Guide to Records Relating to Science and Technology in the National Archives of India: A RAMP Study,* New Delhi, 1983.

Tsukahara, T., *Affinity and Shinwa Ryoku: Introduction of Western Chemical Concepts in Early Nineteenth Century Japan,* Amsterdam, 1993.

Visvanathan, G., *Masks of Conquest: Literary Study and British Rule in India,* New York, 1989.

Visvanathan, S., *Organising for Science,* Delhi, 1985.

Watanabe, M. (tr. Benfry, O.T.), *The Japanese and Western Science,* Philadelphia, 1991.

Worthington, E.B., *Science in Africa,* New York, 1969.

Wurgraft, L.D., *The Imperial Imagination: Magic and Myth in Kipling's India,* Connecticut, 1983.

Zachariah, K., *History of the Hooghly College,* Calcutta, 1936.

Zaidi, A.M. (ed.), *The Encyclopaedia of the Indian National Congress,* I–V, New Delhi, 1978.

Ziadat, A.A., *Western Science in the Arab World: The Impact of Darwinism 1860–1930,* London, 1986.

Zimmerman, F., *The Jungle and the Aroma of Meats: An Ecological Theme in the Hindu Medicine,* Berkeley, 1987.

Name Index

Subject Index